Conservation of Biodiversity in the North Eastern States of India

Ramagopal V. S. Uppaluri · Latha Rangan
Editors

Conservation of Biodiversity in the North Eastern States of India

Proceedings of NERC 2022

Springer

Editors
Ramagopal V. S. Uppaluri
Department of Chemical Engineering
Indian Institute of Technology Guwahati
Guwahati, Assam, India

Latha Rangan
Department of Biosciences
and Bioengineering
Indian Institute of Technology Guwahati
Guwahati, Assam, India

ISBN 978-981-99-0947-6 ISBN 978-981-99-0945-2 (eBook)
https://doi.org/10.1007/978-981-99-0945-2

This Springer imprint is published by the registered company Springer Nature Singapore Pte Ltd.
The registered company address is: 152 Beach Road, #21-01/04 Gateway East, Singapore 189721, Singapore

Foreword

It is a matter of great satisfaction for me that the Indian Institute of Technology Guwahati successfully hosted North-East Research Conclave (NERC) 2022 on May 20–22, 2022. The NERC 2022 was conducted on the theme "Sustainable Science and Technology". Concurrently, Assam Biotech Conclave (ABC) was also organized on May 21–22, 2022. Both events attracted huge participation from policy-makers, researchers, industrialists, the army, and students. Even the participation of schoolchildren was overwhelming.

NERC and ABC had many events including panel discussions, exhibitions, keynote lectures, competitions and paper presentations. Presentation of technical papers forms the core of any research conference. NERC attracted 879 research papers on various themes covering science, technology and humanities. Out of these, some selected papers have been published by Springer Nature in the form of 15 volumes. These papers have been peer-reviewed and thoroughly edited by IIT Guwahati faculty members. I am sure that these volumes will prove to be excellent resource material for research. Most of the papers presented in these volumes highlight the special needs and aspirations of eight states of North-East India. I congratulate and thank authors, reviewers, editors, and publishers for bringing out proceedings.

The motivation for organizing NERC came from none other than Honourable Minister of Education, Government of India, Shri Dharmendra Pradhan Ji. It helped to bring policy-makers, researchers, industrialists, academicians, students, and children into one forum. It is perhaps the rarest Conclave covering almost all possible research themes. For better readability, the Proceedings have been divided into 15 volumes, but each volume reflects diversity in terms of topics and researchers. The only common thread is the sustainable development of North-East India. Invariably, Sustainable North-East India is a prerequisite for sustainable India and the whole world. In that sense, these 15 volumes will serve guiding and stimulating light for all the stakeholders of the development. I am pleased to dedicate these volumes to the nation as a part of Azadi ka Amrit Mahotsav.

T. G. Sitharam
Director
Indian Institute of Technology Guwahati
Guwahati, Assam, India

Preface

North-East India is a well-known region of biodiversity. The abrupt alterations in the altitude and the presence of deep valleys and mountains in the region translate into diverse ecological situations. In addition, altering climatic conditions also influence and determine the vegetative pattern of the region. With its extensive and rich flora and fauna being contributed to the climatic and vegetative diversity, the region is well known in recent times as one of the world's few biodiversity hotspots. However, ongoing and continued emphasis on the urbanization-driven developmental pressures on the environment and the global losses in biodiversity did renew research investigations and emphasis on biodiversity conservation.

It is well known that urbanization-oriented development strategies critically detriment harmony in the physiographic, geographical and eco-climatic conditions that sustain the region's endemic flora and fauna. Further, the region's medical plants and endangered taxa need preservation and conservation. This has been emphasized by several national and international agencies and with site-specific conservation strategies for the indigenous species of the region. Thereby, the North-eastern region of India can mark its rich resources for the unleashing of unexplored interspecific biodiversities and eventually promote a better harmonization of urbanization-driven development, biodiversity conservation and sustainable utilization of its natural resources. The diverse agro-climatic condition in the region varies between alpine to tropical environment. Along with this, the soil pH of 4.5–5 is highly favourable in the region for the abundance of various species. Witnessing the implementation of such strategies, the region's rich conservation of biodiversity can serve as a model for other regions in India and abroad.

Notably, biodiversity conservation shall address rising concerns with respect to the conservation of certain hotspots. This region is not only considered a homogenous entity, but a highly diverse mosaic of ecological, social and physiological landscapes that needs intensive analysis and attention for its conservation. With global warming and climate change posing threats to vegetation, plant and faunal diversity, the importance of conserving biodiversity is gaining momentum. Hence, the ongoing research should engage in risk distribution agronomy that can ensure food security in an era of climate change to make this region self-sufficient in all aspects. The human endeavour to orient towards economic security based on ecological sustainability has a strong role in the conservation of biodiversity. Hence, research strategies encouraging such sustainable economic security for the community's well-being also become a part of the biodiversity conservation research.

With these purposes and sub-objectives, the authors and editors have contributed sixteen book chapters in several sub-themes that elucidate upon the broader perspectives of biodiversity. These sub-themes are (a) conservation of plant biodiversity for sustainable species utilization and ecosystem, (b) restoration and conservation of diversified microbial and animal biodiversity, (c) value-added product development and benefit sharing in biodiversity for the bioprospecting perspective and (d) conservation of biodiversity through critical environmental factors and its sensitivity towards urbanization and holistic methodologies. In these sub-themes, the reader gains useful insights with a knowledge-sharing platform with respect to the diverse backgrounds of the wildlife, ecological and environmental variation with flora and fauna in the North-Eastern region of India. Needless to convey, the potential conservation measures must be conventionally applied to encourage native landraces, intensify cropping by land configuration and sustainability in agriculture.

In the first sub-theme of plant biodiversity conservation, the volume first addresses the plant biodiversity research framework for bottle gourd and okra germplasms in the Garo hills of Meghalaya. Thereafter, the sub-theme elucidates upon the utility of waste precursors to enrich common bean plants through a vivid farming system. Further, the conservation of *citrus indica*, distribution mapping and diversity assessment of *Ilex venulosa,* above-ground biomass and carbon stock estimation using an allometric equation and on-farm conservation of small millets have been delineated in the sub-theme. Thereby, the role of conservation for its germplasm extinction and genetic erosion have been elaborated. The second sub-theme devotes to microbial and animal biodiversity and emphasizes the genotypic and phenotypic characterization based on microbial diversity conservation and Mithun husbandry practices as relevant animal biodiversity conservation practices. The third sub-theme aims to discuss factors affecting the diversity of black rice, value-added product development such as non-leafy vegetable soup formulation and its organoleptic properties, antimicrobial potential of *Callistemon viminalis*, biopharmaceutical potential of *Alpinia Nigra* and post-harvest management of temperate fruit crops in a northeast state of India. In the final sub-theme of conservation of biodiversity through critical environmental factors and its sensitivity towards urbanization and holistic methodologies, the eco-sensitive zone Deepor Beel has been mapped and quantified for its urbanization impact. Thereafter, floristic diversity and its significance in bio-cultural aspects have been elucidated as the last book chapter and thereby, prompt upon holistic methodologies for carbon sequestration.

In summary, this volume on 'Conservation of Biodiversity of North-Eastern States: Proceedings of NERC 2022' targeted research investigations from areas such as land and lake biodiversity, plant and animal biodiversity and tourism prospects. The critical role of agro-climatic conditions to foster biodiversity and the region's emphasis on rural existence can be witnessed in all articles. The beneficial outcome of the volume has been summarized as follows. Firstly, this volume will help all the researchers in their unique field to develop holistic research culture and thereby prompt the exchange of information and thoughts and eventually stimulate new ideas through such holism. Secondly, a sustainable system can be developed to integrate various thoughts and rights of farmers, breeders or village communities in the NE region of India. Such strategies must value the altered agro-climatic and socio-economic conditions of the regional and native communities. Thirdly and most importantly, a conceptual approach can be enabled based on the generalized guidelines provided by the authors in several chapters. Such guidelines will be useful to protect and preserve the co-existence and biological dependence of many species including humankind in the world.

In summary, this volume will provide an immense opportunity for researchers to exhibit a strong willingness to harmonize biodiversity conservation and modern thinking-based development so as to reflect upon the sustainable utilization of ecological resources and thereby ensure livelihood security in the north-east region of India. Thereby, betterment of improved agro-techniques and harmonious entrepreneurship development will provide holistic approaches and can better the self-sufficient status of the region in all possible natural resources. All in all, the volume conceptualizes the indispensable role of biodiversity for healthy human sustenance to eventually prompt the criticality of human dependence on a balanced ecosystem.

Editorial Assistance: Prabhat Kumar Patel, Tinka Singh, Udarauta Bhattacharjee, Sneha Singh, Kumudhini Akasapu, Paushali Mukherjee, Rubeka Idrishi, Aishwarya Jain, Nuruzzaman Choudhury

Guwahati, India

Prof. Ramagopal V. S. Uppaluri
Prof. Latha Rangan

About IIT Guwahati

Indian Institute of Technology (IIT) Guwahati, established in 1994, has completed 25 years of glorious existence in 2019. At present, the Institute has eleven departments, seven interdisciplinary academic centres and five academic schools covering all the major engineering, science, healthcare, management and humanities disciplines, offering B.Tech., B.Des., MA, M.Des., M.Tech., M.Sc., and Ph.D programmes. The institute presently offers a residential campus to 435 faculty members and more than 7,500 students at present. Besides its laurels in teaching and research, IIT Guwahati has been able to fulfil the aspirations of people of the North-East region to a great extent since its inception in 1994. The picturesque campus is on a sprawling 285 hectares plot on the north bank of the Brahmaputra, around 20 km from the heart of Guwahati city.

IIT Guwahati is the only academic institution in India that occupied a place among the top 100 world universities—under 50 years of age—ranked by the London-based Times Higher Education (THE) in the year 2014 and continues to maintain its superior position even today in various International Rankings. IIT Guwahati received 37th global rank in the 'Research Citations per Faculty' category and overall 384 rank in the QS world University Ranks 2023 that were released recently IIT Guwahati has retained the 7th position among the best engineering institutions in the country in the 'India Rankings 2021' declared by the National Institutional Ranking Framework (NIRF) of the Union Ministry of Education. Also, IIT Guwahati has been ranked 2nd in the 'Swachhata Ranking' conducted by the Govt. of India. Recently, IIT Guwahati has been ranked as the top-ranked University in 2019 for IT developers by HackerRank in the Asia-Pacific region.

Among other frontier areas of research and innovation, IIT Guwahati is working towards augmenting critical science research initiatives in Genomics, Developmental Biology, Health Care and Bioinformatics, Flexible Electronics, Advanced Functional Materials, Sustainable Polymers, Rural Technologies, Renewable Energy, Artificial Intelligence, Disaster Resilience and Risk Reduction and Water Resources and Management. In its silver jubilee year, IIT Guwahati is poised to scale newer heights through all-round growth and development.

Indian Institute of Technology Guwahati has dedicated itself to the cause of improving and empowering north-east India through cutting-edge research, region relevant projects, innovations, individual and multilateral collaborations and special initiatives. Being the only IIT in the entire north- eastern region, IIT Guwahati has an immense amount of responsibility to develop the region and empower the people of the region.

While the entire country celebrating the "Azadi ka Amrit Mahotsav"—75 glorious years of Independence, and the great pride with which our nation of more than a billion people has been steadily growing today, IIT Guwahati is strongly committed to support that pace of growth for the entire NE so that we can keep pace along with the rest of the country. The specific areas of focus where IIT Guwahati has been contributing immensely to the region are:

(a) Infrastructure development across multiple sectors.
(b) Providing solutions for multiple natural disasters such as recurring floods, landslides, earthquakes, cyclones, hailstorms and other natural calamities.
(c) Improving the education sector and creating opportunities for employment.
(d) Internet, telecommunication and cultural integration.
(e) Technological intervention in interdisciplinary areas.
(f) Healthcare services and education.
(g) renewable energy generation (solar, wind, biomass, hydro, geothermal).
(h) overall industrialization, refining fossil fuels and setting up biorefineries.

Besides bringing in state-of-the-art technical know-how for most of the above sectors, the institute has been partnering with the local governments and enhancing the technological and educational interactions such that the next generation youth are empowered with knowledge, skills and necessary entrepreneurial ability. These measures in Assam as well as other north-east states will usher in a new era of growth. Further, it will provide numerous opportunities for interaction with the ASEAN countries as a part of the Act East Policy of the Government of India and will thereby bring prosperity in the region.

Prof. Parameswar K. Iyer
Dean, PRBR, IIT Guwahati

North-East Research Conclave 2022: Towards Sustainable Science and Technology

It is extremely important and imperative to have knowledge-driven growth based on innovation in the case of academic higher education institutes of high repute. The North- Eastern region endowed with rich biodiversity comprises eight states. However, the climatic conditions, limited connectivity, lack of research infrastructure/ institutes, territorial conflicts and the mountainous terrain of these regions are major impediments to the research ecosystem in the North-East. Quality higher education focusing on industry-academia collaboration and translational research is extremely beneficial for society. It has also been rightly pointed out by the Hon'ble Prime Minister Sh. Narendra Modi that, "*India cannot develop till Eastern India develops*".

With this idea and as India marks 75 years of Independence, the Indian Institute of Technology Guwahati organized "The North-Eastern Research Conclave" from

20th–22nd May 2022. This grand event was jointly conducted with Science, Technology and Climate Change Department and the Department of Education, Govt. of Assam at IIT Guwahati Campus.

The mission behind the conclave was to showcase the best R&D activities from educational and research institutions across North-East India and to create an environment, conducive to the development of local indigenous technologies and innovations, creating the scope and laying the foundation for entrepreneurship.

In order to attract people and spread awareness about the event, a roadshow was initiated from IIT Guwahati on 7th May 2022 in order to reach all the partnering academic institutes and make them an integral part of the mega event. The Director, IITG, waved the NERC 2022 flag and sent off the road show vehicle from the institute. More than 400 students, staff and faculty participated actively in the roadshow.

A huge response was received from participants throughout the country. The total no. of Participating institutions in this conclave included 7 IITs, 10 NITs, 5 IIITs and other CFTIs, 23 Research Labs, 17 Central Funded Universities and 47 other Universities/Institutes along with about 100 schools. Eminent personalities from industries, start-ups, research councils, and PSUs also joined in.

The presence of dignitaries from important Ministries was observed such as Shri Dharmendra Pradhan, Hon'ble Union Minister of Education and Minister of Skill

Development and Entrepreneurship, GOI; Dr. Himanta Biswa Sarma Hon'ble Chief Minister of Assam State; Dr. Ranoj Pegu, Hon'ble Minister of Education, Government of Assam; Dr. Rajkumar Ranjan Singh, Hon'ble Minister of State for Education, GOI; Dr. Subhas Sarkar, Hon'ble Minister of State for Education, GOI; Shri Keshab Mahanta, Hon'ble Minister of Science Technology & Climate Change, Govt. of Assam and many more.

The inauguration ceremony of the conclave was followed by the signing of an MoU between IIT Guwahati and the Government of Assam to establish 'The Assam Advanced Health Innovation Institute (AAHII)'. This MoU would prove to be a unique partnership between the Government of Assam and IIT Guwahati in order to set up a Research Institution to leverage advanced technologies to transform medical science. This joint venture company will be able to invite participation from intending parties including corporates/businesses/research institutions and philanthropic organizations.

The third edition of Assam Biotech Conclave 2022 was also held as part of NERC 2022. It brought together Biotech Entrepreneurs, industry leaders, researchers, academicians, Government Representatives, policy-makers, innovators and investors together on one platform to explore the possibilities of Biotechnology in North-East India and to discuss the new opportunities in the transition.

Officers from the Indian Army also actively participated in the Conclave. A talk on "Atmanirbhar Bharat—Indian Army Initiatives towards Self Reliance" was delivered by Lt. Gen. D. S. Rana AVSM, YSM, SM General Officer Commanding, Gajraj Corps on 21st May 2022. The talk was aligned with the vision of the apex leadership of the Government of India and initiatives undertaken by the Indian Armed Forces with a focus on the integration of civil-military establishment in the field of self-reliance. He also elucidated that institutions such as IIT Guwahati which has many running research projects and elaborate student exchange and joint collaboration setup with a large number of Countries have the wherewithal to take up defence-related R&D and also facilitate delivery with Industry Partners. He also invited IIT Guwahati to participate in EAST TECH Symposium planned at Kolkata in July 2022. This led to the signing of an MoU between the Indian Army Eastern Command and IIT Guwahati on 7th July 2022 during East Tech 2022. This would further impetus to Indigenization and Raksha Atmanirbharta.

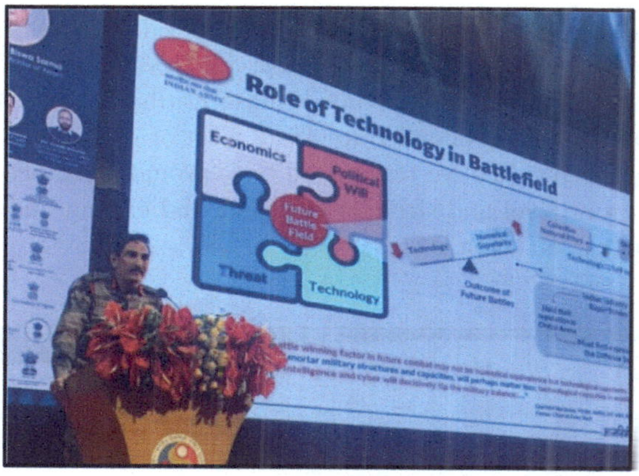

Royal Society of Chemistry, Global battery experiment was performed by more than 1300 students in three sessions starting from 20th May to 22nd May at IIT Guwahati. Along with the Global Battery Experiment, Creating Skilful Educators (Teacher training programme) was also conducted in parallel sessions. Students had arrived from various schools across Assam and other North-Eastern states.

The Guwahati Declaration was launched at the valedictory ceremony of the conclave by Shri Lok Ranjan, Secretary, Ministry of Development of North-Eastern Region (DoNER), in the presence of Shri Kailash Karthik, Deputy Commissioner, Kamrup. The Declaration is intended to create a set of guidelines, through which individual as well as a collective responsibility to promote and encourage innovation at the grass-root level and strive to stimulate and execute indigenization and entrepreneurship, can be taken up.

Science, education, research and innovation are the four pillars on which the development, as well as the work culture of a nation, rests. This was well articulated by the promising number of Exhibitors being seen participating from all across the NE states in the NERC 2022. All the NITs, CFTIs and CFIs were allocated two stalls each, where the delegates showcased the working models of their inventions. Distinctive pavilions were arranged for IIT, NIT, CFIs and CFTIs. Excellent response was obtained from Start-Ups all across the NE states. The Federation of Industry Commerce of North-Eastern Region (FINER) had partnered with NERC 2022 as an Industry Partner and they showcased 50 start-ups as a part of the Exhibition under the FINER Pavilion. Other significant organizations that came forward to showcase their allied R&D start-ups were the Oil and Natural Gas (Oil and Natural Gas Pavilion), Indian Army (Defense Pavilion) and NE-Railway (NE-Railway Pavilion).

Multifarious research work on topics of societal relevance was presented by researchers from different organizations/institutes. The presentations were conducted in oral and poster presentation modes. The thematic areas for these presentations were part of some of the Sustainable Development Goals (SDGs) such as SDG-3: Good health and well-being; SDG-7: Affordable and Clean Energy; SDG-9: Industry, Innovation and Infrastructure; SDG-11: Sustainable cities and communities and SDG-12: Responsible consumption and production. Some of the papers highlighted environmental sustainability, efficiency and management issues, which are important to be presented in the case of North-East regions. Two awards were given under each technical category for these presentations. Overall, the technical sessions were a grand success due to the active cooperation from editors, chairpersons of all the sessions and student volunteers of IITG.

The government of India has taken various steps to encourage women in the field of science and technology. In this line, the IIT Guwahati Woman Researcher Award was approved to recognize the contribution of women Faculty members of IIT Guwahati fraternity. This prestigious award was conferred to Dr. Latha Rangan, who is a Senior Professor in the Department of Biosciences and Bioengineering, Indian Institute of Technology Guwahati, India. Prof. Rangan has played a key role in Plant Biotechnology and Sustainable development, especially in the areas of energy security, food security and medicinal crops.

The Conclave paved the way for creating mass awareness of Research and Innovation for developing a sustainable society. There was knowledge exchange and dissemination that led to the establishment of Centres of Excellence in Translational Collaborative Research and Innovation. This mega event led to the bridging of the gap between Industry–Academia and Creating Hand holding Pathways for setting up long-term collaboration for R&D innovations towards the goal of establishing sustainable NE India. The Conclave brought together over 8000 participants including Hon'ble Ministers, Official Bureaucrats, Eminent Professors, Scientists, Renowned Industrialists, School Children/Teachers and Others delegates. This revolutionized the R&D road map of all the NE states through various dissemination of policies that will benefit the sustainable development of all NE states in near future.

It is an honour and a moment of extreme pride for getting the NERC proceedings published in the prestigious Springer volumes. We would like to thank and acknowledge the globally active publisher Springer for helping us being able to publish the articles in 15 broad areas. We would also like to thank all the authors for their contribution to the grand success of NERC 2022 and wish them great success in all of their future endeavours.

Prof. Vimal Katiyar
Dean, R&D
Department of Chemical Engineering
Centre for the Sustainable Polymer
Indian Institute of Technology Guwahati
Guwahati, India
vkatiyar@iitg.ac.in

Prof. Subhendu Sekhar Bag
Associate Dean, R&D
Department of Chemistry
Centre for the Environment
Indian Institute of Technology Guwahati
Guwahati, India
ssbag75@iitg.ac.in

From the Desk of Chairman of Technical Committee of NERC 2022

North-East Research Conclave 2022 was successfully organized on May 20–22, 2022 with the participation of thousands of delegates. A total of 879 oral and poster papers were presented at the conference on 16 different tracks. The theme of the Conclave was Sustainable Science and Technology, which is very pertinent in the modern era of globalization. Science and technology had to address economic, environmental and social problems of the world. Technology and sustainability are not incompatible. In fact, technology can achieve the goal of sustainability, which also includes preserving our rich cultural heritage. Concurrently with North-East Research Conclave (NERC), Assam Biotech Conclave 2022 was also organized on May 21–22, 2022. These mega events were organized at the Indian Institute of Technology Guwahati (IITG) in physical mode after two years of the pandemic period. Along with IITG, Science, Technology and Climate Change Department & Department of Education, Government of Assam, were also organizers of these events under the patronage of Shri Dharmendra Pradhan Ji, Honourable Minister of Education and Minister of Skill Development and Entrepreneurship in the Government of India, and Shri Himanta Biswa Sarma Ji, Honourable Chief Minister of Assam.

It is a matter of great pleasure that Springer Nature is publishing the select papers from the conclave in 15 volumes. These are Advanced Functional Materials, Low Cost Manufacturing Technologies, Agro and Food Processing Technologies, Artificial Intelligence and Data Science based R&D interventions, Conservation of Biodiversity in the North-Eastern States of India, Disaster Management, Healthcare Research and Related Technologies, Innovative Design for Societal Needs, Policies for Research and Innovation, Research and Innovation for Sustainable Development Goals, Sustainable Environment, Sustainable Energy Generation and Storage, Sustainable Transportation and Urban Development, Teaching and Learning Technologies, Technologies for Rural Development. These volumes are useful archival and reference materials for policy-makers, researchers and students.

As the Chairman of the Technical Committee, I am thankful to all Editors, reviewers and student volunteers who have put tireless efforts to review, select and edit the papers of respective divisions, overcoming the time-constraint. Support provided by Convener, Professor Vimal Katiyar, Dean R&D, IITG, and Co-Conveners Prof.

Subhendu Sekhar Bag, Associate Dean R&D, IITG and Shri Kailash Karthik N, IAS is commendable. It is difficult to express words of gratitude for the Director, IITG, Prof. T.G. Sitharam who has been motivating and guiding all the teams of NERC 2022 and ABC 2022.

Uday S. Dixit
Professor
Department of Mechanical Engineering and
Head Center for Indian Knowledge Systems

Contents

Part I
Conservation of Plant Bio-Diversity for Sustainable Species Utilization and Ecosystem

Chapter 1
Morphological Characterization of Bottle Gourd *[Lagenaria siceraria.(Mol). Standl.]* Germplasms in Garo Hills of Meghalaya

Susmita Chakraborty and A. K. Chaurasiya

1.1 Introduction

With fresh and edible portions, vegetables are considered to be an important class of horticultural products. As key food ingredients, they offer promising benefit to humankind in terms of health protection and disease prevention. Among many palatable vegetables, the gourds are often regarded as the oldest and prime cultivars. Belonging to the Cucurbitaceae family, the commonly grown gourds can be associated with the *Cucurbits* and *Lagenaria* genera. Usually, with their color and shape features, the cucurbit genera are popular. However, in Asia, the *Lagenaria* genera are widely cultivated. In this category, the gourds are known by the name of dipper or bottle gourds. With a diploid chromosome number of 22, the bottle gourd [*Lagenaria siceraria* (Mol.) Standl.] belongs to a member of the Cucurbitaceae family and is often grown during the summer and rainy seasons in south-east Asia. The vegetable origins are in Africa. With varying shapes and sizes, the vegetable can be either long, round, or oblong, depending upon the variety. In the Garo Hills, Meghalaya, they are about 15–100 cm in length and in green color. The fruit is often cultivated in a warm environment.

The wild fruit forms exist in India and South African Gulf countries and have huge potential for sustainable agriculture. With about 13% of world vegetable production, India is only next to China. While Indian farmers cultivate various gourd cultivars being made famous by several organizations, their satisfactory yield has not been ensured to date. To circumvent, bypass, and mitigate this issue, cultivars with divergent performance can be targeted. To this effect, many varieties have been investigated by Morimoto and Mvere [1], Morimoto et al. [2]. All these species investigated to date were those that are accustomed to the climate of Asia, Africa, and the New

S. Chakraborty (✉) · A. K. Chaurasiya
Department of Horticulture, NEHU, Tura Campus, Tura, Meghalaya, India
e-mail: jstsusmy@gmail.com

R. V. S. Uppaluri and L. Rangan (eds.), *Conservation of Biodiversity in the North Eastern States of India*, https://doi.org/10.1007/978-981-99-0945-2_1

World. From a crop improvisation perspective, the Cucurbitaceous class of horticultural products are often considered to be a special class in cross-pollinated crops. A clear distinction can be made for the bottle gourd cultivar in terms of its characteristic white flowers, fruit shape, seed shape, and leaf shape [3]. With wider sexual forms and expressions that favor breeding, the bottle gourd cannot be fully cross-pollinated. Hence, natural pollination on its own could take place despite being at a lower level. Also, a promising feature of the group is that they don't get affected by inbreeding depression. Further, even if such an effect takes place, it will be highly marginal in comparison to other species that underwent cross-pollination. Thereby, this feature customizes cultivar's improvisation through the single plant-based pure line inbreeding. To check and verify the influence of inbreeding, several objectives need to be assessed. These include higher yield, larger fruit weight, earliness, high female to male flower ratio, desired shape features such as long cylindrical or round fruits, skin persisted with bare hairs, edible stage flesh being non-fibrous in nature, non-bitter tasting fruit, attractive texture of fruit without spots, and good resistance for pests and diseases. With good combinations of genetic diversity and morphological parameters, the dipper gourd plants are highly suitable to adapt, grow well, and provide high yields in altered environmental conditions and ecological diversities [4].

Bottle gourd fruits can be consumed as vegetables or used to make sweets such as *halva, kheer, pedha,* and *barfi*, as well as pickles. Even patients can easily digest it as a vegetable. Leaf decoction is an excellent treatment for jaundice. The fruit has a cooling effect having diuretic and cardiotonic properties. The pulp can be used to treat constipation, coughing, and night blindness, as well as an antidote to certain poisons. The plant's extract is used as a cathartic, whereas the seeds are used to treat dropsy. The shapes and sizes of the fruits vary. They are light greenish in color and range in length from 15 to 100 cm. Furthermore, the fruits' dry hard shells have been used to make a variety of everyday items such as bowls, bottles, ladles, containers, fishing net floats, pipes, and musical instruments. Additionally, the seeds and seed oil are edible. Bottle gourd can be grown in dry conditions as well. Depending on the variety, bottle gourd fruits can be long, oblong, or round. Immature fruits can be consumed in a variety of ways. Kofta is one of the most popular preparations. It contains a healthy amount of minerals and vitamins. Its fruit has 95.54% moisture, 10.1 g of vitamin C, 16 IU of vitamin A, thiamine (0.029 g), riboflavin (0.022 g), niacin (0.320 g), carbohydrates (3.39 g), fats (0.02 g), and potassium (150 mg)/100 g. Indian farmers have used various local cultivars and had released (via various organizations) various varieties. However, their output is insufficient. Varietal performance may be useful in solving this issue. Many authors have studied bottle gourd variability, including Morimoto and Mvere [1], Morimoto et al. [2], and others.

Juice is also a popular product. Bottle gourd juice is preferred because it is a natural reservoir of micronutrients and nutritional elements. Additionally, it has been noted for having high concentrations of vitamin C, vitamin B complex, β-carotene, minerals, polyphenols, and carbohydrates. The demand for bottle gourd juice has recently increased due to its reputation as a health drink owing to its nutritional and phytochemical composition. Owing to the health benefits of bottle gourd juice, the extraction of high-quality juice with the greatest retention of bioactive components is a difficult task for processors. Reduced pigment loss during processing is of utmost importance to the processors because the color is a crucial quality determinant. The majority of vegetable juices, including bottle gourd juice, not only turn brown after extraction, but also lose nutrients. Enzymatic activity encourages chlorophyll losses by reacting with the free hydroxyl groups of chlorophyll, which raises the concentration of hydrogen ions and replace the magnesium ions in the center of the pigment with hydrogen thereby, color loss occurs or the production of pheophytin, a yellow-colored substance takes place. Blanching, for example, inactivates enzymes and destroys vegetative microflora, making it an important step in the production of juices that improves color, cloud stability, and nutritional retention. Polyphenol oxidase (PPO), chlorophyllase, β-glucosidase, elastase, and peroxidases are the main enzymes found in bottle gourd. Except for peroxidase, these enzymes are primarily responsible for nutritional and sensory quality loss in vegetables. Peroxides are the most heat-resistant enzymes, and the lack of residual peroxidase activity would imply that other less heat-resistant enzymes have been destroyed as well. Peroxidases are thus utilized as marker enzymes to assess the effectiveness of blanching. The bottle gourd plant has a variety of purported therapeutic qualities. The bottle gourd is a particularly helpful vegetable in reducing digestive issues like constipation due to its dietary fiber. Consuming fiber has been linked positively to a decline in the incidence of diabetes and coronary heart disease. Studies conducted in India revealed a significant regional variation [5]. Without variability, selection may not be efficacious in a population, in terms of variability, it is the genetic portion of the observable variation that provides indicators of transmissibility and responds to selection. The link between genotypic variability and observed variability represents heritability. This is only because the heritability of quantitative traits is large [6]. Bottle gourd cultivation covers 187 thousand hectares in India, with a production of 3079 thousand metric tons (NHB database, 2018–19). It is critical to conserve agricultural genetic resources in order to meet the world's food and feed demand. Crop improvement is primarily based on extensive germplasm evaluation. The greater the genetic resources, the better the chances of capturing specific traits. As high-yielding varieties spread, this genetic variability comprised of landraces is gradually eroding. Thereby, it resulted in a large-scale depletion of variability. This situation necessitates immediate action to preserve the eroding germplasm.

Bottle gourd is a popular cucurbitaceous vegetable crop grown in Meghalaya's Garo Hills. Numerous native cultivars are available with a wide range of variations in fruit size, shape, and color in the region, but few studies have been conducted to develop this crop in this region. With the aforementioned considerations in mind, a research proposal titled "Morphological Characterization of Bottle Gourd [*Lagenaria siceraria* (Mol.) Standl.] Germplasms in Garo Hills, Meghalaya" was undertaken with the goal of studying the morphological characteristics of bottle gourd and identifying superior genotypes.

1.2 Materials and Methods

A field experiment was conducted in 2018 at the horticulture research farm of North Eastern Hill University in Tura district, Meghalaya, India, to morphologically assess the bottle gourd [Lagenaria siceraria (Mol.) Standl.] germplasm in the Garo Hills of Meghalaya. Tura is located at latitude 25° 31′ N and longitude 90° 13′ E with an average elevation of 527 m above the mean sea level. Table 1.1 summarizes the climatological factors that prevailed during the crop growth period as observed at the DAO Research Office in Sangsanggre, Tura, Meghalaya. The following is a description of the materials used, the experimental plan, and the statistical methods deployed for the experimental data.

1.2.1 Experimental Materials

A total of 30 bottle gourd germplasm lines were included as the experimental material that was gathered from locations in the Garo Hills Districts of Meghalaya, India.

Table 1.1 Monthly weather report during the cropping season (2018)

Month	Rainfall	No. of days	Humidity (%)		Temperature (°C)	
			Max	Min	Max	Min
January	0.00	0	91	71	28	13
February	10.00	1	92	65	29	10
March	61.70	2	83	63	30	13
April	161.30	10	86	65	31	13
May	383.00	17	90	61	38	19
June	297.60	10	90	69	39	19
July	131.00	8	90	70	39	19
August	81.50	6	90	70	39	18
September	107.80	7	90	60	39	18
October	6.00	1	82	52	32	17
November	27.00	1	80	59	30	18
December	0.00	0	80	50	30	17
Total	1265.90	63	92	50	39	10

Fig. 1.1 Nursery raising of seedlings

1.2.2 Design of Experiments

The experimentation was conducted in 2018 and was set up by employing a randomized block design (RBD) using three replications. Within each row there were two rows spaced two meters apart. The seedlings (20 days old) were raised in a polyhouse (Fig. 1.1), and were relocated to the field (Fig. 1.2). Nine plants were kept for each germplasm. The crop was raised using a rain-fed system. Thereby, the plants under field conditions were healthy and yielded good quality fruits (Fig. 1.3). The first harvest was made on July 22nd, 2018, after the seeds were sowed on May 19th, 2018. During the cropping period, for proper growth and sustainability of the crop, all ICAR-recommended cultural practices were adopted. The observations were made in accordance with the bottle gourd NBPGR descriptor. In terms of characteristics, data were collected on nine randomly selected plants. Table 1.2 delineates the evaluated characteristics.

1.2.3 Statistical Analysis

The genotype mean values for horticultural traits were used in the statistical analysis. The subsequent statistical variables were computed.

Analysis of Variance

For analysis, the genotype mean values from each replication were used. RBD was used to evaluate the data [7].

For the analytical phase, the following mathematical model was used:

$$X_{mn} = \mu + G_m + R_n + \epsilon_{mn}$$

Fig. 1.2 Field preparation and transplanting

where

m	1, 2, 3,……G
n	1, 2, 3,…….R
X_{mn}	Performance of mth genotype in nth replication
μ	Population mean
G_m	Effect of mth genotype
R_n	Effect of nth replication
ϵ_{mn}	Random error coupled with mth genotype and nth replication

The variance study is depicted as follows:

Fig. 1.3 Plants under field condition

Analysis of Variance (ANOVA)

Source of variation	Degrees of freedom	Mean square	Expected mean square (MS_{ex})	F-value
Replication, R	$R - 1$	MS_R	–	–
Genotype, G	$G - 1$	MS_G	$\sigma^2_\epsilon + R\,\sigma^2_G$	MS_G/MS_ϵ
Error, ϵ	$(R - 1) * (G - 1)$	MS_ϵ	σ^2_ϵ	
Total	$(R * G - 1)$			

where

R	replication counts
G	genotypic counts
MS_R	mean square for replication
MS_G	mean square for genotype
MS_ϵ	mean square error
Genotypic variance (σ^2_G)	$(MS_G - MS_\epsilon)/R$

Table 1.2 Descriptive information of various leaf, plant, and fruit characters considered in the experimental investigations

Characteristics	Classification	Observation
Plant	(a) Vine length, VL (cm)	At the peak of the fruiting period, the VL was noted from the soil surface to the tip of the main stem
	(b) Internode length, IL (cm)	It is the average distance between the fourth and fifth nodes at full foliage development in nine randomly selected plants
	(c) Number of primary branches, NOPB	The primary branch is the one that grows from the main vine or stem. It was measured as the average the of same 9 plants by the end of the flowering period
	(d) Node number when the first female flower blooms, NFFFB	It was recorded during the first appearance of the female flower
	(e) Days till 50% flowering, DTFPF	The number of days from the plantation date till the day the first female flower blooms on at least 50% of the plants was examined and noted
	(f) Sex ratio	The ratio of female (including hermaphrodite) to male flowers at the flowering period was noticed and noted as the sex ratio
	(g) Days until the harvest of the first fruit, DTHFF	It was observed as the days when seeding is done until the first harvest of sellable fruit
	(h) Yield, Y (T/ha)	It was observed and noted as the average of all pickings in the same random 9 plants
Leaf	(a) Petiole length, PL (cm)	At the full leaf stage, the PL was recorded and found to be an average of nine randomly chosen leaves in the mid-region of the vine
	(b) Leaf count, LC	At the sellable fruit harvest stage, the leaf count was recorded as an average of 9 random plants
Fruit	(a) Number of sellable fruit harvest, NOSFH	The total number of fruit pickings was recorded
	(b) Days until the last fruit harvest, DTLFH	It was measured and noted as the number of days from the time of sowing till the DTLFH

(continued)

Table 1.2 (continued)

Characteristics	Classification	Observation
	(c) Fruit length, FL (cm)	Fruit length was measured at the sellable fruit harvest stage as the average of 9 random fruits
	(d) Peduncle length, PDL (cm)	The peduncle length was measured as an average of 9 random fruits during the sellable fruit harvest period
	(e) Fruit width, FW (cm)	At the time of the sellable fruit harvest period, fruit width was examined and measured as the average of 9 random fruits
	(f) Fruit weight, FWT (g)	The fruit weight was evaluated based on fruit output and the number of fruits produced by each plant
	(g) Fruit count per plant, FC/P	It was calculated as the average of the same 9 random plants
	(h) Harvest of sellable fruits, HOSF (kg/plant)	It was observed and noted as the cumulative total yield of all harvests from the same randomly selected 9 plants
	(i) Number of seeds per fruit, NS/F	The average of 9 randomly matured fruits was used to calculate NS/F
	(j) Weight of 100 seeds, WHS (g)	It was calculated as the average weight of 100 randomly selected dry seeds
	(k) Seed length-breadth ratio, SLBR	The ratio of the breadth and thickness of the seed to its width and length

Phenotypic variance ($\sigma^2{}_P$)	$\sigma^2{}_G + \sigma^2{}_\epsilon$
Error variance ($\sigma^2{}_\epsilon$)	MS_ϵ

The significance of genotype-related variance at 5% or 1% level of probability was assessed using the F-test. The critical difference (CD), which was derived to evaluate the importance of the difference between treatment means when the F-value was found to be significant and was determined as follows:

$$\text{CD} = \text{Standard error (SE)} \times t \text{ value at errored degrees of freedom}$$

where

t	table value at 5% or 1% probability levels for error degrees of freedom
Z	Difference in SE between two treatment means
Z	$\frac{\sqrt{2MS_\epsilon}}{R}$

If the mean difference between any two genotypes was larger than CD, the difference was considered to be significant.

Grand mean (GM) can be calculated as

$$\text{GM} = \frac{\sum(G_m + R_n)}{G + R}$$

The coefficient of variation (CV) can be computed by dividing the MS_ϵ by the GM (and then multiplying the result by 100%).

$$\text{CV} = \frac{\sqrt{MS_\epsilon}}{\text{GM}} \times 100$$

1.3 Results and Discussion

With three replications for each genotype in the observed nine plants, the fruit yield and constituent characteristics were determined and reported in terms of the mean value. Thereby, with respect to each combination of genotype and replication, the average value of evaluations being carried out through a random measurement for the nine numbered plants was obtained. Thereafter, a similar methodology was followed for all other replications. Tables 1.3, 1.4, and 1.5 summarizes the average values for respective parameters and for alternate genotypes. In these tables VL, IL, NOPB, NFFFB, DTFPF, SR, DTHFF, Y of plant and PL and LC of leaf genotypes have been delineated in Tables 1.3 and 1.4 respectively. Further, Table 1.5 summarizes average findings for the fruit genotypes, namely FL, PDL, FW, FWT, FC/P, HOSF (kg/p), NS/F, WHS, SLBR, NOSFH, and DTLFH, respectively.

Table 1.3 Data summary of eight evaluated plant characteristics of thirty bottle gourd plant genotypes

Treatment	VL (cm)	IL (cm)	NOPB	NFFFB	DTFPF	SR	DTHFF	Y (T/ha)
GHA-1	455. 40	8. 08	23. 60	26. 04	39. 80	0. 53	89. 73	57. 93
GHA-2	439. 73	6. 87	28. 60	24. 71	38. 73	0. 47	86. 93	48. 06
GHA-3	460. 47	6. 60	26. 60	20. 66	42. 53	0. 71	85. 67	66. 55
GHA-4	455. 53	7. 37	24. 33	21. 32	38. 47	0. 53	79. 47	63. 28
GHA-5	416. 00	6. 57	27. 20	22. 46	40. 53	0. 22	83. 20	63. 54
GHA-6	474. 12	5. 77	21. 87	26. 42	38. 60	0. 81	94. 67	42. 09
GHA-7	496. 00	9. 20	22. 47	25. 33	39. 20	0. 53	78. 40	45. 98
GHA-8	521. 13	8. 50	33. 53	27. 33	36. 47	0. 67	70. 33	157. 63
GHA-9	374. 11	5. 83	23. 47	22. 20	34. 27	0. 63	94. 73	68. 44
GHA-10	421. 37	8. 43	24. 20	24. 40	36. 73	0. 55	69. 60	81. 10
GHA-11	460. 27	5. 73	28. 60	23. 55	29. 33	0. 53	81. 00	46. 31
GHA-12	418. 12	7. 27	24. 07	24. 44	34. 53	0. 85	80. 53	60. 40
GHA-13	430. 00	6. 60	23. 27	22. 46	41. 27	0. 63	69. 60	70. 40
GHA-14	403. 57	5. 60	20. 93	30. 01	38. 33	0. 53	92. 73	28. 89
GHA-15	443. 23	7. 30	24. 53	26. 29	37. 13	0. 93	91. 87	53. 77
GHA-16	422. 83	5. 80	30. 40	29. 49	39. 33	0. 41	77. 13	62. 73
GHA-17	475. 77	5. 13	25. 87	28. 33	35. 33	0. 32	78. 40	90. 90
GHA-18	395. 93	6. 67	24. 40	25. 40	33. 33	0. 15	81. 98	22. 74
GHA-19	428. 86	7. 30	23. 93	23. 46	35. 07	0. 15	92. 47	35. 51
GHA-20	469. 10	8. 17	30. 93	26. 46	38. 60	0. 22	86. 67	117. 96

(continued)

Table 1.3 (continued)

Treatment	VL (cm)	IL (cm)	NOPB	NFFFB	DTFPF	SR	DTHFF	Y (T/ha)
GHA-21	494. 37	6. 80	31. 07	25. 20	34. 60	0. 34	76. 60	133. 83
GHA-22	391. 33	5. 70	25. 27	26. 33	32. 47	0. 34	87. 20	57. 27
GHA-23	442. 03	6. 47	29. 60	28. 40	34. 73	0. 75	90. 60	125. 61
GHA-24	438. 13	5. 10	28. 30	27. 15	35. 33	0. 75	78. 53	80. 91
GHA-25	487. 92	7. 27	29. 33	24. 33	40. 84	0. 43	81. 13	107. 92
GHA-26	433. 97	6. 57	28. 00	25. 31	38. 47	0. 36	77. 13	63. 40
GHA-27	382. 73	7. 27	27. 73	22. 36	29. 07	0. 48	86. 87	63. 68
GHA-28	452. 53	6. 17	24. 27	23. 29	35. 80	0. 25	70. 40	53. 57
GHA-29	408. 53	5. 60	29. 40	25. 24	33. 33	0. 17	89. 80	99. 24
GHA-30	504. 93	8. 63	30. 07	21. 40	36. 93	0. 27	80. 53	154. 36
GM	443. 27	6. 81	26. 53	24. 99	36. 64	0. 48	82. 80	74. 13
CV (%)	11. 12	15. 57	11. 37	10. 41	10. 51	12. 10	10. 27	29. 28
MS_{ϵ}	23. 07	0. 50	1. 41	1. 22	1. 80	0. 03	3. 98	12. 51
CD (5%)	64. 94	1. 40	3. 97	3. 43	5. 07	0. 10	11. 20	35. 42
Range(min)	374. 11	5. 10	20. 93	21. 32	29. 07	0. 15	69. 60	22. 74
Range(max)	521. 13	9. 20	33. 53	30. 01	42. 53	0. 93	94. 73	157. 63
Best case (Rank 1)	GHA-8	GHA-7	GHA-8	GHA-14	GHA-3	GHA-15	GHA-9	GHA-8
Second best case (Rank 2)	GHA-30	GHA-30	GHA-21	GHA-16	GHA-13	GHA-19	GHA-6	GHA-30
Third best case (Rank 3)	GHA-9	GHA-24	GHA-14	GHA-4	GHA-27	GHA-6	GHA-13	GHA-18

Table 1.4 Data summary of two evaluated leaf characteristics of thirty bottle gourd plant genotypes

Treatment	PL (cm)	LC
GHA-1	10. 43	169. 13
GHA-2	11. 43	149. 67
GHA-3	8. 60	171. 00
GHA-4	11. 67	163. 33
GHA-5	10. 73	136. 67
GHA-6	7. 70	186. 00
GHA-7	8. 77	196. 67
GHA-8	7. 40	227. 07
GHA-9	8. 90	135. 87
GHA-10	10. 03	164. 47
GHA-11	6. 93	201. 73
GHA-12	7. 70	196. 87
GHA-13	8. 57	204. 47
GHA-14	9. 30	197. 27
GHA-15	8. 23	186. 00
GHA-16	7. 73	212. 67
GHA-17	9. 50	193. 60
GHA-18	10. 37	188. 33
GHA-19	8. 63	178. 27
GHA-20	9. 30	201. 93
GHA-21	8. 83	195. 80
GHA-22	7. 57	176. 07
GHA-23	8. 83	154. 33
GHA-24	7. 93	204. 47
GHA-25	10. 30	208. 93
GHA-26	10. 50	234. 33
GHA-27	8. 67	170. 53
GHA-28	8. 57	245. 67
GHA-29	9. 43	210. 07
GHA-30	9. 17	232. 33
GM	9. 06	189. 78
CV (%)	15. 96	10. 89
MS_{ϵ}	0. 68	9. 68
CD (5%)	1. 90	27. 24
Range(min)	6. 93	135. 87

(continued)

Table 1.4 (continued)

Treatment	PL (cm)	LC
Range(max)	11. 67	245. 67
Best case (Rank 1)	GHA-4	GHA-28
Second best case (Rank 2)	GHA-2	GHA-30
Third best case (Rank 3)	GHA-11	GHA-9

1.3.1 Plant Characteristics

(a) **Vine length, VL**

The maximum VL in genotype GHA-8 was 521.13 cm, and the minimum VL in genotype GHA-9 was 374.11 cm. The average mean was 443.27 cm. Mishra [8] investigated and reported a maximum VL of 6.7 m in genotype B11 and a minimum VL of 4.7 m in genotype B6. Kandasamy et al. [9] tested 20 bottle gourd genotypes and found that genotype LS12 (720.82 cm) had the longest vines and genotype LS2 had the shortest (465.90 cm). Also, based on the VL, the best, second-best, and third-best genotypes have been found to be GHA-8, GHA-30, and GHA-9, respectively.

(b) **Internode length, IL**

IL recorded a maximum, i.e., 9.20 cm of internode length in GHA-7, whereas a minimum internode length of 5.10 cm was recorded in GHA-24. The average mean was recorded at 6.81 cm. Tirumalesh et al. [10] reported maximum internode length in variety APBG-3 (15.60) and minimum in Pusa Santushti (4.13) among 27 bottle gourd genotypes. Based on the IL, the best, second-best, and third-best genotypes have been found to be GHA-7, GHA-30, and GHA-24, respectively.

(c) **Number of primary branches, NOPB**

The maximum NOPB was found in genotype GHA-8, at 33.53, while the lowest NOPB was found in genotype GHA-14, at 20.93. The average mean was calculated to be 26.53. Kumar et al. [11] investigated and reported the highest NOPB in genotype IBG-7 (10.37) and the lowest in genotype IBG-45 (2.53). Singh and Singh [12] also found that genotype SEL-6 had a higher NOPB (11.5), while genotype SEL-10 had the lowest (4.5). The best, second-best, and third-best genotypes for NOPB have been found to be GHA-8, GHA-21, and GHA-14, respectively.

(d) **Node number when the first female flower blooms, NFFFB**

The maximum NFFFB in GHA-14 was 30.01, while the minimum NFFFB in GHA-4 was 21.32. The overall average was 24.99. Kumar et al. [11] conducted research and discovered that the mean of the first NFFFB was 25.37. Kandasamy et al. [9] found that genotype LS15 had the highest NFFFB (30.20) and genotype LS3 had the lowest (14.72). Significant differences in NFFFB in the current experiment could be

Table 1.5 Data summary of eleven evaluated fruit characteristics of thirty bottle gourd plant genotypes

Treatment/replication	FL (cm)	PDL (cm)	FW (cm)	FWT (g)	FC/P	HOSF (kg/p)	NOSFH	DTLFH	NS/F	WHS (g)	SLBR
GHA-1	52. 77	8. 63	13. 17	1773. 33	13. 07	23. 17	2. 33	144. 93	348. 67	11. 53	2. 02
GHA-2	58. 60	12. 73	14. 60	1676. 67	11. 47	19. 23	3. 33	155. 73	412. 00	12. 83	2. 05
GHA-3	24. 73	9. 57	12. 17	1976. 67	13. 47	26. 62	2. 00	152. 27	461. 33	13. 10	2. 02
GHA-4	50. 97	9. 20	9. 43	1898. 33	13. 33	25. 31	3. 67	148. 47	311. 67	10. 20	1. 70
GHA-5	42. 03	12. 50	12. 57	2823. 80	9. 00	25. 41	3. 33	144. 33	318. 67	10. 90	2. 25
GHA-6	25. 53	8. 37	13. 37	1640. 00	10. 27	16. 84	2. 33	150. 93	465. 33	12. 63	2. 15
GHA-7	14. 67	11. 47	10. 90	1550. 00	11. 87	18. 39	2. 67	153. 53	424. 00	11. 17	1. 70
GHA-8	12. 83	10. 93	14. 17	3616. 67	17. 43	63. 05	4. 53	152. 13	375. 67	11. 13	2. 10
GHA-9	9. 40	7. 93	7. 67	2666. 67	10. 27	27. 38	3. 60	144. 17	255. 67	9. 93	2. 43
GHA-10	13. 07	9. 83	15. 37	2630. 33	12. 33	32. 44	3. 73	144. 13	253. 67	10. 80	2. 10
GHA-11	14. 13	9. 20	12. 53	1526. 67	12. 13	18. 52	4. 33	155. 00	431. 33	12. 87	1. 93
GHA-12	22. 23	8. 90	12. 17	1858. 33	13. 00	24. 16	2. 87	155. 33	454. 33	13. 67	2. 20
GHA-13	24. 13	9. 63	9. 30	2560. 00	11. 00	28. 16	2. 67	145. 00	296. 00	10. 03	2. 44
GHA-14	25. 00	9. 20	10. 83	1507. 53	7. 67	11. 56	3. 47	141. 13	251. 33	9. 63	2. 52
GHA-15	30. 70	13. 00	16. 00	1946. 53	11. 05	21. 51	2. 00	146. 33	409. 33	13. 50	2. 14
GHA-16	22. 40	9. 83	11. 57	1881. 80	13. 33	25. 09	3. 67	169. 60	386. 00	11. 43	2. 39
GHA-17	10. 80	8. 43	13. 93	2870. 40	12. 67	36. 36	3. 00	157. 67	267. 33	9. 07	2. 58
GHA-18	16. 03	13. 20	15. 57	1705. 33	5. 33	9. 10	2. 13	143. 87	344. 33	10. 53	2. 08
GHA-19	12. 23	10. 03	11. 23	1521. 93	9. 33	14. 20	2. 27	149. 99	361. 67	11. 70	2. 15
GHA-20	27. 10	9. 33	16. 03	2973. 87	15. 87	47. 19	4. 87	161. 67	410. 67	12. 73	1. 93
GHA-21	24. 50	9. 30	18. 13	3100. 33	17. 27	53. 53	4. 27	168. 13	524. 67	13. 33	2. 05

(continued)

Table 1.5 (continued)

Treatment/replication	FL (cm)	PDL (cm)	FW (cm)	FWT (g)	FC/P	HOSF (kg/p)	NOSFH	DTLFH	NS/F	WHS (g)	SLBR
GHA-22	23. 03	9. 67	12. 73	2045. 33	11. 20	22. 91	2. 67	152. 33	445. 67	12. 97	1. 82
GHA-23	13. 40	7. 83	16. 37	3088. 77	16. 27	50. 24	3. 00	147. 00	512. 33	13. 57	2. 21
GHA-24	37. 37	9. 17	10. 40	2682. 13	12. 07	32. 36	4. 93	144. 87	436. 00	12. 83	2. 08
GHA-25	13. 23	8. 43	12. 50	3068. 75	14. 07	43. 17	4. 87	158. 13	413. 67	11. 57	2. 20
GHA-26	11. 50	8. 30	12. 47	2173. 60	11. 67	25. 36	3. 67	151. 67	276. 67	9. 50	2. 15
GHA-27	18. 90	8. 50	10. 97	2635. 00	9. 67	25. 47	2. 67	147. 00	314. 90	10. 67	2. 40
GHA-28	12. 90	9. 30	11. 70	1495. 00	14. 33	21. 43	4. 00	146. 73	307. 13	10. 93	2. 57
GHA-29	27. 00	8. 40	14. 93	2238. 47	17. 73	39. 70	4. 87	136. 53	394. 67	10. 63	2. 23
GHA-30	9. 87	10. 10	15. 10	2921. 67	21. 13	61. 74	5. 07	149. 60	416. 33	11. 53	2. 17
GM	23. 37	9. 70	12. 93	2268. 46	12. 64	29. 65	3. 43	150. 61	376. 03	11. 56	2. 16
CV (%)	13. 71	13. 30	16. 46	18. 69	36. 10	29. 28	24. 44	7. 88	2. 14	9. 01	10. 86
MS_{ϵ}	1. 50	0. 60	1. 00	198. 51	2. 14	5. 00	0. 39	5. 60	3. 76	0. 49	0. 11
CD (5%)	4. 22	1. 70	2. 80	558. 67	6. 01	14. 17	1. 10	15. 77	10. 59	1. 37	0. 31
Range(min)	9. 40	7. 83	7. 67	1495. 00	5. 33	9. 10	2. 00	136. 53	253. 67	9. 07	1. 70
Range(max)	58. 60	13. 20	18. 13	3616. 67	21. 13	63. 05	5. 07	169. 60	524. 67	13. 67	2. 58
Best case (Rank 1)	GHA-2	GHA-18	GHA-21	GHA-8	GHA-30	GHA-8	GHA-30	GHA-16	GHA-21	GHA-12	GHA-17
Second best case (Rank 2)	GHA-1	GHA-15	GHA-9	GHA-21	GHA-29	GHA-30	GHA-24	GHA-21	GHA-23	GHA-23	GHA-13
Third best case (Rank 3)	GHA-9	GHA-23	GHA-23	GHA-28	GHA-18	GHA-18	GHA-3	GHA-29	GHA-10	GHA-17	GHA-4

attributed to soil fertility, hospitable environmental conditions, and location. Harika et al. [13] also reported that the NFFFB is highest in Sarika (13.83) and lowest in Louki (6.83). Also, based on the NFFFB, the best, second-best, and third-best genotypes have been found to be GHA-14, GHA-16, and GHA-4, respectively.

(e) **Days to 50% flowering, DTFPF**

DTFPF were recorded at a maximum of 42.53 days in GHA-3 and a minimum of 29.07 days in genotype GHA-27, with a general mean value of 36.64 days. Kumar et al. [11] investigated and documented a maximum of 77.67 days in genotype IBG-55 and a minimum of 47 days in genotype IBG-69 in 69 genotypes of the bottle gourd. Pandiyan et al. [14] found that the maximum DTFPF in genotype 2012/BOGVAR-3 (53.7) and the minimum in NDBG-104(C) (50.0) were similar. Moreover, the best, second-best, and third-best genotypes have been found to be GHA-3, GHA-13, and GHA-27, respectively.

(f) **Sex Ratio, SR**

SR ranged from 0.93 for the genotype GHA-15 to 0.15 for the genotypes GHA-18 and GHA-19, with an average mean of 0.48. Harika et al. [13] reported a maximum SR of 24.40 in Anand Bottle Gour-1 and a minimum SR of 11.87 in NS-443. Based on the SR, the best, second-best, and third-best genotype have been found to be GHA-15, GHA-19, and GHA-6, respectively.

(g) **Days until the harvest of the first fruit, DTHFF**

The maximum DTHFF was recorded in GHA-9 (94.73) and the minimum DTHFF was recorded in GHA-10 (69.60). The average mean values recorded were 82.80 days. In an earlier experiment, Kumar et al. [11] reported a maximum DTHFF of 88.36 days in genotype IBG-23 and a minimum of 62.84 DTHFF in genotype IBG-3 in 69 bottle gourd germplasm. Kandasamy et al. [9] published similar results for 20 bottle gourd genotypes. The best, second-best, and third-best genotypes based on the DTHFF have been found to be GHA-9, GHA-6, and GHA-13, respectively.

(h) **Yield, Y (T/ha)**

The maximum yield for genotype GHA-8 was 157.63 t/ha, while the minimum yield for genotype GHA-18 was 22.74 t/ha. The average mean was 74.13 T. Kumar et al. [11] investigated genotype IBG-61 and found a maximum yield of 433.3 q/ha and a minimum yield of 123.70 q/ha. Pandiyan et al. [14] reported similar results of maximum yield in genotype 2012/BOGVAR-5 (253.6 q/ha) and minimum yield in genotype 2012/BOGVAR-4 (195.0 q/ha). Further, the best, second-best, and third-best genotypes based on the yield factor have been found to be GHA-8, GHA-30, and GHA-18, respectively.

1.3.2 Leaf Characteristics

(a) **Petiole length, PL**

The maximum PL was 11.67 cm in genotype GHA-4 and the minimum PL was 6.93 cm in genotype GHA-11. The overall average was 9.06 cm. Leela and Singh [15] found that genotype V14 had the longest petiole (16.70 cm) and genotype V19 had the shortest (16.70 cm) (8.35 cm). PL can differ significantly depending on soil fertility, environmental conditions, and location. For PL, the best, second-best, and third-best genotypes have been found to be GHA-4, GHA-2, and GHA-11, respectively.

(b) **Leaf count, LC**

The GHA-28 area had the most leaves (245.67), whereas GHA-9 had the fewest leaves (135.87). The recorded average mean value was 189.78. Location, environmental factors, and soil fertility may all have a significant impact on overall LC. Also, for LC, the best, second-best, and third-best genotypes have been found to be GHA-28, GHA-30, and GHA-9, respectively.

1.3.3 Fruit Characteristics

(a) **Number of sellable fruit harvest, NOSFH**

The maximum NOSFH per plant was observed (5.07) in genotype GHA-30 and the minimum NOSFH per plant (2.00) was observed in genotype GHA-15. The NOSFH per plant ranges between 2.0 and 5.07 and had an average mean of 3.43. Also, based on the NOSFH, the best, second-best, and third-best genotypes have been found to be GHA-30, GHA-24, and GHA-3, respectively.

(b) **Days until the last fruit harvest, DTLFH**

The genotype GHA-16 had the longest DTLFH (169.60) and the genotype GHA-29 had the shortest DTLFH (136.53). The overall average was 150.61 days. According to Kaur and Dhillon [16], the last fruit harvest in melon ranged from 116.36 to 140.50 days. Significant differences in the DTLFH in the current experiment could be attributed to late flowering and the appearance of the first flower node. Further, for DTLFH, the best, second- best, and third-best genotypes have been found to be GHA-16, GHA-21, and GHA-29, respectively.

(c) **Fruit length, FL (cm)**

The maximum FL in genotype GHA-2 was 58.60 cm, while the minimum FL in genotype GHA-30 was 9.87 cm. The average mean measured was 23.37 cm. Harika et al. [13] found a maximum FL of 58.92 cm and a minimum FL of 9.18 cm in twenty-five bottle gourd genotypes. Among the eleven bottle gourd lines, Uddin et al. [17] found FL ranging from 18.1 cm in genotype L-3 to 54.9 cm in genotype L-11. According to Ilyas et al. [18], the genotype NS550f1 has the longest FL (0.49 m), and the variety Globe has the shortest (0.38 m). Kandasamy et al. [9] also reported a maximum FL in genotype LS4 (49.28 cm) and a minimum in genotype LS14 (19.20 cm). Based on FL, the best, second-best, and third-best genotypes have been found to be GHA-2, GHA-1, and GHA-9, respectively.

(d) **Peduncle length, PDL**

Maximum PDL was 13.20 cm in GHA-18 and minimum PDL was 7.83 cm in GHA-23. The average mean value was 9.70 cm. For PDL, the best, second-best, and third-best genotypes have been found to be GHA-18, GHA-15, and GHA-23, respectively.

(e) **Fruit width, FW**

The maximum FW was 18.13 cm in genotype GHA-21 and the minimum FW was 7.67 cm in genotype GHA-9. The overall mean measured was 12.93 cm. Uddin et al. [17] investigated and reported the maximum fruit diameter from L9 (18.5 cm) and minimum fruit diameter from L10 (9.7 cm) in eleven bottle gourd germplasm. Kandasamy et al. [9] also reported that genotype LS 15 had the greatest fruit breadth (65.55 cm) and genotype LS2 had the smallest (25.12 cm). Moreover, based on FW, the best, second-best, and third-best genotypes have been found to be GHA-21, GHA-9, and GHA-23, respectively.

(f) **Fruit weight, FWT**

The maximum FWT recorded was 3616.67 g in GHA-8 and the minimum FWT recorded was 1495.00 g in GHA-28. The average mean value was 2268.46 g. Iqbal et al. [19] also reported a maximum FWT of 1112 g for the variety Anmol and a minimum FWT of 926 g for the local cultivar Louki. Kandasamy et al. [9] reported similar findings of maximum average FWT in genotype LS4 (2.92 kg) and minimum in LS2 (0.51 kg). For the FWT, the best, second-best, and third-best genotypes have been found to be GHA-8, GHA-21, and GHA-28, respectively.

(g) **Fruit count per plant, FC/P**

The FC/P varied significantly between bottle gourd genotypes. The genotype GHA-30 produced the most fruits per plant (21.13), and the genotype GHA-18 produced the fewest fruits (5.33). The recorded average mean was 12.64. Mishra [3] investigated and found that genotype B11 had the highest FC/P of 14.2 and genotype B3 had the lowest FC/P of 9.2. Kandasamy et al. [9] observed and reported the highest FC per vine in genotype LS18 (16.86) and the lowest in genotype LS4 (5.13). Based on FC/P, the best, second-best, and third-best genotypes have been found to be GHA-30, GHA-29, and GHA-18, respectively.

(h) **Harvest of sellable fruits, HOSF (kg/plant)**

The maximum HOSF per plant was 63.05 kg in GHA-8 and the minimum harvest was 9.10 kg in GHA-18. The overall mean was 29.65. Kandasamy et al. [9] reported similar findings of maximum HOSF per vine in genotype LS12 (18.80 kg) and minimum in LS2 (8.51 kg). Also, based on the NFFFB, the best, second-best, and third-best genotypes have been found to be GHA-8, GHA-30, and GHA-18, respectively.

(i) **Number of seeds per fruit, NS/F**

NS/F was found to be a maximum of 524.67 in genotype GHA-21 and a minimum NS/F of 253.67 was found in genotype GHA-10. The average mean was 376.03. For NS/F, the best, second-best, and third-best genotypes have been found to be GHA-21, GHA-23, and GHA-10 respectively.

(j) **Weight of 100 seeds, WHS (g)**

The maximum WHS in genotype GHA-12 was 13.67 g, while the minimum WHS in genotype GHA-17 was 9.07 g. Kumar et al. [11] found a maximum WHS of 23.09 g in genotype IBG-31 and a minimum WHS of 10.07 g in genotype IBG-13. Kandasamy et al. [9] also reported a maximum WHS of 21.95 g in genotype LS15 and a minimum weight of 10.96 g in genotype LS10. The best, second-best, and third-best genotypes for WHS have been found to be GHA-12, GHA-23, and GHA-17, respectively.

(k) **Seed length-breadth ratio, SLBR**

SLBR achieved a maximum value of 2.58 in genotype GHA-17. The minimum value of 1.70 was recorded for genotype GHA-4. The average mean value was 2.16. Moreover, based on SLBR, the best, second-best, and third-best genotypes have been found to be GHA-17, GHA-13, and GHA-4, respectively.

1.4 Conclusions

Plant genetic resources are the most significant and necessary basic raw materials for crop development programs. Encoding the genetic information using plant genomes, biological processes translate such information into phenotypes. The structure and level of biological organization by which genes, phenotypes, or groups of phenotypes persist collectively are referred to as plant genetic resources or plant germplasm. Due to the significant variation in fruit size, weight, and length, it was considered important to assess and characterize bottle gourd germplasm lines gathered from the West Garo Hills District of Meghalaya, India. In addition, relevant genetic parameters were calculated in order to effectively use superior germplasm lines in future bottle gourd breeding programs.

From the conducted investigations, it can be observed that there have been a lot of variabilities among the different germplasm lines of bottle gourd and between the bottle gourd genotypes for various traits and genetic diversity analyzed by ANOVA. On the basis of yield and quality attributes, superior genotypes were identified. The best 3 superior genotypes were identified in the study as GHA-8, GHA-30, and GHA-29. This variability among the different germplasm lines can be used as an improved cultivar or as an inbred line for different hybrid breeding programs for the improvement of prevailing variety or development of a new cultivar. The demonstrated research methodology can be followed for similar inbred crops of commercial value in North-East India. Therefore, human consumption-driven best germplasm can be identified through active research in the hybrid breeding programs. Needless to convey, the role of climatic and ecological conditions are to be effectively considered in such studies to foster the importance of regional factors in fostering the conservation of bio-diversity in North-East states of India.

References

1. Morimoto, Y., Mvere, B.: *Lagenariasiceraria* (Molina) Standley. In: Grubben, G.J.H., Denton, O.A. (eds.) Plant Resources of Tropical Africa 2. Vegetables, pp. 353–358 (2004)
2. Morimoto, Y., Maundu, P., Fujimaki, H., Morishima, H.: Diversity of landraces of the white-flowered gourd (*Lagenariasiceraria*) and its wild relatives in Kenya: fruit and seed morphology. Genet. Resour. Crop Evol. **52**, 737–747 (2005)
3. Cutler, H.C., Whitaker, T.W.: Cucurbits from the Tehuacan caves. In: Byers, D.S. (ed.) Prehistory of the Tehuacan Valley, pp. 212–219. University of Texas Press, Austin, Texas, USA (1967)
4. Koffi, K.K., Anzara, K.G., Malice, M., Dje, Y., Bertin, P., Baudoin, J.P., Zoro, B.A.I.: Morphological and allozyme variationin a collection of *Lagenariasiceraria*(Molina) Standl. from Cote d'Ivore. Biotechnol. Agron. Soc. Environ. (Base) **13**, 257–270 (2009)
5. Sivaraj, N., Pandravada, S.R.: Morphological diversity for fruit characteristics in bottle gourd germplasm from tribal pockets of Telangana region of Andhra Pradesh, Indian. Asian Agri-Hist. **9**, 305–310 (2005)
6. Sharma, A., Sengupta, S.K.: Genetic diversity, heritability and morphological characterization in bottle gourd [(*Lagenariasiceraria*) (Mol.) standl.]. The Bioscan **8**(4), 1461–1465 (2013)
7. Panse, V.G., Sukhatme, P.V.: Statistical Methods for Agricultural Workers, p. 359. ICAR, New Delhi (1954)
8. Mishra, M.: Agronomic quality of *Lagenariasiceraria* (bottle gourd). IJSER **6**(9) (2021)
9. Kandasamy, R., Arivazhagan, E., Bharati, S.: Evaluation of growth and yield characters in bottle gourd (*Lagenariasiceraria* (Mol.) Standl.). Int. J. Pharmacogn. Phytochem. **8**(3), 4653–4655 (2019)
10. Tirumalesh, M., Mandal, J.: Evaluation of bottle gourd [*Lagenariasiceraria* (Mol) Standl.] genotypes during post rainy season in red and laterite zone of West Bengal. J. Crop Weed **14**(3), 185–188 (2018)
11. Kumar, L., Trivedi, J., Sharma, D., Janghel, A.K., Ansari, M.: Genetic diversity, heritability and agromorphological characterization in bottle gourd [*Lagenariasiceraria* (Mol.) Standl.]. Int. J. Curr. Microbiol. Appl. Sci. **6**, 264–271 (2018)
12. Singh, N., Singh, H.: Evaluation of bottle gourd [(*Lagenariasiceraria*) (Mol.) standl.] genotypes for various growth and yield. Theasian J. Hortic. **15**(1), 4–6 (2020)
13. Harika, M., Gasti, V.D., Shantappa, T., Mulge, R., Shirol, A.M., Mastiholi, A.B., Kulkarni, M.S.: Evaluation of bottle gourd genotypes (*Lagenariasiceraria* (Mol.) Standl.) for various horticultural characters. Karnataka J. Agric. Sci. **25**(2), 241–244 (2012)

14. Pandiyan, R., Pugalenthi, L., Sathyamurthy, V.A.: Evaluation of vegetative growth and yield performance of bottle gourds [*Lagenariasiceraria* (Mol.) Standl.]. Int. J. Genet. **11**(5), 594–596 (2019)
15. Leela, N.C., Singh, D.: Evaluation trial of bottle gourd [*Lagenariasiceraria* (Mol.) Standl.] under Prayagraj Agro climatic conditions. Int. J. Agric. Environ. Biotechnol. **13**(4), 517–520 (2020)
16. Kaur, B., Dhillon, N.P.S.: Characterization of various genotypes of *Cucumismelo* var. *acidulous* L. Veg. Sci. **35**, 212–214 (2008)
17. Uddin, A.F.M.J., Tahidul, M.I., Chowdhury, M.S.N., Shiam, I.H., Mehraj, H.: Evaluation of bottle gourd (*Lagenariasiceraria*) to growth and yield. Int. J. Biosci. **12**(5), 7–11 (2014)
18. Ilyas, M., Nabi, G., Ali, S., Anjum, M.M., Ali, N., Zaman, W., Jan, S., Samar, I.V., Sadiq, M., Akber, S.: Evaluation of bottle gourd varieties in agro climatic condition of Peshawar Valley. Int. J. Environ. Sci. Nat. Resour. **3**(1) (2017)
19. Iqbal, M., Usman, K., Arif, M., Jatoi, S.A., Munir, M., Khan, I.: Evaluation of bottle gourd genotypes for yield and quality traits. Sarhad J. Agric. **35**(1), 27–35 (2018)

Chapter 2
Diversity Analysis of Indigenous Germplasm of *Abelmoschus Essculentus L.* Moench (Okra) Under Garo Hills Condition of Meghalaya

Arindam Barman, Neha M. Sangma, and Richmond Marboh

2.1 Introduction

Okra or *Abelmoschus esculentus (L)* Monoech, an African-native plant from the Malvaceae family, is well known in India with the name "bhindi" or 'Lady's finger" [1]. In the Garo Hills of Meghalaya, it is called "dorai" locally. There are nine species of Abelmoschus, including *A. tuberculatus, A. ficulens, A. Manihot, A. angulosus, A. crinitus, A. esculentus, A. moschatus, A. tetraphyllus, and A. callei.* Only one of these nine species, Abelmoschus esculentus, which has the chromosomal number $2n = 2x = 130$, is commonly cultivated [2–4]. The crop is abundantly distributed around the world and is grown in warm temperate, tropical, and sub-tropical climates [5, 6]. With its use as buds, blooms, pods, stems, dry stems, fresh leaves, and seeds [1, 7–10], this crop is considered for numerous uses. Due to its superior nutritional value and palatable flavour, it is commonly farmed as a vegetable crop in India and on a commercial scale [11].

Okra is a herbaceous annual crop with bisexual flowers. It has erect vegetative growth with or without branches. Under some favourable conditions, the plant reaches 60 to 80 m in height with profuse branching. Thereafter, it becomes woody at maturity. The okra fruit is a capsule in either red or green or light green in colour. For the edible purpose, the fruits become ready to harvest during the fourth to sixth day after flowering. The cooking quality of the fruit remains high during this stage [12]. It is primarily an autogamous crop. However, 8.75 percent undergo cross-pollination [13].

Characterization of crops is an initial step in any crop improvement programme [14]. As a result, characterization refers to genotypes that have been identified and differentiated based on their nature or quality (traits) [15]. It offers data on genotype collection diversity and makes it possible to identify the particular gene bank

A. Barman (✉) · N. M. Sangma · R. Marboh
Department of Horticulture, North-Eastern Hill University, Meghalaya 794002, India
e-mail: arindamnehu@gmail.com

R. V. S. Uppaluri and L. Rangan (eds.), *Conservation of Biodiversity in the North Eastern States of India*, https://doi.org/10.1007/978-981-99-0945-2_2

accessions that are necessary for breeding and germplasm preservation [16]. Thereby, plant biodiversity conservation can be explored for enhanced productivity and human metabolism centric cultivation.

Due to the availability of such diverse germplasm, any breeding programme and successful crop improvement are mandatory. Such diverse germplasm is crucial for productivity, profitability, sustainability, and bolstering access to food and nutrition for the population pressure. Hence, characterization, evaluation, utilization, and conservation of germplasm is necessary. Several landraces and germplasm of indigenous okra are available in the Garo Hills region of Meghalaya. However, very less work has been conducted to characterize these okra germplasm on a scientific basis. To explore the genetic potential of indigenous germplasm lines of okra from Meghalaya (West Garo Hills), the undertaken research work reported in this book chapter has been formulated for the selection of superior genotypes in the okra crop improvement programme.

2.2 Materials and Methods

Ten native okra germplasm lines were obtained from various locations in the Tura region of the West Garo Hills District of Meghalaya, India, for the basic experimental framework. A randomized block design (RBD) field experiment with three replications was carried out. The okra seeds were planted in April 2019. The plants inside the rows were maintained with a 45 cm gaping, and the gaping between the rows was also maintained at 45 cm. From each germplasm and every replication, five plants were chosen randomly. A total of 150 plants were planted in a well-prepared plot with scientific cultural practices for proper growth and stand of the crop during the growth period.

The indigenous germplasm lines of okra were evaluated for 23 horticultural traits and seven biochemical traits viz total phenol, protein, carbohydrates, moisture, dry matter, fibre, and total chlorophyll, respectively. Among the twenty-three horticultural traits, thirteen traits have been quantitative, viz. plant height, H_p (cm), node of first blossoming (FB_n), node producing first fruit (FFP_n), fruit count per plant (FC/P) in days to 50% blossoming (DB_{50}), diameter of fruit in cm, (D_f), fruit's length in cm (L_f), days to 80% maturity (DM_{80}), seeds count per fruit (SC/F), nodes count per plant (NC/P), weight of 100 seed, SW_{100} (gm), internodes count per plant (INC/P), and fruit harvested per plant, FH/P (gm). Ten traits have been qualitative in nature, viz. plant growth habit, early plant vigour, epicalyx segment counts, branching habit, morphology of epicalyx segments, ridge counts per fruit, immature and mature fruit colour, fruit pubescence, and seed morphology. The observations were recorded according to NBPGR, New Delhi descriptors. Using SAS version 9.2 and the Panse and Sukatme approach, the mean data were statistically analysed using ANOVA [17]. Variance coefficient of the phenotypic and genotypic characters, respectively (CVP and CVG), heritability (h^2), and predicted mean percent of genetic advance (GAPM) were determined with the help of the Genres programme [18] and by applying the relevant formula suggested by the researchers [19]. According to a few researchers' reports, the h^2 was calculated using the method and expressed as a percentage [20]. Methods suggested by a few authors [21] were used to estimate known genetic advances in units (GA) and GAPM. The correlation coefficient was analysed using SAS software among the quantitative characters. Using Mahalanobis D^2 statistics, the genetic divergence in the germplasm was calculated. Details of the same have been presented elsewhere [22, 23].

2.3 Results

2.3.1 Variance Analysis

Tables 2.1 and 2.2, respectively, present the variance analysis for thirteen quantitative traits, namely H_p (cm), FB_n, FFP_n, FC/P, DM_{80}, D_f (cm), L_f (cm), DF_{50}, SC/F, NC/P, SW_{100} (gm), NI/P, and FH/P (gm and total phenol, protein, carbohydrates, moisture, dry matter, fibre, and total chlorophyll levels as seven biochemical parameters on ten indigenous germplasm lines of okra. For each estimated parameter, significant variations existed between the genotypes. It was evident that the genotypes chosen for the study had a substantial variation for each attribute.

Similarly, it was found that genotype G1 recorded the highest phenol content of 19.5 mg/100 g, whereas genotype G6 showed the lowest phenol of 1.3 mg/100 g with a grand mean of 8.10 mg/100 g, respectively. The protein content altered from 1.592 to 3.257% with a grand mean of 2.224%. Genotype G7 had the highest protein content (3.257%) and genotype S10 had the lowest protein level (1.592%). The carbohydrates altered from 6.335 to 14.506% with a grand mean of 11.23%. The maximum carbohydrate content was recorded in genotype G4 (14 506%), and the lowest was recorded in genotype G1 (6.335%). Genotype G5 had the highest moisture content (90.6%), followed by genotype G3. Genotype G1 had the lowest moisture content (87.2%). The grand mean for the case was (89.7%). The highest dry matter content was recorded in genotype G1 at 12.8%, while the lowest was recorded in genotype G5 at 9.4% with a grand mean of 10.47%. Fibre content ranged from 26.3 to 37.5% with a grand mean of 31.3%. In terms of fibre content, genotype G7 was noted to have the highest level (37.5%) and genotype G2 was noted to have the lowest value (26.3%). The total chlorophyll content among the genotypes ranged from 0.047 to 0.103 mg/gm, with a grand mean of 0.067 mg/gm. Genotype G4 had the highest chlorophyll concentration (0.103 mg/gm), while genotype G9 had the lowest chlorophyll content (0.047 mg/gm).

Table 2.1 Analysis of variance data for thirteen quantitative horticultural types in okra

Source of variation	Degree of freedom	Value of Mean squares												
		DB_{50}	FB_n	FFP_n	L_f	D_f	FC/P	H_p	DM_{80}	SC/F	SW_{100}	NC/P	INC/P	FH/P
Replication	2	7.6	0.5	0.5	0.07	0.01	3.0	1.19	2.4	29.6	8.8	2.4	2.4	1812.0
Genotypes (Treatment)	9	397.0	4.4	4.4	16.9	0.1	53.1	2302.0	2.8	347.5	377.3	2.8	2.8	110,195.3
F- value		7.42	3.6	3.6	37.8	6.0	37.8	1403.2	4.0	3.9	102.1	4.0	4.0	260.8

DB_{50}-Days to 50% blossoming; FB_n-node of first blossoming; FFP_n-node producing first fruit; L_f- fruit's length (cm); D_f-fruit diameter of fruit (cm); FC/P-fruit count per plant; H_p-plant height (cm); DM_{80}-days to 80% maturity; SC/F- seeds count per fruit; NC/P- nodes count per plant; SW_{100}-weight of 100 seed (gm); INC/P- internodes count per plant and FH/P-fruit harvested per plant (gm)

Table 2.2 Data of seven horticultural biochemical characters of okra and their variance analysis

Source of variation	Degree of freedom	Value of Mean squares								
		Phenolic (mg/100 gm)	Protein (%)	Carbohydrates (%)	Moisture (%)	Dry matter (%)	Fibres (%)	Chlorophyll (mg/g)		
								(a)	(b)	Total
Replication	2	0.6	0.01	1.1	0.7	0.8	2.0	0.000009	0.00003	0.00002
Genotypes (Treatment)	9	66.6	0.1	20.9	4.9	3.7	48.9	0.0007	0.0007	0.0010
F- value		90.4	9.8	4.9	6.0	3.9	16.6	56.3	38.3	55.3

DB_{50}-Days to 50% blossoming; FB_n- node of first blossoming; FFP_n-node producing first fruit; L_f-fruit's length (cm); D_f- diameter of fruit (cm); FC/P-fruit count per plant; H_p-plant height (cm); DM_{80}-days to 80% maturity; SC/F-seeds count per fruit; NC/P- nodes count per plant; SW_{100}-weight of 100 seed (gm); INC/P- internodes count per plant and FH/P-fruit harvested per plant (gm)

2.3.2 *Qualitative Traits*

The qualitative morphological characters among the okra accessions were estimated and have been depicted in Fig. 2.1. All the genotypes were found to exhibit strong early plant vigour as a significant characteristic. For plant growth habits, all the genotypes exhibited erect plant growth habit except genotype G7 which exhibited medium growth habit. Also, the genotypes G1, G5, and G7 exhibited low branching habit. The genotypes G1, G2, G3, G4, G5, G7, G8, G9, and G10 exhibited 8 to 10 epicalyx segment counts, while the genotype G6 showed 5 to 7 epicalyx segment counts. Lanceolate morphology of epicalyx segment counts was observed in genotypes G2, G3, and G6. The Genotype G7 showed dark green as an immature fruit colour. The Genotype G7 showed 5–7 ridge counts per fruit, whereas the rest of the genotypes exhibited 8–10 ridges. Therefore, 8 to 10 showed prominent characters for ridge counts per fruit. Downy showed as a prominent character for fruit pubescence. Genotypes G1, G2, G4, G5, and G10 showed the yellowish green colour of mature fruits, whereas genotypes G3, G6, and G9 showed green colour with red patches of mature fruits. Other genotypes such as G7 and G8 showed the green colour of mature fruits. Round seed shape was observed in all investigated genotypes estimated.

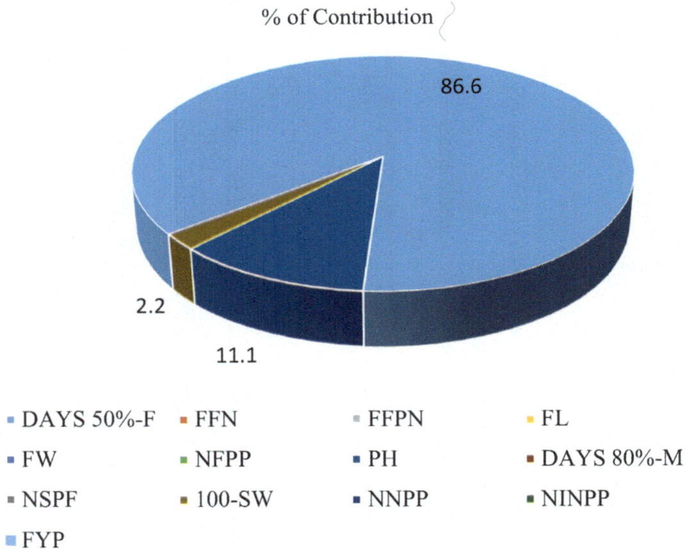

Fig. 2.1 Pie chart depicting percentage of morphological characters' contribution towards divergence

Table 2.3 Genotypic and phenotypic correlation coefficients data of seven biochemical characters

Characters		Total phenolics	Protein content	Total carbohydrates	Moisture content	Dry matter	Fibre content	Total chlorophyll
Total phenolics	G	1.0	−0.01	−0.8*	−0.7*	0.7*	−0.0	−0.2
	P	1.0	0.0	−0.6*	−0.5*	0.5*	−0.01	−0.2
Amount of protein	G		1.0	0.2	−0.6*	0.7*	0.9*	0.1
	P		1.0	0.2	−0.3	0.3	0.6*	0.1
Total carbohydrates	G			1.0	0.6*	−0.5*	0.4	0.6*
	P			1.0	0.2	−0.1	0.2	0.4
Moisture content	G				1.0	−1.0*	−0.4	0.1
Dry matter	P				1.0	−0.9*	−0.3	0.1
	G					1.0	0.4	−0.1
Fibre content	P					1.0	0.3	−0.1
	G						1.0	0.2
Total chlorophyll	P						1.0	0.2
	G							1.0
	P							1.0

* Significant at 5%

2.3.3 *Coefficient of Variation*

The CVG and CVP were obtained for FH/P and H_p, respectively, while NC/P had the lowest CVG and CVP. On the other hand, the highest CVG and CVP for biochemical parameters were recorded for total phenol and protein content, respectively, and the minimum value of CVG and CVP was attained for moisture content (Table 2.3).

2.3.4 *Heritability (h^2)*

The h^2 values were generally medium to high and altered from 46.4% to 99.7%. FB_n and FFP_n-node recorded the least heritable traits (both traits with 46.4% each). However, H_p (99.7%) exhibited the most heritable traits followed by FH/P (98.8%) (Table 2.3). The related GAPM has shown a wide range of values (10.5 to 74.4%). A higher FH/P (74.4%) recorded a value of GAPM and then H_p (66.2%), whereas minimum values were found for NC/P (10.5%), respectively. However, in biochemical parameters, the h^2 was generally altered from 49.32% to 96.75%. The highest h^2 was exhibited by total phenolics (96.75%) followed by total chlorophyll (94.77%), whereas the lowest h^2 was exerted by dry matter content (49.32%), respectively. As shown in Table 2.3, all characters had medium to high h^2 values, and the maximum GAPM value was found in the total phenolics (116.9%), followed by the protein content (62.5%), total chlorophyll (52.2%), total carbohydrates (32.4%), fibre content (23.3%), and dry matter (13.1%). Moisture content (2.09%) had the least value in the case.

In correlation coefficient analysis for phenotypic correlation (R_p), the DB_{50} has shown a highly significant positive relation with FB_n (0.639), DM_{80} (0.868), D_f (0.614), and FFP_n (0.639). FC/P (0.75) and DM_{80} (0.673) have shown a highly significant positive relation with FB_n and FFP_n. L_f had a significant positive association with H_p (0.578). The D_f had a significant positive relation with DM_{80} (0.550). The reported FH/P (0.688) exhibited a highly significant positive relation with FC/P. H_p had a significant positive association with SW_{100} (0.502) and FH/P (0.632), whereas, for genotypic correlation, DB_{50} have shown a highly significant positive association with FB_n (0.794), DM_{80} (0.956), FFP_n (0.794), and D_f (0.855). FB_n and FFP_n had a significant and positive link with D_f (0.705) and DM_{80} (0.875), but a negative relationship with the NC/P (-0.565) and INC/P (−0.565). L_f had a highly significant positive outcome with H_p (0.604) and SC/F (0.524). D_f exhibited a strongly significant positive relation with DM_{80} (0.911) and SC/F (0.524). FC/P have shown strong positive outcomes with FH/P (0.711) and significant negative relation with NC/P (−0.514) and INC/P (−0.514), respectively. SC/F (0.621), SW_{100} (0.508), and FH/P (0.638) showed a significant positive association with H_p. SC/F has shown a significant positive response with NC/P (0.594) and INC/P (0.594).

The genotypic correlation coefficient (R_g) analysis for biochemical parameters has been presented in Table 2.4. The table conveyed that for the parameter, the total phenolic showed highly positive significant correlation with dry matter (0.769) and highly negative significance correlation with total carbohydrates (−0.857) and moisture content (−0.792). Protein content exhibited a moderate negative correlation with moisture content (−0.618), but a substantial positive correlation with dry matter (0.714) and fibre content (0.900). However, moisture content (0.618) and total chlorophyll (0.651) exhibited a positive significant correlation, whereas total carbohydrates to dry matter content (−0.585) resulted in a negative significant correlation. The total phenolic to dry matter content (0.512) exhibited a positive significant correlation and a negative significant correlation with total carbohydrates (−0.664) and moisture content for the R_p, respectively. Moisture content (0.593) had a highly negative significant correlation with dry matter content (−1.018). Protein content had a highly positive significant correlation with fibre content (0.667). Moisture content showed a highly negative significant correlation with dry matter (−0.969).

Table 2.4 Clustering pattern of ten genotypes data of okra germplasm

Cluster	Total genotypes count	Genotypes
I	9	G_1, G_4, G_6, G_8, G_3, G_9, G_5, G_2, G_7
II	1	G_{10}

Fig. 2.2 Dendrogram depicting the morphogenetic relationships among ten genotypes based on Mahalanobis

2.3.5 *Genetic Divergence Analysis*

Among ten genotypes, genetic divergence for 13 quantitative types was estimated by Mahalanobis D^2 analysis [24]. Ten genotypes were categorized into two groups and findings have been presented in Table 2.4 and Fig. 2.2. Cluster I had the most genotypes (9 genotypes) and Cluster II had the least (1 genotype) value, respectively.

2.3.6 *Cluster Means of Characters*

Cluster II scored the highest DB_{50} (81.33) and cluster I score the lowest DB_{50}(74.07). Cluster II had the highest FB_n and FFP_n readings (8.33 each), and also 7.0 each resulted for cluster I. Cluster II (22.03) and cluster I (22.03) had the maximum L_f (17.33). Cluster II (2.20) had the highest cluster mean for D_f, followed by cluster I (2.16). Cluster II had the highest FC/P (19.0) and highest average H_p value (148.93), while cluster I had the lowest FC/P (17.07) and Hp (78.98). Cluster II (100.60) followed by cluster I (92.2) existed in terms of the highest cluster mean for DM_{80}. While cluster II (66.5) and cluster I (80.2) had the highest SW_{100} cluster mean, cluster II (87.3) and cluster I (80.2) had the highest SC/F with cluster mean (48.5). Moreover, cluster I had the greatest NC/P (11.7) and that of cluster II (11.3), while cluster I also recorded the highest INC/P (10.7) and with a value of 10.3 for cluster II. Cluster II (855.9) and cluster I (490.0) both had the highest FH/F values.

The data recorded on ten germplasm lines for seven biochemical characters, namely total phenolic, protein content, total carbohydrates, moisture content, dry matter content, fibre content, and total chlorophyll were used to classify the okra lines based on Mahalanobis D^2 statistics. Ten genotypes of okra germplasm were clustered according to genetic divergence. There were four groups created from the genotypes. Cluster I had the most genotypes, counting 5, cluster II had 3, and clusters III and IV had the fewest genotypes, counting one each.

Cluster III recorded the highest cluster mean for total phenolics (19.5), followed by cluster IV (8.6), and cluster I (8.08), and cluster II reported the lowest cluster mean (4.16). The highest protein content was observed in cluster III (14.6), then in cluster II (13.8), cluster I (12.2), and the lowest in cluster IV (6.0). For total carbohydrates, the maximum average of the cluster was recorded in cluster II (13.4) and cluster IV (12.6) and lowest for cluster III (6.3), respectively. Cluster II (90.2), cluster I (89.5), and cluster IV (89.4) had the greatest cluster means for moisture content, whereas cluster III had the lowest cluster mean (86.5). Additionally, cluster III (12.8), cluster IV (10.5), and cluster I (10.4) had the greatest cluster averages for dry matter content, whereas cluster II (9.7) had the lowest cluster mean, while the fibre content was highest in cluster III (33.8), then cluster II (33.5) and finally cluster I (30.0) and the minimum in cluster IV (28.3). Maximum cluster means for total chlorophyll were recorded in clusters II (0.08) and IV (0.08), III (0.06), and finally I (0.05).

2.3.7 Intra and Inter-Cluster Distance

Cluster I (0.00) had the lowest inter-cluster D^2 value, whereas cluster II had the highest (37,696.7). Furthermore, cluster II with one genotype had the lowest intra-cluster D^2 value (0.00), whereas the highest intra-cluster D^2 value (6231.8) with nine genotypes was recorded for cluster I. A relative contribution of at least thirteen characters was assessed for the total genetic divergence indicating that the FH/P has exhibited maximum genetic divergence (86.6) with H_p (11.1) and SW_{100} (2.2), respectively. Moreover, the lowest divergence share was observed for DB_{50}, FB_n, FFP_n, L_f, D_f, FC/P, DM_{80}, SC/F, NC/P, and INC/P.

The D^2 values of the inter- and intra-clusters altered from 377.6–1295.6 and 0.0–153.4, respectively. Cluster II had the maximum intra-cluster distance (153.4), Cluster I (122.8), and Clusters III and IV had the lowest. Cluster II and cluster III of 1295.6 value have been observed to have the highest inter-cluster distances, then between clusters II and IV (655.4), and finally between clusters III and IV (509.3), while clusters I and IV had the lowest (377.6), followed by clusters I and II (407.7) and clusters I and III (490.9). The relative contribution of seven biochemical characters was evaluated for the total genetic divergence which shows that the total chlorophyll has the highest genetic divergence (42.2) among all the characters followed by fibre content (28.8) and total phenolics (13.3). Moreover, protein content, total carbohydrates, moisture content, and dry matter content each aided the least in genetic divergence.

2.4 Discussion

Genotype G7 exhibited the earliest flowering, whereas genotype G2 showed late flowering. Early flowering tends to give early fruiting and early marketable harvest in okra studied as confirmed in relevant prior art [25, 26]. The Genotype G7 produced the earliest flowering at the lowermost node among all the genotypes resulting in early flowering and early fruiting nodes. Few authors reported significant variation for FB_n (4.87 to 8.2) [26]. The variation of L_f could be due to genetic factors, as well as a direct or indirect influence by environmental factors such as rainfall, humidity, temperature, etc. during the fruiting period. Conclusively, it was found that the maximum and minimum L_f was observed for G10 and G3, respectively. Few researchers have reported that L_f in okra altered from 12.68 to 25.02 cm [27]. Similarly, significant variation among genotypes of okra for green pod length (11.0–14.7 cm) was also recorded in a relevant prior art [27]. Among the genotypes studied, the variation for D_f may be attributed to temperature, humidity, rainfall, soil fertility, nutritional requirement, etc. Few authors detected an important variation between various genotypes of okra in terms of green pod diameter (1.15–2.34 cm) [26]. FC/P is directly proportional to the yield component. Therefore, plants with more fruits are desirable for crop improvement programme [27]. Few researchers reported a broad range of variation for FC/P (32.84) [25]. Likewise, significant variation for FC/P (10.27 to 19.07) was reported by few researchers [27]. The variation in H_p among various genotypes might be associated with their environmental factors, soil fertility, nutritional requirement, individual genetic makeup, and other factors. Researchers suggested that the H_p of okra ranged between 38–147.65 cm [28]. Few authors have obtained a wider variation for H_p (mean = 80.8 cm) [29]. According to one study, the Arka Anamika okra had the highest H_p value of 90.62 cm [30]. The variation for DM_8 among the genotypes studied might be attributed to environmental factors such as fluctuation in night and day temperature, humidity, rainfall and nutritional factors.

The variation in SC/F among the genotypes might be due to rainfall fluctuation, genetic inconsistency among the genotypes along with inconvenience practices during the harvesting of dried pods. Few authors observed a broad range of variation for SC/F [29]. The SC/F ranged from 35.2–100.6 studied by a few authors [27]. For SW_{100}, the variation among the ten indigenous germplasm of okra might be due to genetic factors of the genotypes. Researchers have also reported similar genetic variability for SW_{100} in okra [27]. The data collected on NC/P and INC/P is noteworthy as it is directly or indirectly contributed to the yield. A study reported that reasonable variation was achieved for NC/P (14.58) [31]. Considerable genetic variability among the genotypes for INC/P was observed [32]. Also, FH/P depends upon the FC/P and individual weight of the fruit. Therefore, a higher FH/P might be associated with higher FC/P and higher individual fruit weight. Moreover FH/P variations might be influenced by environmental factors as well as agronomical practices. Few researchers have reported similar findings for FH/P in okra [25, 26].

Few authors have observed that the phenol content in the okra fruit of the Arka Anamika cultivar was recorded with 167.62 mg/100 g and they found it to be significantly higher than the Sinnova with 112.27 mg/100gm and shagun with 106.26 mg/100 g, respectively [33]. The variation in protein among the ten genotypes might be attributed due to genetic makeup among the genotypes as well as environmental factors, such as pH that influences the enzymes involved in the synthesis of various substances in okra fruit. Researchers have reported a carbohydrate content of 11.17% in okra fruit [34]. The mature fruit of okra had the highest total carbohydrate content (11.17%), that of the leaf (8.83%) and young fruit (7.05%) as studied by a few authors [35].

The okra fruit's moisture content ranged between 87.7 and 90.9% as per a few studies [36, 37]. Few authors reported variation in moisture content in okra fruits might be attributed to maturity stages where the young mature fruits recorded the highest moisture content followed by leaves and lowest in mature fruits [38]. Variation of dry matter content in okra fruits depends upon the advanced growth stages. As a result, as the growth stage advances, the dry matter content rapidly increases. Authors have reported that the fibre content in okra fruits is directly correlated with the advanced growth stages [33]. Similarly, few researchers have reported the highest fibre content in mature okra fruit followed by young fruit and lowest in leaves [35]. Variation of total chlorophyll content in okra might be attributed to the advanced growth stages in which the highest chlorophyll content was recorded in young fruit and the lowest was recorded in mature fruit. According to a few authors, mature fruit had a lower chlorophyll level of 0.11 mg/0.1 gm. However, mature leaves had a higher chlorophyll content of 1.56 mg/0.1 g [35].

The obtained CVP was greater than the other characters associated with their corresponding CVG. Thereby, it indicated that these character variations have been caused by environmental factors in addition to genotypes. Hence, selecting crop improvement based on phenotypes may lead to their expression as it depends more on environmental factors. The CVP were greater than their corresponding CVG denoting environmental factors that influence their expressions to a certain degree, as being delineated by a research group [39]. Few authors have reported that selection could be successful for future crop improvement programmes. In this case, high heritability is combined with high genetic advance and high CVG [40].

The R_g study aided in the identification of the characters that have a significant correlation with yield. As a result, understanding the nature of the correlation of various characters is of great interest in crop improvement programmes, as it aids breeders to determine the relative significance of harvesting components and their relevant parameters. Several studies on R_g and R_p analysis among yield contributing characters including plant growth characters, fruit characters, and quality of okra have been reported in India and abroad by a research group [41]. As a result, in the current study, it can be concluded that the R_g values were maximum than the equivalent R_p value. This pointed towards a strong intrinsic relationship between various character pairs. The authors also reported that the R_g between yield and its attributing traits were stronger than the R_p [26].

The genetic divergence study assisted in the identification of the distant parents among the genotypes for selecting the superior genotypes in the parental material. Genotypes from the same geographical origin were placed in different clusters and eventually resulted in a wider diversity. Hence, this may be due to the frequent exchange of germplasm between different geographical regions. Despite the mean values obtained for the various number of genotypes in each cluster, which could not be compared statistically, they have been compared to get a relative idea of diversity among the clusters. Therefore, it was feasible to identify the characters affecting the divergence based on the range of means for each character.

The maximum inter-cluster D^2 value indicates that the genotypic trait for cluster II was found to be highly divergent from cluster I. Further, they are not closely associated while the lowest inter-cluster D^2 value indicates that the genotypes belonging to cluster I were very close to one another. As a result, the genotypes of okra germplasm within the same cluster appeared to be closely related due to lesser intra-cluster distance than that of inter-cluster. In conclusion, there was a bigger difference in the distance between the inter and intra clusters. According to this assessment, genotypes within a cluster are homogeneous and display less divergence than those occurring in other clusters as they have a similar genetic makeup. In order to achieve transgressive segregation compounds for yield with broad spectrum genetic diversity and other constituent qualities that differentiate high-yielding genotypes in okra genotypes that belong to a larger inter-cluster distance are chosen and could be used in hybridization programmes [42]. The highest relative contribution of the traits to divergence shows that these traits are genetically diverse and should be taken into account in due course of the selection of parents for programmes involving hybridization with genotypes as recommended by a few authors [43].

2.5 Summary

For the characterization of the genotypes, 23 morphological characters (13 quantitative in nature and 10 qualitative in nature) were observed at different growth stages following the descriptors developed for okra by the National Bureau of Plant Genetic Resources Institute (NBPGR), New Delhi. Five plants were chosen arbitrarily from each genotype and for each replication for the collection of data on plant attributes.

Alternate genotypes reported significant variation in all investigated characters. The genotype G10 exhibited the highest H_p (148.93 cm) and G5 had the highest NC/P (14) and INC/P (13), respectively. DB_{50} was highest in genotype G2 with 87.6 days. Genotype G7 produced FB_n at the lowermost node (5). Maximum L_f was recorded in genotype G10 (22.03 cm). Maximum D_f was recorded in genotype G6 (2.53 cm). The best FC/P was achieved in genotype G8 (24.66). Maximum SC/F was found in genotype G4 (99). Maximum SW_{100} was obtained in genotype G10 (66.5). DM_{80} was highest in genotype G3 with 103 days and genotype G10 recorded the highest in FH/P (855.92 gm), respectively.

The early plant vigour, plant growth habit, Immature fruit colour, fruit pubescence, epicalyx counts, ridge counts per fruit, and seed morphology of ten okra germplasm lines were evaluated. Maximum variability was observed in branching habit, morphology of epicalyx segments, and mature fruit colour. Genotype G7 exhibited a medium branching habit with dark green immature fruit colour and showed five to seven ridge counts per fruit. Low branching habit was observed in G1, G5, G7, and G10. Five to seven epicalyx segment counts were observed in G6, G2, G3, and G6 showing the lanceolate morphology of epicalyx segments. Genotypes G6 and G7 showed five to seven ridge counts per fruit. Moreover, the genotypes G3, G6, and G9 showed green but with red patches in the mature fruit colour characteristics.

CVP was highest for FH/P (36.53) and total phenol (58.76), whereas the CVG was maximum for FH/P (36.32) and total phenol (58.80), respectively. FB_n, DB_{50}, L_f, FFP_n, D_f, FC/P, SC/F, SW_{100}, DM_{80}, H_p, NC/P, and INC/P have achieved the lowest CVG and CVP values that altered from 7.20 to 32.24. For biochemical characteristics such as protein content, total carbohydrates, moisture content, dry matter, fibre content, and total chlorophyll, the CVG and CVP altered from 1.30 to 41.09, respectively.

Among the quantitative morphological characters, the h^2 values altered from 46.4% to 99.7%. The highest h^2 was exhibited by H_p (99.7%) followed by FH/P (98.8%), SW_{100} (97.1%), FC/P, and L_f (92.4%). The lowest h^2 was exerted by FB_n and FFP_n with 46.4% each. Thus, all the characters showed medium to high h^2 estimates. However, in biochemical characters, the h^2 altered from 49.32% to 96.75%. The highest h^2 was exhibited by total phenol (96.75%) followed by total chlorophyll (94.77%). The lowest h^2 was achieved for the dry matter content (49.32%). Thus, all the characters showed medium to higher h^2 estimates.

The related GAPM was altered from 10.5 to 74.4. The highest value of GAPM was obtained for FH/P (74.4%), H_p (66.2%), FC/P (47.4%), SW_{100} (44.9), L_f (25.8%), DB_{50} (24.2%), DM_{80} (17.1%), FB_n (20.1%), D_f (17.5%), FFP_n (20.1%), SC/F (16.7%), NIN/P (11.4%), and NC/P (10.5%), respectively. For biochemical characters, the maximum value of GAPM was observed for total phenol (116.9%) followed by protein content (62.5%), total chlorophyll (52.2%), total carbohydrates (32.4%), fibre content (23.3%), dry matter (13.1%), and moisture content (2.09%).

For morphological characters, PC-1, PC-2, PC-3, and PC-4 were described for the relative contribution of each variable which accounts for 87.65% of the total variation. PC-1 reported about 35.50% of total variability and was positively loaded on FB_n (0.94), DB_{50} (0.87), D_f (0.69), FFP_n (0.94), and DM_{80} (0.89). This PC exhibited the important contribution of variables (quantitative traits) towards variability. PC-2 accounts for 25.44% of the total variability and was positively loaded on L_f (0.73), H_p (0.54), SC/F (0.73), NC/P (0.76), and INC/P (0.76). PC-3 accounts for 18.58% of the total variability; PC-4 accounts for 8.12% of the total variability. On the other hand, for biochemical traits, the relative contribution was determined by two principal components (PC-1 and PC-2) that accounted for about 80.45% of the total variation. With total phenolics (0.76) and dry matter (0.96), PC-1 conveyed favourable correlation, and it accounted for 46.34% of the entire variation. This PC showed the important contribution of variables (biochemical traits) toward variability. PC-2 accounts for 34.10% of the total variation.

For phenotypic correlation, DB_{50} had a significant positive association with D_f (0.614), FB_n (0.639), FFP_n (0.639), and DM_{80} (0.868). FB_n and FFP_n have shown a significant positive association with FC/P (0.75) and DM_{80} (0.673). L_f and H_p had a strong positive relationship with a value of 0.578. D_f strongly correlated with DM_{80} (0.550). FC/P showed a significant positive association with FH/P (0.688). H_p has shown a significant positive association with that of SW_{100} (0.502) and FH/P (0.632).

For genotypic correlation, DB_{50} has shown a highly significant positive association with DM_{80} (0.956), FB_n (0.794), D_f (0.855), and FFP_n (0.794) FB_n and FFP_n showed a positive association with D_f (0.705), DM_{80} (0.875) and displayed negative relation with NC/P (−0.565) and INC/P (−0.565). L_f had a significant positive outcome with H_p (0.604) and SC/F (0.524). D_f showed significant positive results with DM_{80} (0.911) and SC/F (0.524). Substantial negative correlations were observed between NC/P (−0.514) and INC/P(−0.514), while significant positive correlations between FC/P and FH/P (0.711) were observed. H_p had shown significant positive results with SC/F (0.621), SW_{100} (0.508), and FH/P (0.638). SC/F showed a significant positive relation with NC/P (0.594) and INC/P (0.594).

Total phenol exhibited a highly significant negative relationship with total carbohydrate (−0.857) and moisture content (−0.792), but a substantial positive correlation with dry matter (0.769). Protein content exhibited a highly significant negative association with moisture content (−0.618); however, dry matter (0.714) and fibre content (0.900) showed a significantly strong positive correlation. Moreover, total chlorophyll (0.651) and moisture content (0.618) resulted in a substantial positive correlation. On the other hand, a negative correlation with dry matter (−0.585) has been exerted in terms of total carbohydrates. Moisture content showed a highly inverse relationship with dry matter content (−1.018). A significant negative correlation with total carbohydrates (−0.664) and moisture content (0 593) was apparent, whereas a substantial positive correlation with dry matter (0.512) was observed for total phenol. Protein content had a positive significant correlation with fibre content (0.667). Moisture content showed a significantly high negative correlation with dry matter (−0.969).

Okra genotypes were classified into two clusters. The highest number of genotypes were classified into cluster I (9) viz. G_1, G_4, G_6, G_8, G_3, G_9, G_5, G_2, and G_7, followed by cluster II (1), viz. S_{10}. The maximum and the least inter- cluster D^2 values were observed in Cluster II (37,696.7) with nine genotypes and Cluster I (0.00) respectively. Therefore, the maximum inter-cluster D^2 value indicates that the genotypes belonging to cluster II were found to be highly divergent from cluster I and are not so closely related, whereas the lowest inter-cluster D^2 value indicates that the genotypes belonging to cluster I were very close to each other, respectively. While cluster I (6231.8) with nine genotypes had the highest intra-cluster D^2 value, cluster II (0.00) with one genotype had the lowest. Moreover, in biochemical parameters, all the genotypes have been classified into four clusters. A total count of 5 genotypes for Cluster I and Cluster II with 3 genotypes had the maximum genotype counts. However, Cluster (III) and Cluster (IV) had the lowest genotype counts with only one genotype. The inter and intra-cluster D2 values altered from 377.6 to 1295.6 and 0.0 to 153.4, respectively. Clusters II (153.4) and I (122.8) had the maximum intra-cluster distance, while clusters III and IV had the minimum intra-cluster distance. Additionally, the distance between clusters was maximum between clusters II and III (1295.6), clusters II and IV (655.4), and then clusters III and IV (509.3), whereas clusters I and IV (377.6), clusters I and II (407.7), and finally cluster I and III recorded the lowest inter-cluster distance (490.9). Moreover, comparing the intra-cluster distance conveyed that the distances have been higher. Hence, it can be concluded that the genotypes within the cluster had a comparable genetic makeup and were probably homogeneous and less diverging than those found in other clusters.

2.6 Conclusions

Greater variability exists for economically important characters. Henceforth, they provide a good scope for genetic improvisation of the crop. However, the mean performances for harvested fruit and its component characters (DB_{50}, FC/P, L_f, D_f, DM_{80}, and FH/P) and different biochemical characters (phenol, protein, carbohydrates, fibre, and total chlorophyll) of seven germplasm lines, namely G4, G5, G6, G7, G8, G9, and G10 (presented in Table 2.4) were selected as the superior and most promising genotypes for the okra crop improvement programme. This investigation furnished a futuristic understanding of the advancement of research in okra for both agronomical and biochemical characteristics. The diversity present among the okra genotypes for various characters can be used to enhance desirable traits through the selection and hybridization breeding programme. The existing cultivars can be improved in a hybrid breeding programme by the superior genotypes identified in the present investigation. The genotype G7 exhibited a unique leaf structure, possessed tolerability, and was resistant to YVMV. Therefore, for further crop improvement programme, genotype G7 can be used as a YVMV-resistant source. The seeds collected from the okra genotypes can be used for germplasm conservation. Research on scientific cultivation methods and integrated management of diseases of okra may be adopted in order to reduce loss, maximize the yield and increase the economic return. Thereby, the demonstrated research directions in the article can be adopted for inline breeding research for similar crops of commercial value.

Acknowledgements The authors express their sincere thanks to the Department of Horticulture, North-Eastern Hill University, Tura campus, Meghalaya, for their tremendous support.

References

1. Schippers, R.R.: African indigenous vegetables. In: An Overview of the Cultivated Species. *Natural* Resources Institute/ACP-EU Technical Centre for Agricultural and Rural Cooperation, Chatham, UK, pp 103–118 (2000)
2. Joshi, A.B., Gadwal, V.R., Hardas, M.W.: In: Hutchinson, J.B. (ed.) Evolutionary Studies in World Crops, Diversity and Change in the Indian Subcontinent, pp. 99–105. Cambridge (1974)
3. Singh, H.B., Bhatnagar, A.: Chromosome number in okra from Ghana. Indian J. Genet. PlantBreed **36**, 26–27 (1975)
4. Kumar, S., Dagnoko, S., Haougui, A., Ratnadass, A., Pasternak, D., Kouame, C.: Okra (*Abelmoschus* spp.) in West and Central Africa, Potential and Progress on its Improvement. Afr. J. Agric. Res. **5**(25), 3590–3598 (2010)
5. Tindal, H.D.: Vegetables in the Tropics. Mac- millan Education Limited Houndmills Hampshire, p. 533 (1983)
6. Hammon, S., Van Sloten, D.H.: Characterization and evaluation of okra. In: The Use of PlantGenetic Resources, pp. 173–174 (1989)
7. Martin, F.W.: A second edible okra species and its hybrid with common okra. Ann. Botany **50**, 277–283 (1982)
8. Kalra, C.L., Pruthi, J.S.: Chemistry and technology of okra (*Hibiscus esculents* L.). Central Food Technology Research Institute, Indian Food Packer, pp. 7–57 (1984)
9. Markose, B.L., Peter, K.V.: Review of research on vegetables and tuber crops. Directorate of extension, KeralaAgriculture University, Mannuthy, 43p. (1990)
10. Preston, S.R.: Aibika/Bele-*Abelmoschus Manihot* (L.) Medik. International Plant GeneticResources Institute (IPGRI) Publication, p. 97 (1998)
11. Lamont, W.: Okra a versatile vegetable crop. Hort Technol. **9**(2), 179–184 (1999). http://horttech.ashspublications.org/content/9/2/179
12. Sistrunk, W.A., Jones, L.G., Miller, J.C.: Okra pod growth habits. Proc. Am. Soc. Hortic. Sci. **76**, 486–491 (1960)
13. Purewal, S.S., Randhawa: Genetic of quantitative character in bhindi (*Abelmoschus esculentus* L. Moench). Indian J. Agric. Sci. **17**, 129–136 (1947)
14. De-Vicente, M.C., Guzmán, F.A., Engels, J., Ramanatha Rao, V.: Genetic characterization and its use in decision making for the conservation of crop germplasm. In: The Role of Biotechnology, VillaGualino, Turin, Italy, 5–7, 63 p. (2005)
15. Merriam-Webster: Webster's ninth new collegiate dictionary. Merriam-Webster Inc. publishers. Springfield, Massachusetts, USA (1991)
16. Brown, A.H.D., Marshall, R., Frankel, O.H., Williams, J.T.: The use of plant genetic resources. Cambridge University Press, London (1990)
17. Panse, V.G., Sukhatme, P.V.: Statistical Methods of Agricultural Workers. ICAR Publication, New Delhi (1954)
18. GENRES: Statistical package for genetical researchers. Pascal International Software Solutions, Version 7.01 (1994)
19. Burton, G.W., Devane, E.H.: Estimating heritability in tall fescue (*Festuca aurandinacea*L.) from replicated clonal material. Agron. J. **45**, 478–481 (1953)
20. Weber, C.R., Moorty, B.R.: Heritable and non-heritable relationships and variability of soil content and agronomic characters in F2 generation of soya-bean crosses. Agron. J. **44**, 2 (1952)
21. Johnson, H.W., Robinson, H.F., Comstock, R.E.: Estimates of genetic and environmental variability in Soya-bean. Agron. J. **47**, 314–318 (1995)
22. Mahalonobis, P.C.: On the generalized distance in statistics. In: Proceedings of the National Academy of Sciences (India), vol. 2, pp. 49–55 (1936)
23. Rao, C.R.: Advanced Statistical Methods in Biometrical Research, 1st edn. (Reprinted with correction 1960). Wiley Haffner press, New York (1952)
24. Lush, J.L.: Heritability of quantitative characters in farm animals. Proc. Eight Int. Cong. Gen. **43&49**, 356–375 (1949)

25. Dhankhar, B.S., Dhankhar, S.K.: Genetic variability, correlation and path analysis in Okra (*Abelmoschus esculentus* (L.) Moench). J. Veg. **29**(1), 63–65 (2002)
26. Dhankhar, B.S., Dhankhar, S.K.: Variability studies in okra (*Abelmoschus esculentus* (L.) Moench). Haryana J. Hortic. Sci. 31(1&2), 82–84 (2002)
27. Singh, B., Pal, A.K., Singh, S.: Genetic variability and correlation analysis in Okra. Indian J. Hortic. **63**(3), 281–285 (2006)
28. Guddadamath, S., Mohankumar, H.D., Salimath, P.M.: Genetic analysis of association studies in segregating population of okra. Karnataka J. Agric. Sci. **24**(4), 432–435 (2011)
29. Hazra, P., Basu, D.: Genetic variability, correlation and path analysis in okra. Ann. Agric. Res. **21**(3), 452–453 (2000)
30. Sachan, V.K.: Performance of okra (*Abelmoschus esculentus* L.) varieties in mid hills of Sikkim. Orissa J. Hortic. **34**(2), 131–132 (2006)
31. Alam, A.K.M.A., Hussain, M.M.: Variability of different growth contributing parameters of some okra (*Abelmoschus esculentus* L. Moench) accessions and their interrelation effects on yield. J. Agric. Rural Dev. Gazipur **6**(1/2), 2535 (2008)
32. Kumar, V., Kumar, A., Gayen, R.: Estimation of genetic parameters in okra for quantitativetraits. Indian J. Hortic. **68**(3), 336–339 (2011)
33. Kabir, J., Pillu, N.: Effect of age harvest on fruit quality of okra (*Abelmoschus esculentus* (L.) Moench). J. Environ. Res. Dev. **5**(3), 615–622 (2011)
34. Sreeshma, L.S., Bindu, R.N.: Biochemical changes associated with fruit development in *Abelmoschus esculentus* cv. Arka Anamika. Int. J. Agric. Technol. **9**(2), 373–382 (2013)
35. Nwachukwu, E.C., Nulit, R., Go, R.: Nutritional and biochemical properties of Malaysian okra variety. Adv. Med. Plant Res. **2**(1), 16–19 (2014)
36. Bajaj, K.L., Kaur, P.P., Arora, S.K.: Varietal variation in the chemical constituents of okra (*Abelmoschus esculentus* (L.). Monech). Ind. Food Packer **42**, 32–26 (1988)
37. Chavan, U.D., Adsule, R.N., Kachare, D.P.: Chemical composition and nutritional quality of some promising cultivars of okra (*Abelmoschus esculentus* (L.) Monech*)*. J. Maharashtra Agric. Univ. **16**, 287–288 (1991)
38. Gopalan, C., Sastri, S.B.V., Balasubramanian, S.: Nutritive value of Indian foods. In: National Institute of Nutrition (NIN), ICMR, India, pp. 67–75 (2007)
39. Umrao, V., Sharma, S.K., Kumar, R., Kumar, V., Sharma, A.: Genetic variability and divergenceanalysis in okra(*Abelmoschus esculentus* (L.) Moench). Hort Flora Res. Spectr. **3**(2), 127–132 (2014)
40. Nasit, M.B., Dhaduk, L.K., Vachhani, J.H., Savaliya, J.J.: Variability, heritability and genetic advance in okra (Abelmoschus esculentus (L.) Moench). Asian J. Hortic. **4**(2), 415–417 (2009)
41. Magar, R.G., Madrap, I.A.: Genetic variability, correlations and path co-efficient analysis in okra(*Abelmoschus esculentus* (L.) Moench). Int. J. Plant Sci. **4**(20) 498–500 (2009)
42. Vahab, M.A., Devi, S.N., Mathew, S.K., Prabhakaran, P.V.: Genetic divergence in okra (*Abelmoschus esculentus* (L.)Moench). Hortic. J. 7(2), 117–120 (1994)
43. Singh, D., Dudi, B.S., Dhankhar, S.K., Rajkumar: Genetic diversity analysis of Okra genotypes using morphological markers. Int. J. Curr. Microbiol. App. Sci. 7(1), 1667–1675 (2018)
44. Singh, N., Singh, D.K. Sati, U.C., Rawat, M., Pandey, P.: Genetic Analysis Studies in Okra (*Abelmoschus esculentus* (L.) Moench). Int. J. Pure App. Bio Sci. **5**(4), 361–367 (2017)
45. Singh, A.K., Ahmed, N., Raj, N., Chatoo, M.A.: Genetic variability, correlations and path-coefficient analysis in okra under Kashmir conditions. Indian J. Hortic. **64**(4), 472–472 (2007)
46. Mishra, R.S., Chhonkar, Y.S.: Genetic divergence in okra. Indian J. Agric. Sci. **49**, 247–249 (1979)

Chapter 3
Utility of Cow Urine-Based Bio-enhancer: A Boon Substrate for the Growth Study in *Phaseolus Vulgaris* and Its Variability in the Modified Jeevamrutha Bio-formulation

Udaratta Bhattacharjee and Ramagopal V. S. Uppaluri

3.1 Introduction

The exploitation of soil by the intensive use of chemical fertilizers caused reduced natural soil fertility, deterioration of the ecosystem, pollution, and sustainability. In this regard, soil organic matter has shown depletion in nutrient availability and constant soil productivity. Soil compaction, nitrogen leaching and loss of organic carbon are some of the emerging threats due to the excessive use of chemical fertilizers. These fertilizers harden the soil texture and strengthen bio-pesticides that disrupt the balanced sustainable growth of plants. In general, chemical fertilizers do not aid in soil nutrient replenishment, but, enhance nitrogen, phosphorus, and potassium. These eventually lead to a decline in the soil equilibrium state [1–3]. A balance in the nutrient quantity in the soil is necessary for improving its biological and physio-chemical properties. Thereby, it is important to visualize the applied fertilizers in terms of plant growth-related factors. These growth factors are highly affected by the presence of aggressive microorganisms that cause qualitative and quantitative damage to the plants. In terms of crop economics and yield traits, the causative plant pathogen associated with wilting and damping of several crops are fungi such as *Sclerotium rolfsii and Fusarium oxysporum.* In this regard, chemical agents are often deployed for plant disease control. However, these agents signify pathogen resistance in plants and toxic effects on nom-specific organisms triggering

U. Bhattacharjee
Centre for the Environment, Indian Institute of Technology Guwahati, Assam, India
e-mail: udaratta@iitg.ac.in

R. V. S. Uppaluri (✉)
Department of Chemical Engineering, Indian Institute of Technology Guwahati, Assam, India
e-mail: ramgopalu@iitg.ac.in

R. V. S. Uppaluri and L. Rangan (eds.), *Conservation of Biodiversity in the North Eastern States of India*, https://doi.org/10.1007/978-981-99-0945-2_3

disintegrated microbial consortium [4, 5]. Therefore, natural bio-products such as cow urine provide benefits to the farming system in terms of their availability and eco-friendly attributes. Cow urine comprises good nitrogen, phosphate, potassium, magnesium, calcium, and sulphate with 2.5% urea, enzymes, minerals, and 95% water [6, 7]. This could be sprayed in the crucial growth phase of the plant to address the problems associated with the slow release of organic nutrients for its growth, development, and pest management. The application of cow urine has been proven to kill numerous resistant fungi, bacteria, and viruses. Also, cow urine along with other plant extracts has been reported to serve as a profound biodegradable disinfectant due to its antibacterial property [8, 9].

The allelochemicals or bioactive constitutions of cow urine have not been explored in a wider sense. Aurum oxide, 2-phenyl phenol, halogenated phenols, manganese, and carbolic acid present in it act as an antibiotic, antifungal, antiviral, and pesticide. As an indigenous crop pest control, spraying cow urine is an efficient eco-friendly crop protection measure. Besides enhancing soil texture and improving micronutrient deficiency, cow urine acts as an excellent plant hormone and hence can be used as an efficient bio-fertilizer and bio-pesticide [10]. In the prior art, the application of 125 and 100 kg N ha^{-1} by cattle urine has shown enhanced vegetative, morphological, and yield features along with a 50% substitution of urea by urine. Thereby, the study targeted the efficacy and potential of the alternate cattle urine to nitrogen fertilizers for the improvement in the yield of cauliflower from cattle urine combined with urea [11]. Another study was carried out for the efficacy of two foliar sprays comprising various concentrations of cow urine (2%, 4%, and 6%) for 25 and 40 days after sowing for soybean. The results revealed that 6% urine concentration has shown better combinations of yield, biochemical, morphological, and physiological parameters. This was carried out using a statistical tool called Factorial Randomized Block Design (FRBD) in *kharif* with a spacing of 30 cm $\times$ 10 cm [12]. Plant growth parameters such as the number of branches (4.8), leaf area index (1.3), and plant height (35.7 cm) have shown improvement in comparison with the control group by the application of 10% cow urine during the flowering phase and 15 days after flowering for chickpea. The field application was conducted during *the rabi* season in the dry area of northern Karnataka [13]. Devakumar et al. (2014) and Singh et al. (2014) reported higher grain yield and nitrogen content (18.6 and 17.6 q/ha) with different levels of panchagavya, Nitrogen-Phosphate-Potassium (NPK) fertilizer along with cow urine. The results suggested that the application of nitrogen when used in combination with cow urine at various levels enhanced the fertility status of soil in the case of paddy crops during the *kharif* season. Also, the presence of a beneficial microbial population enhanced phenotypic characteristics in plants to reduce the use of external inputs for the farmers [14, 15]. The application of cow urine recorded higher stem girth (0.64 cm) and plant height (116.2 cm) on buckwheat. This study has been investigated during the *rabi* season to identify the best microbial strain in a hilly ecosystem in North-East India. *Azospirillum* and *Azotobacter* spp. are the strains that thrive well in the soils of Sikkim which are acidic [16]. Twelve different treatments were set up in four RCBD

(Randomized Complete Block Design) and split-plot designs. Each plot consists of four rows comprising 0.25 cm × 0.25 cm (1.75 cm each). Fresh and dry stem, leaf, root, and mass along with stem length and commercial yield were evaluated. Thereby, different concentrations of cow urine (0.00, 0.25, 0.50, 0.75, 1.00, and 1.25%) have increased lettuce productivity up to 28.1% (leaves) and 47.3% (soil) in comparison with the control and suggested that cow urine improves lettuce productivity [17].

Apart from the growth and yield features in plants, cow urine has an excellent nutritional effect on the crop. Better chlorophyll and protein content in *Abelmoschus esculentus* and *Trigonella foenum-graecum* along with better N and K concentration in grass and clover were obtained using cow urine in comparison with the control group, respectively. Cow urine-treated groups have shown fewer symptoms of wilting and damping off disease in the plants. The study demonstrated the inhibitory response and antifungal activity of cow urine at 15% concentration and its use as a bio-pesticide [4, 10, 18]. Additionally, a higher dose of liquid organic manures provides enhanced organic carbon, EC, and pH values. The growth parameters such as the number of panicles, tillers, leaves, leaf area index and plant length and per hectare grain yield as 7.56 ton/ha in comparison with the control case (5.38 tons/ha) [19].

Organic farming solely relies on livestock manure application to strictly limit the use of chemical fertilizers, antibiotics, and genetically modified organisms. However, owing to its pharmacological usage and as a cure for several bacterial and fungal infections, cow urine possesses many beneficial potentialities in the commercial market. Cow urine has been recognized for its medicinal properties granted by the US Patents (No. 6,896,907 and 6,410,059). The anticancer, antibiotic, antifungal, and bio-enhancer properties of cow urine have been mentioned in the patents. It has also been reported that cow urine increases the efficacy of paclitaxel against a human breast cancer cell line (MCF-7) (US Patent No. 6,410,059). In the rural economy, it has a high socio-cultural impact that represents cattle bio-diversity and its importance in traditional medicine. This is due to the presence of certain calcium salts, vitamin A, B, C, D, and E, lactose, creatinine, enzymes, hormones, gold acids, tartaric, maleic, citric acids, and macro and micronutrients like nitrogen, phosphate, manganese, iron, silicon, magnesium, and chlorine with a pH range of 7.4–8.4 and specific gravity of 1.02–1.04, respectively. *Aspergillus malassezia*, *Candida tropicalis*, and *Candida glabrata* growth are usually inhibited by the application of cow urine. It has also been found that neem leaves and lemon juice extract are less effective in terms of various diseases in crops in comparison with cow urine. However, well-organized studies are required to gather information about its potential to be utilized in field applications [20]. Besides having therapeutic properties, cow urine distillate acts as a booster agent for B- and T-cell proliferation and phagocytic activity of macrophages, thereby aiding in the control of infections [21]. Additionally, the synthesis of IgM, IgG, and IgA antibody titers and interleukin-1 and 2 are largely facilitated by cow urine. A large number of poisonous chemicals can be treated by the use of cow urine which acts as an anti-toxin [8]. Certain rural areas confine the use of cow urine as an efficient raw source for wound healing, bathing, and curing other diseases. Other utilities of cow urine involve the rearing of honeybees which do serve as a holistic and eco-friendly measure for conservation [22]. Owing to the numerous medicinal applications, the rational utility of cow urine is necessary to eradicate its substantial use in agricultural systems. The traditional practices and considerate insights for this animal by-product are essential to promote its wider applicability in pharmacological sections [20].

Panchagavya, an excellent cow excreta-based organic bio-fertilizer, has profound utility in agricultural land. This is a concoction made up of five random cow products, namely, cow dung, cow urine, curd, ghee, and milk, respectively. Similarly, Sanjibani and Amritpani were prepared using cow dung, cow urine, and water along with ghee and honey and used in the field after sowing the seed. It can offer application as both a fertilizer and a bio-pesticide [5, 23, 24]. Another cow excreta-based bio-fertilizer, Kunapajola, contained rice husk, sesame seed, molasses, Bombay duck fish, and cow urine, and was left to decompose. However, cost-effectiveness is necessary for the due course of designing any fertilizer bio-formulation. This justifies the wider utility of Jeevamrutha [25]. It was termed by Subhash Palekar, one of the progressive farmers of Maharashtra, India [26]. It contains growth-promoting beneficial microorganisms such as *Azospirillum*, *Acetobacter*, *Azotobacter*, *Pseudomonas,* and *Bacillus* species [19, 26–28]. Jeevamrutha, an organic bio-fertilizer, provides enrichment to the soil through the presence of a diverse microbial population [26]. This bio-formulation comprises cow dung, cow urine, soil, jaggery, and gram flour. Cow dung comprises rich microflora predominantly bacteria, certain beneficial fungi, actinomycetes, and yeast [29, 30]. Similarly, cow urine enhances the N content due to its rich constitution of amino acids. Also, pest resistance in bio-formulation improves due to its bio-pesticidal properties [1]. Soil, an effective bio-inoculant and rich in NPK, provides microbial consortium for nitrogen fixing and phosphate solubilizing bacteria [5, 31, 32]. On the other hand, jaggery is a good carbon source and supplies adequate potassium content for daily plant growth requirements. Gram flour contains amino acids aiding in a healthier root system [33]. This bio-formulation is relatively inexpensive due to the rural availability of raw precursors. Therefore, it will be interesting to observe the plant growth characteristics in the bio-formulation with and without the addition of cow urine and its potential for soil fertility status.

Common beans, also known as *Phaseolus vulgaris* are an economically and nutritionally significant food crop across the globe. The plant is well-known to adapt to temperate, subtropical, and tropical zones [34]. It provides essential nutrients like minerals, vitamins, dietary fiber, protein, carbohydrates, and a wide range of polyphenolic compounds that target the decrease in the rate of degenerative diseases in humans [35]. Numerous health-related properties like antidiabetic, anticancer, anti-inflammatory, anti-obesity, and antioxidant are provided by beans [21, 36–38]. The present study provides useful insights with respect to the indigenous Jeevamrutha bio-formulation and its modification by the elimination of cow urine on the *Phaseolus vulgaris* growth profile. The significance of viable microbe during the fermentation process will briefly comprehend the criticality of individual precursors in the formulation and elaborate the scope of the research theme by reducing cow urine as a potent bio-enhancer [12].

3.2 Materials and Methods

Jeevamrutha solid formulation (unmodified) and Jeevamrutha solid formulation without cow urine (modified) was evaluated for their nutrient composition and growth profile study for *Phaseolus vulgaris*. In this study, a modified formulation without cow urine was considered to observe the variation and growth patterns in the model plant and its respective root health. Also, the microbial count for each treatment was investigated in CFU/mL.

3.2.1 Chemicals and Instruments

Sodium hydroxide, di-sodium hydrogen phosphate dehydrate, sodium bicarbonate, sodium nitroprusside solution, sodium salicylate, sodium hypochlorite, sodium–potassium tartrate, 2,6-dichlorophenolindophenol, hydrazine sulphate, and ammonium molybdate (analytical grade) were purchased from Himedia, India and Merck, USA. For analytical instruments and apparatus, the method is similar to the one described by Bhattacharjee and Uppaluri (2022) [39].

3.2.2 Sample Preparation and Seed Collection

Unmodified Jeevamrutha solid formulation was prepared by adding 1 kg cow dung, 0.02 L cow urine, 0.03 kg jaggery and gram flour, and 0.36 kg soil with an initial pH of 6.76 [26, 39]. The modified Jeevamrutha formulation contains a similar compositional set without cow urine. The temperature and humidity during the growth phase were 27–31 °C and 56–71%, respectively. Both formulations were selected for the growth study of *Phaseolus vulgaris* along with their macronutrient composition. The experimental plant seeds of *Phaseolus vulgaris* (common bean) were procured from a commercial market in Guwahati, India. The seeds were of equal weight, size, and shape and were sterilized using distilled water before being sown. Figure 3.1 illustrates the procedure used to evaluate the seed germination and the growth features of *Phaseolus vulgaris.*

3.2.3 Microbial Growth Study of the Jeevamrutha Bio-fertilizer Formulations with and without Cow Urine

After the growth study of the common bean, the soil samples were analyzed for microbial analysis (in duplicates). The microbial diversity was obtained between the 9th and 12th days after preparation. Enriched consortia in the bio-formulation

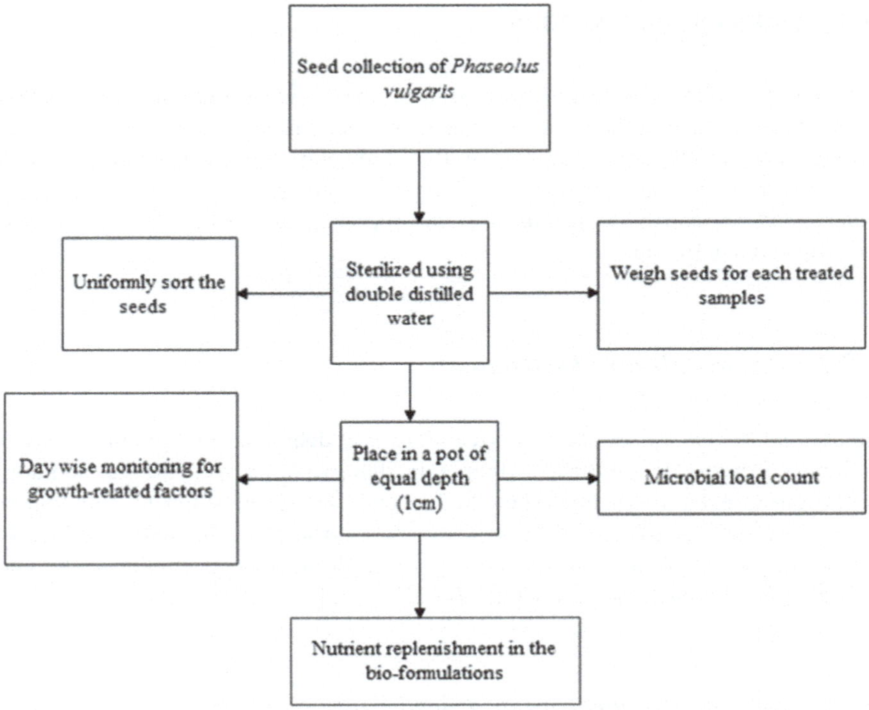

Fig. 3.1 Schematic of the procedure adopted for seed collection and preparation of pot experiments in *Phaseolus vulgaris*

signified a better growth profile for the microorganisms [40]. The total microbial count was conducted by culturing the plates for bacteria, fungi, actinomycetes, N-fixers, and P-solubilizers. Nutrient agar was used for growing and counting bacteria, potato dextrose agar for fungi, starch M-protein agar for actinomycetes, Jensen's medium for N-fixers, and Pikovskaya's agar for P-solubilizers. In this study, the highest CFU count of bacteria, fungi, and actinomycetes was 14×10^5, 17×10^3, and 2×10^3 CFU/mL, respectively [41].

3.2.4 *Physio-Chemical Characterization of the Jeevamrutha Bio-fertilizer Formulations with and without Cow Urine*

The unmodified and modified Jeevamrutha bio-formulations were analyzed for their pH, Electrical conductivity (EC), and Organic Carbon (%OC) using a pH and conductivity meter along with a TOC analyzer. The freshly prepared bio-formulation is acidic (5.6). Toward the progressing end of the storage time, a high alkaline pH was observed (9.7). This variation is due to the alcoholic and lactic acid fermentation due to the presence of ethanol, butanol, propanol, and methanol as fermentation by-products [2, 42–44].

3.2.5 *Macronutrient Constituents of the Jeevamrutha Bio-fertilizer Formulations with and without Cow Urine*

Ammonium Nitrogen (AN) and %NPK of the bio-formulations were evaluated using standard methods. These are similar to those adopted by Bhattacharjee and Uppaluri (2022) [39]. The moisture content (%) of the soil samples was assessed by the AOAC method and dried at 105 °C [45]. However, it was found that the nitrogen content in the bio-formulations does not vary in a significant way due to subsequent biochemical interactions and volatile constituents during the fermentation process (0.3–0.8%) [46, 47]. The potassium range was reported to increase during storage, while phosphate content gradually reduced with storage time. Other micronutrients like calcium and magnesium do not vary due to the aging of the bio-formulations. However, sulphur content varied during the storage period. The presence of zinc and manganese was also detected in the range of 0.4–7.5 and 1.7–26.9 ppm respectively [2, 11].

3.2.6 *Evaluation of Growth and Yield Parameters in* Phaseolus Vulgaris

The growth characteristics in *Phaseolus vulgaris* for each treatment were studied for their number of shoots, leaves, flowers, fruits, leaves, and root length after 45 days. The uprooted plant has shown its root health with and without the addition of cow urine. The yield parameters such as fruit weight and root length were also assessed for both unmodified and modified samples. The fresh and dry weight of the fruit samples obtained after 45 days was measured using a weighing balance. The dry weight of the fruit samples was taken after oven drying at 80 °C for 24 h [48–51].

3.3 Results and Discussion

In general, the fertility content in soil is likely to reduce unless a sufficient amount of nutrients are applied to it. This is necessary to achieve a higher grain yield and optimum plant growth profile. This work aims to investigate the effect of cow urine on plant growth and yield parameters and to reduce its utility in organic farming techniques owing to its medicinal properties in the commercial aspects.

Table 3.1 Physio-chemical and macro-nutritional components data of unmodified (with cow urine) and modified (without cow urine) Jeevamrutha bio-fertilizer formulations

SI. no.	pH	AN (mg/L)	N (%)	P (%)	K (%)	OC (%)	EC (ds/m)	Moisture content (%)
Control	6.58	158.67	0.24	0.47	0.35	0.31	3.4	20.4
Unmodified	7.21	426.58	0.63	0.58	0.51	0.68	2.4	19.6
Modified	7.09	289.54	0.57	0.55	0.48	0.57	2.6	18.7

3.3.1 Macronutrients and Physio-Chemical Composition in the Bio-formulations

In the Jeevamrutha bio-formulation, a higher macronutrient content was obtained, i.e., 426.58 mg/L AN, 0.63% N, 0.58% P, and 0.51% K, respectively (Table 3.1). On the other hand, the modified bio-formulation recorded 289.54 mg/L AN, 0.57% N, 0.55% P, and 0.48% K, respectively. However, the control group achieved lower macro-nutritional constituents (158.67 mg/L AN, 0.24% N, 0.47% P, and 0.35% K). Also, organic carbon content was higher (0.68% and 0.57%) for the unmodified and modified bio-fertilizers, respectively, than those obtained for the control case (0.31%). The bio-formulations have shown a neutral pH level (7.21 and 7.09). The EC and moisture content of the unmodified and modified bio-fertilizers were significantly lower (2.4 and 2.6 and 19.6% and 18.7%, respectively) in comparison with the control group (3.4 and 20.4%). The pH for all samples was favorable for plant growth (6.5–7.5) [52].

3.3.2 Microbial Population in the Bio-formulations

The amount of cultivable microbial cultures is represented using the plate count method (Table 3.2). It was observed that maximum bacteria, actinomycetes, N-fixers, and P-solubilizers were observed for the unmodified followed by modified bio-formulations. However, fungal counts were highest for the control group followed by modified bio-formulations (15×10^3 and 9×10^3). The fungal growth in both these samples could be justified by the fact that cow urine acts as a pest control agent and has fungicidal properties. For each treated sample, the microbial growth pattern has shown its significance in the fermentation process during the preparation of the bio-formulations (Fig. 3.2).

Table 3.2 Microbial analysis data of unmodified (with cow urine) and modified (without cow urine) Jeevamrutha bio-fertilizer formulations

Sl. no.	Bacteria	Fungi	Actinomycetes	N-fixers	P-solubilizers
Control	71×10^5	15×10^3	4×10^2	17×10^4	1×10^4
Unmodified	169×10^5	4×10^3	8×10^2	59×10^4	31×10^4
Modified	157×10^5	9×10^3	7×10^2	27×10^4	26×10^4

3.3.3 *Effect of the Bio-formulations on Growth and Yield Parameters of* **Phaseolus Vulgaris**

After transplanting bean seedlings into the potting media, growth parameters along with pest attacks were continuously monitored. Figure 3.3a–c depicts the growth profile of the bean plant 30–40 days after transplantation. The nutrient depletion in the control group is evident from the growth pattern study. Also, during the 10–20 days period, yellow leaves were observed for the control plant suggesting the lack of an adequate supply of nutrients during their initial growth period. Furthermore, the root health of the unmodified and modified bio-formulations is illustrated in Fig. 3.4a–c. The unmodified treatment has shown better root conditions in comparison with the modified and the control group. This is due to the presence of a higher proportion of phosphorus content in the unmodified treatment. Tables 3.3, 3.4 and 3.5 summarizes the growth characteristics of *Phaseolus vulgaris* from 0–40 days time period. The shoot and leaf length has shown remarkable development (8 and 10.4 cm) for the unmodified treatment in comparison with the other two sample studies. In the case of unmodified treatment, the observation for days 20–30 reveals a higher growth yield in terms of the number of leaves, branches, flowers, and shoot and root length followed by modified bio-formulations. The better flowering and heading are due to the enhanced uptake of nutrients, water retention, and photosynthesis causing higher root expansion. Day 30–40 infers maximum fruit yield (23) and root length (8.2 cm) for the unmodified sample followed by modified treatment (Fruit 17, root length 7.4 cm). Hence, cow urine had positive effects on the growth and fruit yield of the bean plant. Nevertheless, modified bio-formulation did have a better impact on the flowering and fruiting process in comparison with the control. Table 3.6 represents the fruit length for all samples and their respective dry and wet weight (in g). Figure 3.5 shows the bean samples (*Phaseolus vulgaris*) collected after 45 days of growth.

3.4 Conclusions

Many challenges faced by sustainable agriculture could be addressed by using cow excreta-based fertilizers as a better supplement for nutrient management in the organic farming system. In the agricultural sector, the benefit of using cow urine

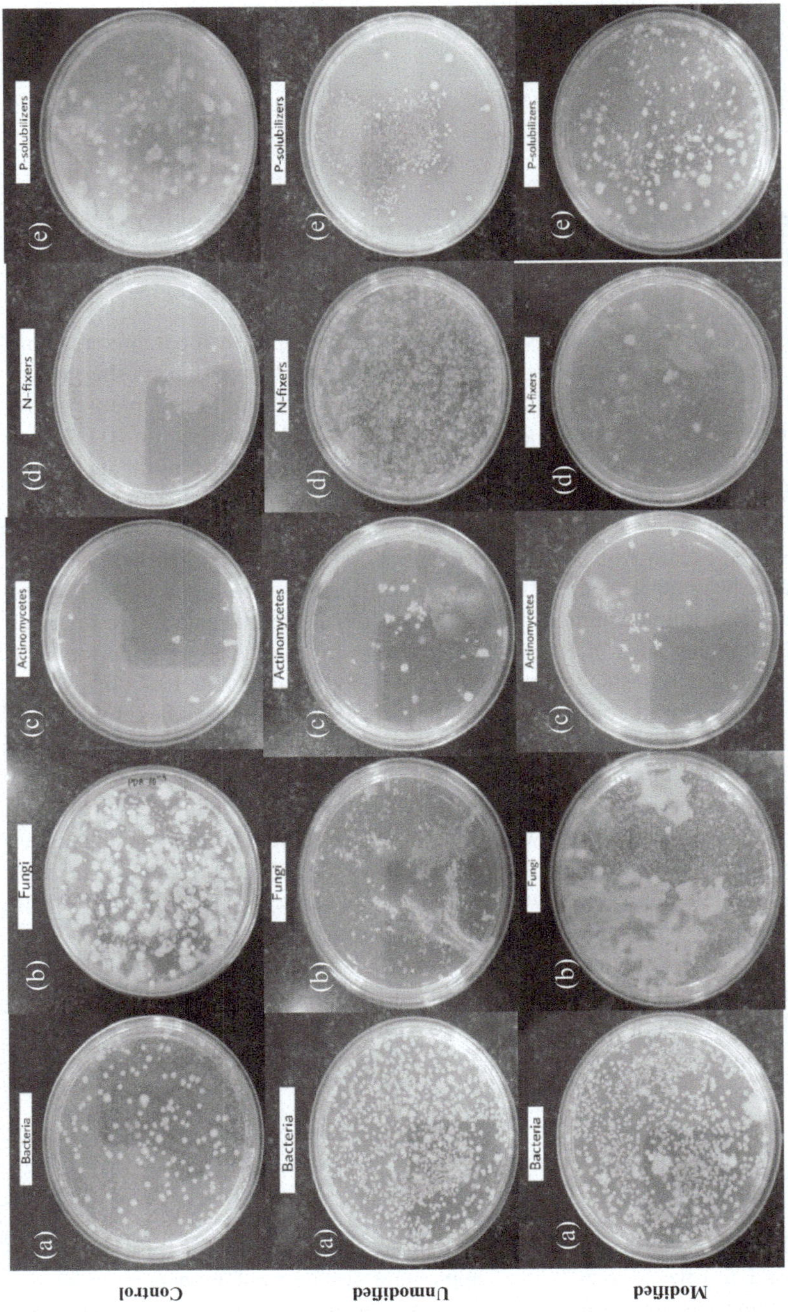

Fig. 3.2 Images depicting microbial load of control (**a-e**), unmodified (with cow urine) (**a-e**), and modified (without cow urine) (**a-e**) Jeevamrutha bio-fertilizer formulations

Fig. 3.3 **a–c** Photographs depicting growth patterns of *Phaseolus vulgaris* in unmodified (with cow urine) and modified (without cow urine) Jeevamrutha bio-fertilizer formulations

Fig. 3.4 **a–c** Photographs depicting root growth of *Phaseolus vulgaris* in unmodified (with cow urine) and modified (without cow urine) Jeevamrutha bio-fertilizer formulations

Table 3.3 Day summary of 0–10 days growth parameters of *Phaseolus vulgaris*

SI no.	Leaf length (cm)	Shoot length (cm)	No. of leaves
Control	3	7	2
Unmodified	8	10.4	2
Modified	6.7	9.8	2

Table 3.4 Day summary 20–30 days growth parameters of *Phaseolus vulgaris*

SI no.	Leaf length (cm)	Shoot length (cm)	No. of leaves	No. of branches	No. of flowers
Control	8	16.4	11	8	14
Unmodified	11.4	24.8	14	10	23
Modified	9.4	13.89	13	8	17

Table 3.5 Day summary 30–40 days growth parameters of *Phaseolus vulgaris*

SI no.	Leaf Length (cm)	Shoot length (cm)	No. of leaves	No.of branches	No.of Fruits	Root Length (cm)
Control	8	16.4	11	8	14	6.6
Unmodified	11.4	24.8	14	10	23	8.2
Modified	9.4	13.89	13	8	17	7.4

Table 3.6 Day summary of 45th-day yield parameters for *Phaseolus vulgaris*

SI no.	Fruit length (cm)	Fruit weight in g (wet)	Fruit weight in g (dry)
Control	6.7	4.9	4.5
Unmodified	8.4	6.2	5.9
Modified	9.6	5.7	5.4

Fig. 3.5 Photograph of 45th-day fruit sample achieved with unmodified (with cow urine) and modified (without cow urine) Jeevamrutha bio-fertilizer formulations

is a conventional way to integrate it for improved crop profitability and productivity. This bio-product has certain distinctive compounds that contribute to soil sustainability and health. The presence of metabolites, enzymes, and antimicrobials requires further investigation for its noble applications and in a greener way. In other words, cow urine has fungicidal and bio-pesticidal properties and they do enhance the biochemical constituents in plants. The present study is based on the benefits of using cow urine-based bio-formulations for the growth characteristics in *Phaseolus vulgaris* and its enrichment in the microbial consortium.

Better shoot and leaf length along with other growth and yield parameters were achieved using cow urine-based jeevamrutha bio-formulations. Also, the flowering and ripening process was rapid in the case of the addition of cow urine into the formulation. However, the modified Jeevamrutha bio-fertilizer formulation has shown better yield attributes (no. of fruits 17, fruit 5.7 g, and root length 5.7 g) in comparison with the control sample (number of fruits 14, fruit weight 4.9 g, and root length 6.6 g). Similarly, the incorporation of cow urine signifies the least growth of fungi in comparison to the modified (without cow urine) and the control group. The current investigation assures the minimal applicability of cow urine in farming systems due to the limitations of its generous general accessibility. This could be instead, harnessed for usage in traditional medicine and industries. In this regard, available waste precursors could be incorporated into the bio-formulation for improving the macro and micronutrient availability in the soil. The addressed research pedagogy affirms the integrated bio-diversity conservation being driven by microbes and nutrients for good plant health and crop yield. The role of cow urine in bio-diversity conservation in agriculture or farm environment has been addressed in this chapter. Thus, cow dung and/or cow urine fertilizer have strong relevance for such bio-diversity in the agricultural sector of North-East India.

Acknowledgements We are grateful to the Centre for the Environment and School of Agro and Rural Technology, IIT Guwahati for providing the essential instruments to conduct this work.

References

1. Pradhan, S.S., Verma, S., Kumari, S., Singh, Y.: Bio-efficacy of cow urine on crop production: A. IJCS **6**(3), 298–301 (2018)
2. Rameeza, E.: Shelf life of liquid organic formulations, Department of Agronomy, College of Horticulture, Vellanikkara (2014)
3. Mondal, T., Datta, J.K., Mondal, N.K.: Chemical fertilizer in conjunction with biofertilizer and vermicompost induced changes in morpho-physiological and bio-chemical traits of mustard crop. J. Saudi Soc. Agric. Sci. **16**(2), 135–144 (2017)
4. Jandaik, S., Thakur, P. and Kumar, V.: Efficacy of cow urine as plant growth enhancer and antifungal agent. Adv. Agric. (2015)
5. Kulkarni, S., Gargelwar, A.: Production and microbial analysis of Jeevamrutham for nitrogen fixers and phosphate solubilizers in the rural area from Maharashtra. IOSR J Agric. Veterinary Sci. **12**, 85–92 (2019)
6. Randhawa, G.K., Sharma, R: Chemotherapeutic potential of cow urine: a review. J. Intercult. Ethnopharmacol **4**(2), 180 (2015)
7. Panneerselvam, P., Sharma, L., Nayak, A.K., Senapati, A., Bhutia, J.T., Prabhukarthikeyan, S.R., Kumar, U., Kumar, A., Govindharaj, P.P., Mitra, D.: Development of microbial consortium for plant growth promotion, nutrient and disease management in rice-horticulture based cropping system in Sikkim (2020)
8. Mohanty, I., Senapati, M.R., Jena, D., Palai, S.: Diversified uses of cow urine. Int J Pharm Pharm Sci. **6**(3), 20–22 (2014)
9. Mazumdar, K., Das, S.: Phytoremediation of Pb, Zn, Fe, and Mg with 25 wetland plant species from a paper mill contaminated site in North East India. Environ. Sci. Pollut. Res. **22**(1), 701–710 (2015)

10. Ledgard, S., Sheath, G., Gillingham, A.: Influence of some soil and pasture components on the growth of hill country pastures 1. Winter and spring production. New Zealand J. Exp. Agric. **10**(3), 239–244 (1982)
11. Khanal, A., Shakya, S., Sharma, M.: Utilization of urine waste to produce quality cauliflower. J. Agric. Environ. **12**, 91–96 (2011)
12. Deotale, R., Kalamkar, V., Banginwar, A.: Effect of foliar sprays of cow urine and NAA on morpho-physiological, chemical, biochemical parameters and yield of soybean (Glycine max L.). J. Soils Crops **21**(2), 332–337 (2011)
13. Patil, S., Halikatti, S.I., Hiremath, S.M., Babalad, H.B., Sreenivasa, M.N., Hebsur, N.S., Somanagouda, G.: Effect of organics on growth and yield of chickpea (Cicer arietinum L.) in vertisols. Karnataka J. Agric. Sci. **25**(3), 326–331 (2012)
14. Singh, M.K., Singh, R., Rai, S.: Effect of nitrogen levels and cow urine on soil N status, growth and yield on paddy (Oryza sativa L.). Environ. Ecol. **32**(4), 1277–1281 (2014)
15. Devakumar, N., Subha, S., Rao, G.G.E., Khan, I.J.: Studies on soil fertility, cow urine and panchagavya levels on growth and yield of maize. Build. Org. Bridg. **2**, 627–630 (2014)
16. Raghavendra, S., Babu, S., Avasthe, R.K., Yadav, G.S., Chettri, T.K., Phempunadi, C.D., Chatterjee, T.: Bacterial inoculation effect on soil biological properties, growth, grain yield, total phenolic and flavonoids contents of common buckwheat (Fagopyrum esculentum Moench) under hilly ecosystems of North-East India. Afr. J. Microbiol. Res. **9**(15), 1110–1117 (2015)
17. de Oliveira, N.L.C., Puiatti, M., Cecon, P.R., Santos, R.H., Rodrigues, P.H.: Soil and leaf fertilization of lettuce crop with cow urine. Hortic. Bras. **27**, 431–437 (2009)
18. Ignacimuthu, S., Vendan, S.: Botanical pesticides in insect pest management. Emerg. Trends Res. Insect Pest Manag. Environ. Saf. **I**, 141–154 (2008)
19. Andreeilee, B.F., Santoso, M., Maghfoer, M.D.: The effect of organic matter combination and azola dosage (Azolia pinnam) on growth and the production of paddy (Oryza sp.) Ciherang variety. Res. J. Agron. **9**(1–6), 1–4 (2015)
20. Gulhane, H., Nankanekar, A., Mahakal, N., Bhople, S., Salunke, A.: Gomutra cow urine): a multidimensional drug review article. Int. J. Res. Ayurveda Pharma **2017**(8), 1–6 (2017)
21. Dhama, K., Chauhan, R., Singhal, L.: Anti-cancer activity of cow urine: current status and future directions. Int. J. Cow Sci. **1**(2), 1–25 (2005)
22. Chand, A., Tiwari, R.: Efficacy of cow urine against bacterial disease, European foulbrood, in honey bee, Apis mellifera (L.) colonies at different locations of Uttarakhand-an eco-friendly and novel approach. Int. J. Basic Appl. Sci. **1**, 179–189 (2012)
23. Kaur, A.: Jeevamrutham: an effective activator of soil microorganisms. Just Agric. Multidiscip. E-Newsl. (2020)
24. Veeresh, S.J., Narayana, J., Silva, J.T.: Influence of jeevamrutha (biodynamic formulation) on agro-industrial waste vermicomposting. Vermitechnol **II**(4), 96–99 (2010)
25. Sutar, R., Sujith, G., Devakumar, N.: Growth and yield of Cowpea [Vigna unguiculata L.) Walp] as influenced by jeevamrutha and panchagavya application. Legume Res. Int. J. **42**(6), 824–828 (2019)
26. Palekar, S.: Text book on shoonya bandovalada naisargika krushi, published by swamy anand. Agri Prakashana, Bangalore (2006)
27. Anandan, R., Priya, L., Rajendran, P.: Dynamics of organic biofertilizers on Oryza sativa ADT-43. Int. J. Curr. Microbiol. App. Sci. **5**(4), 902–908 (2016)
28. Veeresh, S.J., Narayana, J.: Earthworm density, biomass and vermicompost recovery during agro industrial waste treatment. Int. J. Pharma Bio. Sci. **4**(2), 1274–1280 (2013)
29. Swain, M., Ray, R.: Biocontrol and other beneficial activities of Bacillus subtilis isolated from cowdung microflora. Microbiol. Res. **164**(2), 121–130 (2009)
30. Boraiah, B., Devakumar, N., Shubha, S., Palanna, K.B.: Effect of Panchagavya, Jeevamrutha and cow urine on beneficial microorganisms and yield of capsicum (Capsicum annuum L. var. grossum). Int. J. Curr. Microbiol. App. Sci. **6**(9), 3226–3234 (2017)
31. Anthony, P., Malzer, G., Zhang, M., Sparrow, S.: Soil nitrogen and phosphorus behaviour in a long-term fertilization experiment. Agron. J. **104**(5), 1223–1237 (2012)

32. Anandaraj, B., Delapierre, L.: Studies on influence of bioinoculants (Pseudomonas fluorescens, Rhizobium sp., Bacillus megaterium) in green gram. J. Biosci. Technol. **1**(2), 95–99 (2010)
33. Shilpa, V., Sangita, S.: A study of nutritional profile and antioxidant properties of chick pea (Cicer arietinum L) and Horse Gram (Macrotyloma Uniflorum). Magnesium (mg/100g) **160**, 162.21 (2016)
34. Meza, N., Rosas, J.P., Ortiz, J.M.: Biodiversity of common bean (Phaseolus vulgaris L.) in Honduras, evidenced by morphological characterization. Genet. Resour. Crop Evol. **60**(4), 1329–1336 (2013)
35. Hayat, I., Ahmah, A., Masud, T., Ahmed, A., Bashir, S.: Nutritional and health perspectives of beans (Phaseolus vulgaris L.): an overview. Critical Rev. Food Sci. Nutr. **54**(5), 580–592 (2014)
36. Ganesan, K., Xu, B.: Polyphenol-rich dry common beans (Phaseolus vulgaris L.) and their health benefits. Int. J. Molecular. Sci. **18**(11), 2331 (2017)
37. Aktar, M.W., Sengupta, D., Chowdhury, A.: Impact of pesticides use in agriculture: their benefits and hazards. Interdiscip. Toxicol **2**(1), 1 (2009)
38. El Naim, A.M., Jabereldar, A.A.: Effect of plant density and cultivar on growth and yield of cowpea (Vigna unguiculata L. Walp) (2010)
39. Bhattacharjee, U., Uppaluri, R.V.: Screening and scoping of precursors associated to the production of Jeevamrutha bio-fertilizer. In: Materials Today: Proceedings (2022)
40. Basavaraj, K., Devakumar, N., Sheshadri, T.: Influence of farm yard manure, jeevamrutha and panchagavya on growth and yield of french bean (Phaseolus vulgaris L.). Mysore J. Agric. Sci. **50**(2), 279–283 (2016)
41. Chakraborty, B., Sarkar, I.: Quality analysis and characterization of Panchagavya, Jeevumrutha and Sasyamrutha. Int. J. Curr. Microbiol. App. Sci. **8**(5), 2018–2026 (2019)
42. Panwar, N.: Studies on physicochemical characteristics and fertility of soil by addition of banana peels-waste management. Int. J. Sci. Res. Dev. **3**(1), 121–125 (2015)
43. Patel, K., Patel, M.: Improving bioremediation process of petroleum wastewater using biosurfactants producing Stenotrophomonas sp. S1VKR-26 and assessment of phytotoxicity. Bioresour. Technol. **315**, 123861 (2020)
44. Ogbo, F., Odo, M.: Potential of rice husk and cassava peel as carriers for bio-fertilizer production. Nigerian J. Biotechnol. 23 (2011)
45. Horwitz, W.: Official methods of analysis of AOAC International. Volume I, agricultural chemicals, contaminants, drugs/edited by William Horwitz. 2010: Gaithersburg (Maryland): AOAC Int, 1997
46. Faugno, S., Pindozzi, S., Infascelli, R., Okello, C., Ripa, M.R., Lorenzo, B.: Assessment of nitrogen content in buffalo manure and land application costs. J. Agric. Eng. **43**(2), e13 (2012)
47. Liu, Y., von Wirén, N.: Ammonium as a signal for physiological and morphological responses in plants. J. Exp. Bot. **68**(10), 2581–2592 (2017)
48. Lalitha, S., Santhakumari, R.: Improving and effect of bio fertilizer on enhancement of the growth and bio chemical characteristic of photosynthesis on the Blackgram (Vigna mungo L.). J. Biol. Chem. **11**, 1–10 (2020)
49. Jariwala, H.J., Syed, H.S.: Study on use of fruit peels powder as a fertilizer. Research gate (2017)
50. Isleib, J.: Pros and cons of granular and liquid fertilizers. Michigan: Michigan State University Extension (2016)
51. Khairnar, M.D., Nair, S.S.: Study on eggshell and fruit peels as a fertilizer (2019)
52. Lucas, R., Davis, J.F.: Relationships between pH values of organic soils and availabilities of 12 plant nutrients. Soil Sci. **92**(3), 177–182 (1961)

Chapter 4
A Scientific Study of the Probable Reasons Causing the Loss of *Citrus Indica* from the Ecosystem

Upasana Deb and Sheena Haorongbam

4.1 Introduction

Citrus fruits, with their distinct flavour and aroma, are one of the most popular fruits in the world. In 2020, around 75.46-million-hectare tonne of oranges were produced globally, making it the sixth most-produced fruit in the world [1]. They are known for their antioxidant properties with the goodness of Vitamin C and phytochemicals like flavonoids, carotenoids and limonoids. Besides these components, it is rich in dietary fibres but low in proteins and fat contents. It is the storehouse of various nutrients like Vitamin B complex and carbohydrates. These play a major role in boosting immunity and help in lowering cholesterol and gastrointestinal ailments [2].

Citrus family is a big and diverse family spread all over the world. The fruits of *Citrus* family mainly consist of *Citrus sinensis* Osbeck (Sweet orange), *Citrus reticulata* Blanco (Mandarin), *Citrus aurantifolia* Swingle (lime), *Citrus limon* (L) Burm. f (lemon), *Citrus paradisi* Macf. (grapefruit) and *Citrus grandis* (L.) Osbeck (pummelo). They are reported to be diversely grown and spread out in over 114 countries with cultivation commercially occupying around 53 countries [2, 3]. They belong to the family Rutaceae under the order Sapindales, with certain species such as *Citrus indica* (Indian wild orange), *C. aurantifolia* (acid lime), *C. pseudolimon* (hill lemon also called as Galgal), *C. karna* (Karna Khatta), *C. assamensis*, *C. latipes*, *C. ichangensis*, *C. maderaspatana* (Kichili), *C. pennivesiculata* (Gajirima), *C. rugulosa* (Attanni), *C. jambhiri* (rough lemon), *C. medica* (citron), *C. limonia* (Rangpur lime), being reported to have their parent precursor generations originating in India [2]. Indigenous species like *C. macroptera* and *C. indica* are found in the North Eastern region of India and *C. pennivesiculata* and *C. maderaspatana* in the south India. *C. megaloxycarpa* and *C. pseudolimo* are found in the foothills of the Himalayas. The distribution of commercial species of *Citrus* in India is in the lower Pulney Hills of

U. Deb (✉) · S. Haorongbam
North Eastern Hill University, Tura Campus, Tura, Meghalaya, India
e-mail: upasanadeb123@gmail.com

R. V. S. Uppaluri and L. Rangan (eds.), *Conservation of Biodiversity in the North Eastern States of India*, https://doi.org/10.1007/978-981-99-0945-2_4

Tamil Nadu, Maharashtra, Punjab, Himachal Pradesh, Coorg region of Karnataka, Madhya Pradesh, Rajasthan and Northeast India for the *C. reticulate*. Similarly, Andhra Pradesh, Maharashtra, Telangana, Punjab, Madhya Pradesh and Mizoram host *C. sinensis*. Also, Karnataka, Maharashtra, Odisha, Punjab, Andhra Pradesh, Telangana, Himachal Pradesh, Madhya Pradesh, Tamil Nadu, Bihar, Chhattisgarh and Uttarakhand host *C. aurantifolia*, Uttarakhand and Northeast India host *C. limon* [2]. As for *C. grandis*, the distribution is reported from some gardens in the homestead of the Northeast region in India and Bihar. However, the cultivation of *C. paradisi* has been reported to be limited in India [2]. Among all these, *Citrus indica* Tanaka, commonly known as Indian Wild Orange can be asserted to be the most substratal primordial species and feasibly the ancestral progenitor of all cultivated *Citrus* fruits [4].

The first mention of *Citrus* was done in Sanskrit literature named Vajasaneyi Samhita, where it was known as Jamhila [5]. However, not many people may know that the *Citrus* roots originated in the Nokrek region of the West Garo Hills of Meghalaya, India (which is entitled by the United Nations Educational, Scientific and Cultural Organization (UNESCO) as a Biosphere Reserve in May 2009). It existed in the form of *Citrus indica* Yu Tanaka or the Indian Wild Orange [6]. The plant was discovered by Tanaka in 1928 in the Nowgong district of Khasi hills and in Manipur (at that time was part of Assam), and was recorded as a wild species of orange [7]. According to Bhattacharya and Dutta (1965), since no vernacular names were present at that time, it was given the identity of Indian Wild Orange [8]. Thus, the Indian Wild orange was included among all *Citrus* fruit crops in the Order Sapindales and Family Rutaceae, christening it with the name 'the progenitor of oranges' [1]. The locals of Garo Hills call it 'Memang Narang' or the fruit of the spirits, the Khasis call it Soh Kumphlair, while the Jaintias call it Sa Kymphrai [9].

The plants of *Citrus indica* were once abundant in Meghalaya and its adjacent states like Assam (Kaziranga National Park) and Nagaland (Naga Hills) [4]. In ancient times, the plant was found almost in every home garden or backyard garden of the Garo people. The plants of *Citrus indica* were once abundant in Meghalaya in India and its adjacent states. Unfortunately, in recent times, the number of plants had drastically decreased and as per International Union for Conservation of Nature (IUCN) norms, the plant fell under the list of endangered species and became endemic to the Garo Hills region of Meghalaya. Presently, the plants, due to their rarity, are found non-uniformly and in isolated regions only in the zone of the buffer of Nokrek Biosphere Reserve [6].

Hence, various initiatives had been taken by the Government of India to save the plant from extinction. One of them was the establishment of the National Citrus Gene Sanctuary in the Nokrek Biosphere Reserve in 1984 [10]. Also, the North-eastern Regional Agricultural Marketing Corporation Limited in 2011, applied for its Geographical Indicator (GI) tag from the Government of India, which was eventually granted. Thus, it became the second plant from Meghalaya after the Khasi mandarin to be given such protection [11].

However, not many efforts were done by the common people to save the plant, especially by the people of Garo Hills, where the fruit belongs. Small indirect efforts by the local authorities such as limiting the practice of jhum cultivation could be a big saviour of the plant [12]. Under the PMEGP (Prime Minister's Employment Generation Programme) scheme, bee boxes were distributed to locals. Though the bee boxes were given to set up an apiary, the bees would fly towards the *Citrus* plants and pollinate them, thereby indirectly helping the plants in reproduction [13]. Local brands of Meghalaya like Zizira make products using indigenous plants. They also write blogs to popularize local plants like *C. indica*, and thereby create awareness among people about the plant [14]. However, a direct effort by the people to save the plant was not seen except in Manipur. Here, the Rongmei tribe converted their forest into sacred grooves to protect some indigenous plants including *Citrus indica* [15].

It should be noted that these efforts were not enough. Even with the Government of India's intervention, the plant remained under the category of endangered species. It is due to the reason that the efforts have been made but blindly. Limited research work had been done on the plant, which mainly included its characterization, establishing its relationship with other *Citrus* species, and identification in other places. There had been no study to understand the cause of its endangerment. Had there been knowledge of why the plant started disappearing, necessary steps could be taken to save the plant.

The disappearance of the progenitor orange from the ecosystem would be a great loss to biodiversity. Without the mother orange, the future of modern oranges will be at stake. Hazarika (2012) believed that the plant possesses hardy genes. Hence, the study of those genes and their incorporation into the modern orange can enhance the rate of survival of the other oranges [16]. The plant is also used by the Garo tribes as an ingredient for traditional medicine. This also needs to be studied to conserve their indigenous knowledge. Therefore, if the plant is lost, the knowledge for the preparation of these lifesaving traditional medicines will also be lost from the world. Therefore, the present study was taken up to study the reasons causing its reduction from the earth and whether the plant can be conserved and grown in areas outside the buffer zone of Nokrek Biosphere reserve.

4.2 Materials and Methods

The present study was divided into five categories: review of literature, data collection, morphological study, biochemical studies, and sensory evaluation.

4.2.1 Critical Literature Review

Published papers and articles related to *Citrus indica* were referred to involve studies on its natural habitat and know the view of other authors regarding its loss from the ecosystem. The articles were accessed through google scholar and shodhganga. The study started on the 1st of August 2020 and it was continued till

May 2022. The study period was for one year and nine months. Research works of Malik et al. (2006), Hazarika (2012), Malik et al. (2013), Borah et al. (2018), Devi et al. (2022), etc. were consulted [6, 15–18]. Several websites like Telegraph (last accessed 2022/05/15), geotsy (last accessed 2022/05/15), weather atlas (last accessed 2022/05/15), meteoblue (last accessed 2022/05/15), etc. were also referred [19–22].

The study conducted by Malik et al. (2006) focussed on the characterization and distribution of *C. indica.* The authors discussed the morphology and the probable reasons that caused the loss of *C. indica* from the ecosystem [6]. Similar work was also done by Hazarika et al. (2012), where they studied the genetic diversity of *Citrus* and the main reasons for genetic diversity depletion from the planet [16]. Malik et al. (2013) studied the socioeconomic importance of *Citrus* through a survey method [17]. Borah et al. (2018) discovered *Citrus indica* in Behali Forest, Assam [18]. Also, Devi et al. (2022) reported the presence of *Citrus indica* outside Nokrek (through survey and morphological characterization) in Dailong Forest, Manipur. They also gave some recommendations to save the plant by citing the example of the Rongmei tribes, who converted the forest area into sacred groves and stopped the practice of jhum cultivation [15].

This methodology for the studies in review of literatures was analysed. Since very few studies have been conducted on the plant itself, studies can be targeted to ensure that the data collection can be centralized. Thereby, local tribals of the region in which the plant is grown can be engaged further to awareness-based biodiversity conservation.

4.2.2 Data Collection

Locals from Tura town and surrounding areas were interviewed about their knowledge of *C. indica.* It was a personal face-to-face interview with 50 respondents (of the age range of 20–50 years) belonging to the Garo community of West Garo Hills, Meghalaya. The mode of communication was in the local Language, i.e. Garo. The prepared semi-structured interview was filled at the spot of the interview and doubts regarding the questions were clarified [23, 24].

The research questions were the following:

1. Were the people aware of *C. indica*?
2. The places where the plants were grown as per their knowledge.
3. Did the plant grow in other places except for Nokrek?
4. Were the locals cultivating the plant?
5. If not, then why were they not cultivating the plant?
6. Was the plant consumed, and if so, then which part?
7. Were they aware of its uses, especially medicinal properties?

4.2.3 Morphological Studies

Fruits (ripe) were collected from the local market of Tura. Eight morphological features that had been taken for the present study were: size of the fruit, colour of the peel and the interior portion of the fruit, number of segments, seed number, seed sizes and seed colours. The method was done in three replicates and the observations were noted [6].

4.2.4 Physio-Chemical Studies

The parameters that had been considered for the present study were the juice content of the fruits, pH and TSS (Total Soluble Solids). The procedures were also performed in replicates of three and the results were noted.

Juice content: Juice content was measured by the method given by Moreno (2003). A total of 10 fruits were used for the extraction of fruit juice. The fruits were weighed, and juice was extracted, strained, and weighed in grams (Fig. 4.1). The results were expressed in % juice content [6].

% Juice content = (Juice Weight /Fruit Weight) x 100 [25]

pH: pH was measured with a pH meter (Hanna's portable pH meter, HI 9813–5) using the 945.27 Association of Official Analytical Chemists (AOAC) official method. The pH meter was calibrated, and then the probe was placed in the juice of *C. indica*. This method was repeated two more times and the readings were noted [26].

TSS: Total Soluble Solids (which mainly indicated sugar) were measured using a TSS meter (KEM Brix Meter BX-1). The brix meter was calibrated using distilled water. A drop of *C. indica* juice was placed using a dropper in the brix meter. The reading was noted. The procedure was done in triplicates [27].

Fig. 4.1 Pictures depicting the procedure adopted to produce juice from *C. indica* fruits

Table 4.1 Scorecard for sensory evaluation by the panellists

Serial no	Judgement	Score
1	Like Very much	9
2	Like much	8
3	Like moderately	7
4	Like slightly	6
5	Neither like nor dislike	5
6	Dislike slightly	4
7	Dislike moderately	3
8	Dislike much	2
9	Dislike very much	1

4.2.5 Sensory Evaluation

Sensory evaluation was carried out for the juice of *C. indica* following a 9-point Hedonic scale (ISO 11136:2014/AMD 1:2020) rating to evaluate the like/dislike in terms of taste, texture, colour, aroma, mouth feel, and general acceptability (Table 4.1) [28]. It was a scale that was given to the consumers to rate the organoleptic properties of the fruit. The maximum point allotted was 9 which meant that they liked the fruit juice very much. Accordingly, average scores would indicate a moderate preference, while lower scores like 2 and 1 indicated that the consumers did not like the relevant property of the fruit.

In the present evaluation, eight panellists in the age range of 20–50 years were selected to participate. They were given the scorecard and asked to rate the taste, flavour, aroma and texture of *C. indica* juice. It was done in a replicate of three. Their scores were noted and analysed.

4.3 Results and Discussion

The findings of the present work had been presented in five sub-categories: Review of literature, Data collection, Morphological study, Physico-chemical study and sensory evaluation.

(i) **Review of literature:** The findings of these sections had been discussed in two categories: A) the cause of loss of *C. indica* and B) reports of occurrence of *C. indica* in places except Nokrek.

A. **The cause of loss of *C. indica***: The plant had a reputation for being a hardy plant as per Hazarika (2012) [16]. However, the plant was not completely resistant to diseases and had several threats that affected its survival and existence. It was observed that nature had very little role as its potential threat, as the main destruction was anthropogenic.

With respect to nature, the significant threat to *C. indica* was the geography of its habitat.

a. Geographical location of *Citrus indica*: The plant resided in the Nokrek Biosphere Reserve (25°32′N 90°7′E), which is a hilly region with lots of rainfall. It made the area prone to soil erosion. Due to heavy rainfall, there would be leaching of the topsoil leading to nutrient deficiency. Also, continuous rainfall and high humidity might enhance the growth of pathogens in *Citrus*. This deficiency in nutrients and pathogen attack would ultimately affect the plant's growth and survival [16].

   According to Marak et al. (2018), the majority of the pathogens were fungi. They discovered nine fungal diseases and one algal disease in *C. indica*. The presence of these pathogens was greatly influenced by the depth and zone of the plant and the season. The topsoil hosted beneficial microbes like *Penicillium digitatum*, *Aspergillus flavus*, *Rhizopus nigricans*, etc. which inhibited the growth of pathogens like *Capnodium citri*, *Alternaria citri*, *Cephaleuros virescens*, etc. However, as the depth of the soil increased, the positive dominant microbes' zone started to decrease, thereby paving way for the growth of pathogenic microbes. It was also observed by Marak et al. (2018) that the microbes like *Drechslera oryzae* which existed below the topsoil supported the growth of pathogens like *A.citri* [29]. Thus, the theory of Hazarika (2012) was supported by the fact that the leaching of the topsoil had a significant effect on the growth of *C. indica* [16, 29].

b. Agricultural pressure: People were more interested in growing crops that were profitable to them. With an increase in population, there had also been an increase in the demand for food. So, people tend to clear the land to produce crops of commercial importance. For that, they practiced jhum cultivation in the forest area which destroyed the wild *Citrus indica* plants [16].

   During earlier times, the practice of shifting cultivation was done after 20–30 years. However, due to agricultural pressure, the practice was done at an interval of 5–7 years. As a result, the wild plants would not have much chance to grow. In addition to that, frequent practice of the slash-and-burn method resulted in the destruction of topsoil and its erosion. The loss of Nitrogen and Potassium from frequent shifting cultivation was 58.9 and 7.1 (Mg ha − 1 yr − 1), respectively [30]. As per Zekri and Thomas, 2003, these nutrients were essential for the growth of *Citrus* plants [31]. Therefore, the loss of these nutrients would hamper the growth of the plant. With the loss of topsoil, the theory of Marak et al. (2018) (from the previous topic of geographical location) also got accepted [29, 31].

c. Urbanization: With modernization and the increase in population, there has been a demand for more space to buildings and roads. Therefore, to fulfil the demands, the forests were cleared. With the clearance of forest, the wild plants of *C. indica* are also lost. It had been an irony that once the plants were removed, the plants were not replanted and hence lost in this process [16].

   According to Mckinney (2008), around 80% of urban areas were usually covered by buildings and pavements and only less than 20% area was left for vegetation. Another problem with urbanization was that during the process of landscaping and maintenance of commercial and residential areas, the wild plants would get removed giving space to ornamental plants [32].

d. Change in climate: Nokrek Biosphere Reserve had been blessed with a pleasant cool climate. So, it was assumed that the Indian Wild Orange preferred a cool climate for its growth and reproduction [3]. With an increase in daily temperature and changing climate conditions, the plant's system might not cope up with the changes and hence failed to flower and produce fruit. The theory was supported by the findings of Nawaz et al. (2019), where they found that climate change had a major role in the yield and quality of *Citrus* plants. The increase in average temperature could trigger infestation by pests, higher incidence of disease and usage of water, thereby causing an overall increase in input cost. Fog, drought and heat stress reduced the yield and quality of the fruits. The processing cost and export risk would also increase with higher average temperature [33].

B. **Reports on the occurrence of *Citrus indica*:** As reported by Malik et al. (2006), the plant *C. indica* was endemic to North Eastern Himalaya and was found in Nowgong district, Khasi hills, Manipur, northeastern Naga hills of Nagaland, Kaziranga Reserve Forest in Assam and Garo hills of Meghalaya. Presently, may be, they only exist in their origin, i.e. Nokrek Biosphere Reserve, Meghalaya [6]. Contrary to their statement, few plants were reported in Behali Reserve Forest (26°54′32″N 93°18′29″E), Assam [18] and a plant was reported in Dailong Forest (25.0080° N, 93.5231° E), Manipur [15]. This has been depicted in Fig. 4.2. Thus, it proved that the plant did not enter its endemic stage. More research could lead to the detection of more *Citrus Indica* plants.

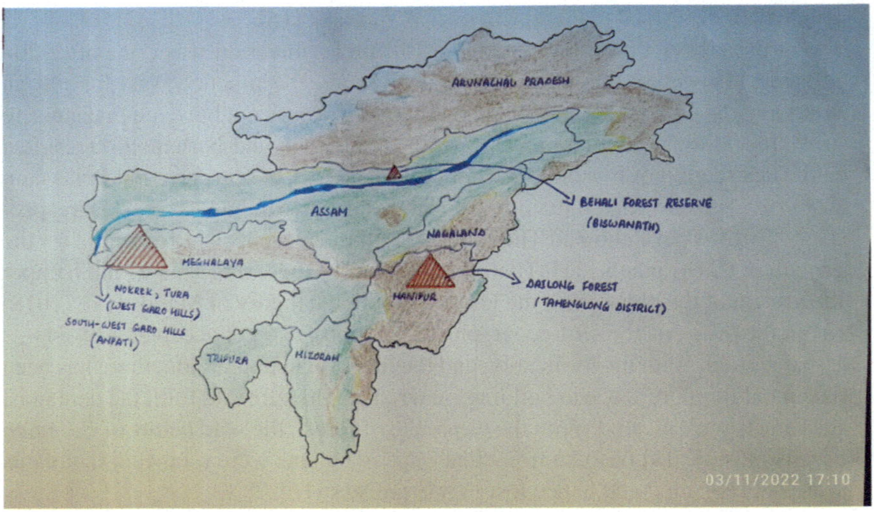

Fig. 4.2 Map of North-East India with triangles depicting the present location of *C. indica*

(ii) **Data collection:** The outcome of these methods gave an idea with respect to the beliefs of the Garo tribes of Meghalaya regarding *C. indica*. The information procured from them was very interesting. This is because, although the plant originated in Garo Hills, very few people knew about it. Mostly the elderly people from the outskirts of Tura town gave a positive response. According to them, the fruit was used in the traditional rituals of the Garos. The fruit was also used for the preparation of their ethnic medicines. However, they did not disclose the method for it. They either did not know it or did not want their secret to getting revealed as it was a process of knowledge passed down from generation to generation in each family. However, evidence of its medicinal properties was obtained from the work of Momin et al. (2016). According to them, the fruit was used for the treatment of jaundice, smallpox and stomach ailments in animals and humans. Chetry et al. (2021) reported that the fruit could also cure hypertension and snake bite. The juice of the fruit was taken for dehydration and fatigue [9, 34].

The respondents of the present study also had limited knowledge with respect to the existence of *C. indica* in the wild. According to the majority, the plant could exist only in Nokrek Biosphere Reserve owing to its need for a particular climate and soil. Even if grown outside Nokrek, the plant might not enter the fruiting stage, and if it fruits, the quality would be inferior to Nokrek's. Thus, their theory satisfied the concept given by Malik et al. (2006) [6].

However, the breakthrough discovery was observed when six respondents reported the presence of *C. indica* in the adjoining areas of Tura (25.5141° N, 90.2032° E) (Fig. 4.4). The areas included Tura peak, Darengre, Chandigre, Duragre of West Garo Hills district of Meghalaya. Some plants were also grown in the two households in Tura town. One plant was spotted in the Katuli village near Ampati (25.4615° N, 89.9347° E) of the Southwest Garo Hills of Meghalaya (Fig. 4.5). The ones reporting the *C. indica* plants were mostly from the outskirts of Tura. Except for the ones having the *C. indica* plants in their place, the residents of Tura had no information regarding the existence of *C. indica* in the wild except Nokrek.

The plantation of *C. indica* was also spotted on the farm of the Department of Rural Development and Agricultural Production, NEHU Tura Campus and on the outskirts of Tura town. The plants grown were tissue cultured plants (Fig. 4.3).

The existence of *C. indica* plants outside Nokrek was an exception to the concept given by Malik et al. (2006) [6]. But it supported the work of Devi et al. (2022) and Borah et al. (2018) that *C. indica* could exist outside Nokrek [15, 18]. It proved that there was a lack of information among the locals. The data from our survey also revealed that almost 90% of the respondents believed that the fruit was sour and unfit for consumption (which has been proved in the section on sensory evaluation). Hence, with modernization and an increase in the demand for other popular fruits, people were least interested in growing the plant. They preferred growing plants of commercial importance.

Fig. 4.3 Plants of *C. indica* growing in RDAP, NEHU, Tura Campus

Another observation was that there was no proper documentation of the usage of *C. indica* among the people of the region. The elders of the Garo society were reluctant to share the information with the authors. Hence, it could be assumed that with modernization, their knowledge might not pass down to the next generation. Hence, due to people being unaware of its uses, they would not be interested to grow these plants. According to Devi et al. (2020), difficult topographic regions and inaccessibility could be a reason that the plant remained unknown to the world [15].

As the hypothesis of Malik et al. (2006) had been disapproved, one clarification was required. This is to know whether the plant really needed a microclimate for its growth. The recent work of Devi et al. (2020) supported that of Malik et al. (2006). The authors stated that the plant preferred a cool shady place [6, 15]. But to know if can survive outside the weather of Nokrek, the average maximum and normal temperature of Nokrek was recorded and compared with the maximum and minimum temperature of the places where the plant was spotted.

Nokrek had the highest temperature of 33 °C in the months of July, August and September, while the lowest temperature was 14 °C (shown in Tables 4.2

Fig. 4.4 Plants of *C. indica* growing in the household of Tura town

Fig. 4.5 Plants of *C. indica* growing near Ampati, Southwest Garo Hills, Meghalaya

and 4.3; Figs. 4.6 and 4.7). The average highest temperature of Nokrek was 28.5 °C and the lowest temperature of Nokrek was 20.58 °C (Table 4.4; Fig. 4.8). The average temperature indicated that Nokrek experienced a pleasant climate which would neither be hot in summer nor too cool in winter.

The maximum temperature of Tura was 32.5 °C during the month of April and the minimum temperature of Tura was 10.8 °C in January (shown in Tables 4.2 and 3; Figs. 4.6 and 4.7). The average maximum temperature of Tura was 29.25 °C, while the average lowest temperature of Tura was 16.56 °C (Table 4.4; Fig. 4.8). The average maximum temperature of Tura town was comparatively higher than Nokrek by almost 1^{0}C, while the minimum temperature was lower than Nokrek by around 4 °C.

In case of Ampati, the maximum temperature was 35 °C in the month of April, while the lowest temperature was 13 °C in the month of January (Tables 4.2 and 3; Figs. 4.6 and 4.7). The average maximum temperature of Ampati was 30.33 °C and the average lowest temperature was 20.83 °C (Table 4.4; Fig. 4.8).

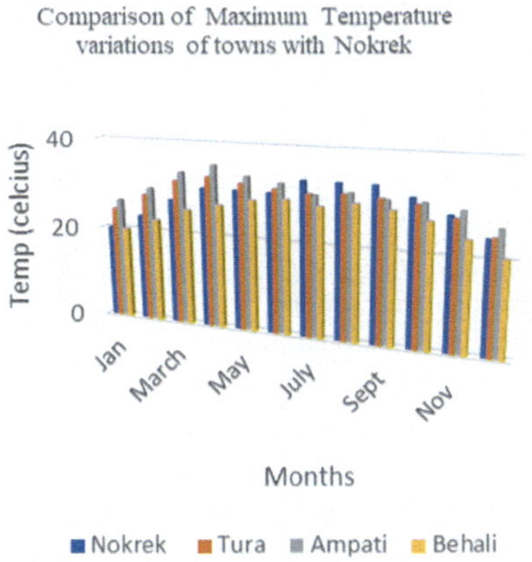

Fig. 4.6 Bar chart for comparative analysis of Maximum Temperature of towns hosting *C. indica*

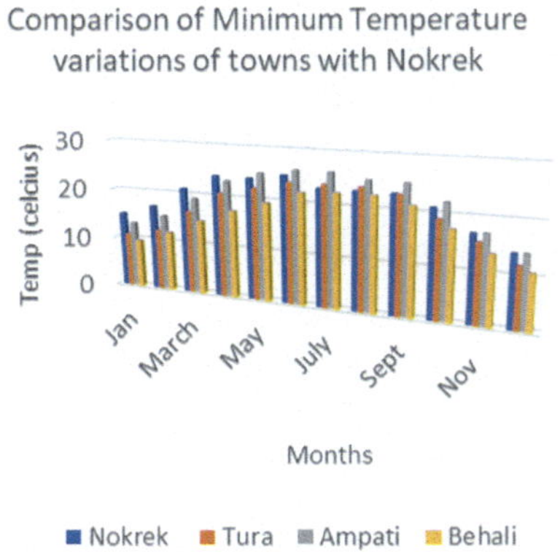

Fig. 4.7 Bar chart for comparative analysis of minimum temperature of towns hosting *C. indica*

Table 4.2 Maximum temperature data of the towns hosting *C. indica*

Months	Nokrek	Tura	Ampati	Behali
Jan	20	24.1	26	19.8
Feb	23	27.7	29	22.3
March	27	31.3	33	25
April	30	32.5	35	26.6
May	30	31.6	33	28.2
June	30	30.9	32	28.8
July	33	30.3	30	27.9
August	33	30.8	31	29
Sept	33	30.4	30	28.2
Oct	31	29.7	30	26.6
Nov	28	27.5	29	23.6
Dec	24	24.3	26	20.3

Table 4.3 Minimum temperature data of the towns hosting C. indica

Months	Nokrek	Tura	Ampati	Behali
Jan	15	10.8	13	9.5
Feb	17	12.1	15	11.8
March	21	16.5	19	14.7
April	24	20.6	23	17.3
May	24	22.1	25	19.4
June	25	23.7	26	21.9
July	23	23.8	26	22.3
August	23	23.8	25	22.4
Sept	23	22.9	25	21.1
Oct	21	19	22	17.4
Nov	17	15.4	17	13.5
Dec	14	11.8	14	10.7

The average maximum temperature of Ampati was higher than Nokrek by 2 °C, while the lower temperature was almost the same.

The maximum temperature of Behali was 28.8 °C in the month of June, while the minimum temperature of Behali was 9.5 °C in January (Tables 4.2 and 4.3; Figs. 4.6 and 4.7). The average maximum temperature of Behali was 25.5 °C which was 3 °C lower than Nokrek and the minimum temperature was 16.83 °C which was approximately 4 °C lower than Nokrek (Table 4.4; Fig. 4.8).

So, it could be stated that marginal temperature variations could be tolerated by the plant. This proved the hypothesis of Hazarika (2012) and that the plant had hardy genes [16]. The observation also supported the statement of Devi

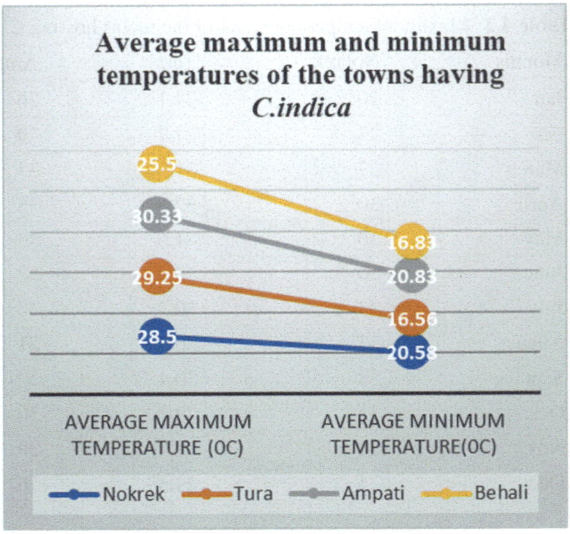

Fig. 4.8 Graph depicting the average temperature of towns hosting *C. indica*

Table 4.4 Average maximum and minimum temperature data of the towns hosting *C. indica*

Towns	Average maximum temperature (°C)	Average minimum temperature (°C)
Nokrek	28.5	20.58
Tura	29.25	16.56
Ampati	30.33	20.83
Behali	25.5	16.83

et al. (2022) which conveyed that the lack of access to topological variations could become an obstacle in spotting the plant [15]. There could be more plants existing in the wild.

Morphological study: The morphological study had some interesting observations. The shape of the fruit was oblate, and the colour of the peel was green in the unripe stage. However, at the ripened stage, both the peel and segment colour were deep orange. The size of the fruit was around 3 cm in diameter (shown in Table 4.5 and Fig. 4.9). The number of segments varied from fruit to fruit. Some fruits had eight segments while others had ten segments. Each segment size was around 2 cm (supported by Fig. 4.10). Another interesting finding was the seed count. Some fruits had one seed, while other fruit had fourteen seeds (presented in Figs. 4.12 and 4.13). The seed measured one cm and the colour was ivory (shown in Table 4.5 and Fig. 4.11). The results of the morphological study are given in Table 4.3.

Thus, the morphological variations proved that the fruit was indeed unique, the observation was supported by the work of Malik et al. (2006) and Upadhaya and Chaturvedi (2013), as they found the fruit to be smaller than the other indigenous

Table 4.5 A summary of morphological features of *C. indica* fruit

Sl.No	Fruit Features	Description
1	Shape	Oblate
2	Colour of the peel	Deep orange when ripe, green while unripe
3	Size	3 cm in diameter
4	Segments	8–10
5	Colour of the segment (pulp)	Orange
6	Segment size	2 cm
7	Seed no	Varies, sometimes one seed in one fruit, while others had 14 seeds in one fruit
7	Seed size	1 cm
8	Seed colour	Ivory colour

Fig. 4.9 Colour and size of *C. indica*

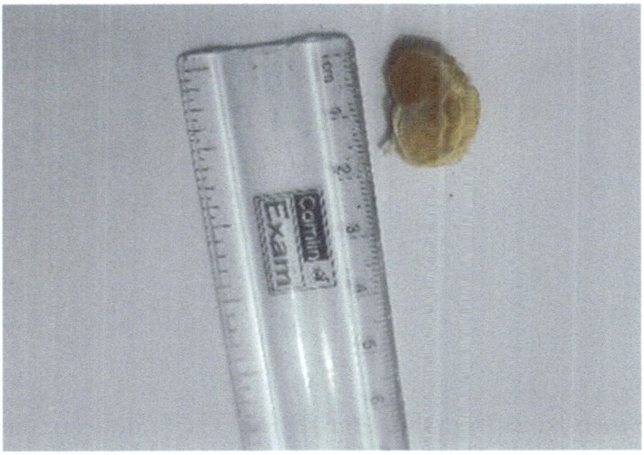

Fig. 4.10 Colour and size of the fruit segment of *C. indica*

Fig. 4.11 Colour and size of *C. indica's* seed

Fig. 4.12 Fruit of *C. indica* having many seeds

Citrus fruits from the same region. The findings of the present study in the morphological aspect had similarity to the work of Malik et al. (2006) and Upadhaya and Chaturvedi (2013), as the colour of the fruit in the ripened stage was the same, i.e. orange [6, 35]. However, the size of the fruit in the present study was bigger than in the previous work. These might be due to the collection of the samples in the

Fig. 4.13 Fruit of *C. indica* with only one seed

different time ranges and from different places. It also gave an indication that with favourable conditions, the size of the fruit could increase by a few millimeters.

There was a correlation with Malik et al. (2006) in terms of the segment number of fruit and the seed colour of the fruit. But there was a significant contradiction in the observation of seed count. While a recent study revealed that the seed number varied from 1 to 14, Malik et al. (2006) reported the seed count could vary from 7 to 12 [6]. It came out to be a major revelation because as per Kumar et al. (2010), the major means of propagation of the plant *C. indica* was through seeds. Also, the dispersion of the seeds in the Nokrek Biosphere Reserve had been a big setback due to its topological condition. The dispersal of the seed had been due to gravity. Due to the attraction of gravity, the fruit would fall and the seeds would get scattered [36]. Hence, in this case, the plant bore fruits with one seed and due to the topological conditions, it could not go far away, thereby it could not reach a place unfavourable for its growth. Hence, the chances to get progeny were also less. With few progenies, the future of the plant would be jeopardized.

(iii) **Physico-chemical analysis:** The analysis gave an impression of the basic characteristics of the fruit in terms of its weight, juice content and taste. The weight of ten fruits was 157 g (shown in Fig. 4.14), indicating that per fruit could weigh around 1.57 g. The juice procured from these ten fruits were 24 mL (expressed in Fig. 4.15), indicating that the juice content of the fruit was 15%. The fruit had an incredible low pH of 2.49, and the TSS was found to be 7.3%.

As per Ketsa (1988), the size of the fruit had direct relationship with juice content. The size of the fruit had inverse relationship with the TSS. Therefore, with respect to size of the fruit, the juice content was found to be satisfactory [37]. The same concept was accepted by Malik et al. (2006) and Upadhaya and Chaturvedi (2013). According to the authors, the juice content of the fruit was

Fig. 4.14 Weight of ten *C.indica* fruits

high [6, 35]. But it should be noted that high fruit content was referred to the fruit size.

Total Soluble Solids (TSS) indicated the sugar content in the juice. Hence, the readings of 7.3% indicated mostly sucrose in 100 mL of the juice. It also satisfied the theory of Ketsa (1988) as being a small sized fruit, the sugar content was not that high [37].

The detailed summary of the physico-chemical analysis is given in Table 4.6.

(iv) **Sensory evaluation**: The extracted juice presented to eight panellists had varied opinions. In case of colour, the maximum point given was 8, and the minimum point given was 5. This conveyed that none of the panellist liked the colour of the juice very much and none disliked it. Out of the eight people, two people gave 8 points which meant that they liked the colour much. Three people gave 7 points indicating that they liked the colour moderately. Two people gave 6 points, and one gave 5 points meaning that they slightly liked it and neither liked nor disliked, respectively. The average score given by the panellists for the colour was 6.75 which meant they moderately liked the colour of the fruit juice.

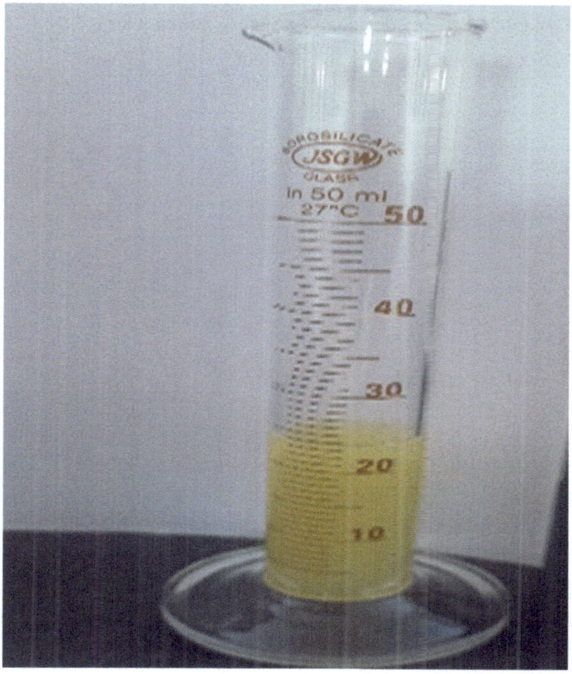

Fig. 4.15 Juice extracted quantity (ml) from ten fruits of *C. indica*

Table 4.6 Physico-chemical parameters of *C. indica* fruit

Sl.No	Parameters	Results
1	Fruit weight (10 fruits)	157 g
2	Juice extracted (10 fruits)	24 ml
3	Total Juice content	15%
4	pH of fruit juice	2.49
5	TSS of fruit juice	7.3%

The flavour of the fruit also had diverse result with a maximum point of 7 and minimum of 2. This conveyed that there was variation in the opinion of the panellists. Some liked the flavour very much and some disliked the flavour very much. Here, two of the panellists gave 7 points which meant that they moderately liked the flavour of the juice. Four of them gave 6 points which reflected moderate likeness of the flavour. One gave 3 points, while the other gave 2 points indicating that they disliked it and disliked the flavour very much, respectively. The average score calculated for the flavour of the fruit juice was 5.37 and affirmed that the panellist neither liked nor disliked the taste.

The aroma of the fruit had significantly higher scores. The maximum score was 8 while the minimum was 6. It showed that all of them liked the aroma. The highest rating of 8 (liked very much) was given by three panellists, three gave 7 points (liked

Table 4.7 Sensory data summary of *C. indica* juice

S.No. of Penalist	Colour	Flavour	Aroma	Texture
1	6	6	7	8
2	5	3	6	8
3	7	2	6	7
4	8	6	8	8
5	8	6	8	8
6	7	7	7	7
7	6	6	7	7
8	7	7	8	7
Overall acceptibility	6.75	5.37	7.31	7.5

slightly) and two gave 6 points (liked moderately). The average score attained by the aroma was 7.35 (liked moderately).

The texture had similar higher ratings like the aroma. The maximum score was 8 and the minimum was 7. Thus, the texture of the fruit juice was highly accepted. Half of the panellist gave 8 points (like much) while the other half gave 7 points (like moderately). The average score was 7.5, meaning that the texture was moderately liked.

The score card summary has been presented in Table 4.7.

Though the flavour was not much liked by the panellist due to extreme sourness, but the aroma was highly appreciated. Hence, the aroma could be used in perfume industry and confectionaries.

4.4 Conclusions

Citrus indica, the mother of oranges had been of great importance to the Garo tribe. The present study was to know the reasons of its disappearance from the planet. However, in the process, many new facts about the fruit were discovered. The first was that the plant and especially its fruit was unique and the fruit smaller than other *Citrus* fruits. With its miniature fruits, it had the potential to be cultivated as an ornamental plant. However, it is unfortunate that the plant had been in a very critical condition with its number decreasing day by day. Literature survey had contrasting views about the plant. Malik et al. (2006) claimed the plant to be an endemic to Nokrek Biosphere Reserve. This is contrary to the findings of Borah et al. (2018) and Devi et al. (2022), who reported the existence of the plant in Behali Forest Resserve and Dailong Forest, respectively [6, 15, 18]. Even the presence of plants in the surrounding of Tura and Ampati (known through data collection) proved that little alterations in climate could be tolerated by the plant. The plant can survive outside its microclimate and might not have entered the endemic stage. There could be more plants in places having similar climate but due to difficult geography and lack of access and knowledge about the other places and the plant, the plants remained a secret. The cause of destruction of the plant was also studied. This reflected that nature (geographic location and pathogens) had a very little role in the plant's disappearance. The major antagonist for the same was human intervention like urbanization, agricultural practices such as jhum cultivation, modernization, lack of cultivation of the plant, etc. Upon further studies with the semi-structured interview schedule to understand its pathetic condition, there was another serious issue as the plant and its fruits were a mystery among many locals of Garos, the people belonging to the place of origin of *Citrus*. This observation directed towards two possible conditions: one in which the locals did not know about the plant, or the other one in which they knew but did not want to reveal it. With cases of reluctance to share knowledge, the knowledge of significance of the plant would get lost as there was not any proper documentation of the same. The ignorance about the plant could be due to modernization and lack of interest in the plant among the locals. The ones who knew about it believed that the fruit was very sour and unfit for consumption. They were interested in growing plants that could draw some profit to them. The sourness of the fruit was confirmed through sensory evaluation and physico-chemical analysis. Morphological study had a major revelation regarding the variable seed counts of the fruit. While some fruit had one seed, other fruit could have around twelve seeds. Since the propagation of the plant is mostly through seeds, due to geographical barriers, if the fruit having only one seed could not disperse into a favourable place, then the chance of its survival is scare, thus limiting the progeny count.

With the loss of the progenitor oranges from the earth, it would be a major loss to biodiversity as the plant had hardy genes that could be incorporated into some susceptible *Citrus* species and medicinal properties known to the Garos. Simply relying on the government to save the plant would not be enough, while the morphological and physico-chemical parameters of the plant were beyond the reach of humans, proper documentation and awareness on the traditional knowledge could be done to help revive the interest of people towards the plant. Awareness of the plant's importance, discontinuation of jhum cultivation and conversion of the place of its occurrence into sacred grooves can be practiced. Since there had been no utilization of the fruits, they were simply wasted. However, with its strong aroma, the peels could be used for the extraction of essential oils for pharmaceutical purposes and confectioneries. Even value addition of the fruits and juices might enhance its taste and make it popular among consumers. With the demand of the fruit, the cultivators might be interested in growing the plant, thereby saving it from getting extinct.

In summary, the book chapter demonstrated relevant research emphasis for the biodiversity conservation of *Citrus Indica*. Similar studies can be targeted for endangered and endemic wild species in the region. Other studies emphasizing complex relationship between *Citrus Indica* and other species can be explored to substantiate upon knowledge-based awareness and prominence for its eco-sustainability in the midst of urbanization.

References

1. Statista Statistics page. https://www.statista.com/statistics/264001/worldwide-production-of-fruit-by-variety/. Accessed 10 Nov 2022
2. Anbalagan, T., Tejkumar, J.P., George, A., Kommu, K.K.: A snapshot of citrus in India. AgroSci. Today **2**(2), 77–80 (2021)
3. Hynniewta, M., Malik, S.K., Rao, S.R.: Karyological studies in ten species of Citrus (Linnaeus, 1753)(Rutaceae) of North-East India. Comp. Cytogenet. **5**(4), 277–287 (2011)
4. Singh, B.: Establishment of First Gene Sanctuary in India for Citrus in Garo Hills, 1st edn. Concept Publishing Company, New Delhi, India (1981)
5. Scora, R.W.: On the history and origin of Citrus. Bull. Torrey Bot. Club **102**(6), 369–375 (1975)
6. Malik, S.K., Chaudhury R., Dhariwal, O.P., Kalia, R.K.: Collection and characterization of *Citrus indica* Tanaka and *C. macroptera* Montr.: wild endangered species of northeastern India. Genet. Resour. Crop. Evol. **53**(7), 1485–1493 (2006)
7. Tanaka, T.: On certain new Species of citrus. Stud. Citrol. **2**(1), 155–164 (1928)
8. Bhattacharya, S.C., Dutta, S.: Classifications of citrus fruits of assam. ICAR Sci. Monogr. **20**(1), 1–110 (1956)
9. Chetry, S., Suresh, C.P., Chaurasiya, A.K., Lyngdoh, J.P.: Morpho-Physico-Chemical Characterization of Indian Wild Orange (Citrus indica Tanaka) grown in Nokrek Biosphere Reserve, Garo Hills, Meghalaya. India. Int. J. Curr. Microbiol. App. Sci. **10**(1), 1910–1922 (2021)
10. Downtoearth in-depth page dated 31st January,2009. https://www.downtoearth.org.in/coverage/local-citrus-goes-global-2951. Accessed 13 July 2020
11. The Telegraph e-paper dated 15th May,2015. https://www.telegraphindia.com/states/north-east/gi-tag-for-citrus-fruit/cid/1530418. Accessed 15 May 2022
12. India Culture Nokrek Biosphere Reserve page. https://indianculture.gov.in/node/2537419. Accessed 11 Oct 2022
13. West Garo Hills District DCIC Tura page. https://westgarohills.gov.in/grt/dcic-tura-gr/. Accessed 11 Oct 2022
14. Zizira Wild Citrus Fruits of Meghalaya - Uses You Never Knew About page. https://www.zizira.com/blogs/plants/wild-citrus-fruits-meghalaya-uses. Accessed 11 Oct 2022
15. Devi, E.J., Labala, R.K., Sanabam, R., Singh, N.S., Modak, R., Devi, H S.: New report of *Citrus indica* Yu. Tanaka, a wild progenitor species of citrus from Dailong Forest, Manipur, and recommendation for its conservation. Genet. Resour. Crop. Evol. **69**(2), 1–14 (2022)
16. Hazarika, T.K.: Citrus genetic diversity of north-east India, their distribution, ecogeography and ecobiology. Genet. Resour. Crop Evol. **59**(6), 1267–1280 (2012)
17. Malik, S.K., Kumar, S., Singh, I.P., Dhariwal, O.P., Chaudhury, R.: Socio-economic importance, domestication trends and in situ conservation of wild Citrus species of Northeast India. Genet. Resour. Crop Evol. **60**(5), 1655–1671 (2013)
18. Borah, D., Kafley, P., Tangjang, S., Das, A.P.: Population structure and conservation of endangered *Citrus indica* Yu. Tanaka (Rutaceae) in Behali Reserve Forest of Assam, India. Pleione **12**(2), 181–186 (2018)
19. Geotsy weather page. https://geotsy.com/en/india/nokrek-national-park-61692/weather-and-climate. Accessed 15 May 2022
20. Weather atlas Tura climate page. https://www.weather-atlas.com/en/india/tura-climate. Accessed 15 May 2022
21. Meteoblue History and climate page. https://www.meteoblue.com/en/weather/historyclimate/climatemodelled/ampati_india_11875610. Accessed 15 May 2022
22. Weather atlas Behali climate page. https://www.weather-atlas.com/en/india/behali-climate. Accessed 15 May 2022
23. Cotton, C.M.: Ethnobotany: Principles and Applications, 1st edn., pp. 347–374. John Wiley and Sons Ltd, Manchester, England (1996)
24. Martin, G.J.: Ethnobotany: A Method Manual, 1st edn., pp. 267–347. Champan and Hall, London, UK (1995)

25. Sanchez-Moreno, C., Plaza, L., De Ancos, B., Cano, M.P.: Vitamin C, provitamin A carotenoids, and other carotenoids in high-pressurized orange juice during refrigerated storage. J. Agric. Food Chem. **51**(3), 647–653 (2003)
26. Horwitz, W., and Latimer, G.W.: AOAC. Official Methods of Analysis of Association of Official Analytical Chemists, 18th Ed. AOAC Internationals, Gaithersburg, USA (2005)
27. Magwaza, L.S., Opara, U.L.: Analytical methods for determination of sugars and sweetness of horticultural products—a review. Sci. Hortic. **184**(1), 179–192 (2015)
28. IS0. Sensory analysis — Methodology — General guidance for conducting hedonic tests with consumers in a controlled area — Amendment 1, ISO 11136:2014/AMD 1:2020. https://www.iso.org/standard/73557.html. Accessed 15 May 2022
29. Marak, L.B.C., Pereira, L.S., Chakraborty, R., Majumdar, D.: Plant pathogens and soil microbes associated with *Citrus indica* in Nokrek biosphere reserve of Meghalaya. Res. Crop. **19**(3), 486–491 (2018)
30. Nath, A.J., Brahma, B., Lal, R., Das, A.K.: Soil and Jhum Cultivation. Encyclopedia of Soil Science CRC Press, pp. 1–9 (2016)
31. Zekri, M., Obreza, T.A.: Plant Nutrients for Citrus Trees, 1st edn., pp. 1–5. University of Florida Cooperative Extension Service, Institute of Food and Agricultural Sciences, EDIS, Gainesville (2003)
32. McKinney, M.L.: Effects of urbanization on species richness: a review of plants and animals. Urban Ecosyst **11**(2), 161–176 (2008)
33. Nawaz, R., Abbasi, N.A., Hafiz, I.A., Khalid, A., Ahmad, T., Aftab, M.: Impact of climate change on kinnow fruit industry of Pakistan. Agrotechnology **8**(1), 1–6 (2019)
34. Momin, K.C., Suresh, C.P., Momin, B.C., Singh, Y.S., Singh, S.K.: An ethnobotanical study of wild plants in Garo Hills region of Meghalaya and their usage. Int. J. Minor. Fruits Med. Aromat. Plants **2**(1), 47–53 (2016)
35. Upadhaya, A., Chaturvedi, S.S.: Biodiversity and Nutritive Value of Semi Wild and Wild Citrus Species of Meghalaya. North Eastern Hill University, Shillong (2013)
36. Kumar, S., Jena, S.N., Nair, N.K.: ISSR polymorphism in Indian wild orange (*Citrus indica* Tanaka, Rutaceae) and related wild species in North-east India. Sci. Hortic. **123**(3), 350–359 (2010)
37. Ketsa, S.: Effects of fruit size on juice content and chemical composition of tangerine. J. Hortic. Sci. **63**(1), 171–174 (1988)

Chapter 5
Distribution Mapping and Diversity Assessment of *Ilex venulosa* from Meghalaya Using Internal Transcribed Spacer Regions, *matK* and *rbcL*

Amilia Nongbet and N. K. Chrungoo

5.1 Introduction

India is one of the most biodiversity-rich countries in the world. Among nearly 45,000 plant species recorded in India, 7500 species have been recognized to have medicinal properties [1]. These plants have been used extensively since time immemorial in traditional (Ayurveda, Unani, Siddha, Homeopathy) as well as modern medicinal practices [2]. While many medicinal plants have been domesticated and are under cultivation, a large number of species are present in the Indian forests as wilds. The historic dependency of different communities on forest plants for their day-to-day needs has always impacted the density, if not the range of distribution, of such species. While many medicinal plants have been domesticated and are under cultivation, a large number are present in the Indian forests as wilds. Amongst the existing land resources in India, North East (NE) India is a 'biodiversity hotspot' and forms a unique biogeographic province harboring major biomes recognized in the world under the Indo-Burma biodiversity ring. The region comprises a wide range of phytogeographical zones. The most prominent medical plants found in the region include *Acorus calamus, Costus speciosus, Dioscorea spp, Hydnocarpus kurzii, Gloriosa superb, Rauvolfia serpentine, Coptis teeta, Gentiana kurroa, Nardostachys jatamansi, Picnorhiza kurroa, Podophyllum, Acer laevigatum, Ardisia disperma, Amenotaxus assamica, Baliospermum micranthum, Embelia ribes, Embelia floribunda, Ilex khasiana, Ilex venulosa, Ilex embeliodes, Diospyros celibica, Ardisia odontophylla,*

A. Nongbet (✉)
University of Science and Technology, Meghalaya, Techno City, 9th Mile, Kiling Road, Baridua, Ri-Bhoi 793101, Meghalaya, India
e-mail: amilianongbet1125@gmail.com; amilia@ustm.ac.in

N. K. Chrungoo
Centre for Advanced Studies in Botany, North Eastern Hill University, Shillong 793022, India

R. V. S. Uppaluri and L. Rangan (eds.), *Conservation of Biodiversity in the North Eastern States of India*, https://doi.org/10.1007/978-981-99-0945-2_5

Dendrobium densiflorum, Pleione maculate and *Piper griffthii* [3]. Over the last decade, the high rates of biodiversity loss have come to the forefront as one of the most important issues that need urgent attention. One of the major hurdles in biodiversity conservation is the near absence of scientific conservation efforts with a special focus on taxonomically and economically important plants that are facing a higher threat of extinction. A major hurdle in devising an appropriate conservation strategy is the limited or unavailable data on population biology and niche characteristics of such plants. While efforts to conserve such plants through habitat conservation, sustainable harvest, and species-specific conservation actions are being accelerated to reduce pressure on plant resources, numerous medicinal plant species are in the process of extinction owing to the anthropological pressures [4, 5].

Dearth of fundamental biological data, such as fine-scaled mapping of individual species and approximations of population figures, makes it difficult to quantify genetic diversity in plants. A vulnerable species' estimation of genetic variation within and across communities is especially crucial because biologically varied populations may be employed to increase a species' extant populations while reducing the likelihood of genetic drift. Molecular profiles of a species have great importance in determining the extent of diversity amongst populations as well as in defining phylogenetic relationships of the species with other species in the same genus, assessment of biologically varied populations for use as source germplasm for conservation activities and confirmation of the taxonomic status of contested taxa [6–10]. A narrow genetic base within the populations is one of the significant intrinsic drivers of species extinction. The possibility of the plant being extinct is owing to the genetic drift elevated with the fragmented population size of these threatened species. Data of their diversity and distribution within and among populations of such plant species is vital for the designing of strategies for their conservation and genetic enrichment.

The use of genetic polymorphism assessed by different biochemical and molecular markers has been applied in many species to resolve genetic relationships for purposes of phylogenetic studies, varietal identification as well as for QTL mapping. Analysis of DNA polymorphism has been proven to be very successful in addressing different issues in the management of plant genetic resources, such as the evaluation of germplasm [11], identification of duplicate accessions [12], detection of misclassifications [13], and genetic improvement [14] and estimation of genetic diversity [15, 16]. Molecular profiles of a species have tremendous significance in determining the extent of diversity amongst populations as well as in defining phylogenetic relationships of the species with other species in the same genus, establishment of the taxonomic identity of uncertain taxa and recognition of genetically diverse populations for their utilization as germplasm for species conservation [17–19]. *Ilex venulosa* is an elegant tree with grey bark, serrate leaves, and red fruits. Due to incomplete and rudimentary embryos, the seed fails to germinate naturally [20]. The prolonged period of dormancy of any viable seed utilization makes them challenging in conventional conservation work. Considering the high rate of disappearance of the species and reduction in species population size, measures to conserve this plant are to be developed. It is in this context that the present study reports on the distribution mapping and genetic diversity in *Ilex venulosa* from Meghalaya.

5.2 Materials and Methods

5.2.1 Plant Materials

Leaf tissues from twenty-three individual plants of *Ilex venulosa* from the populations belonging to Lad Mawphlang and Swer from the Sohra region and Pomlum, Langkyrdem and Mawkajem, locations ranging from 1536 to 1825 m altitudes were collected (Fig. 5.1). Thereafter, the samples were subjected to thorough cleaning with tap water and rinsing with distilled water.

5.2.2 Distribution Mapping

A distribution map of *Ilex venulosa* was created from the data on its location based on a field survey as well as the secondary data available from other sources [21–24]. The data shown in Table 5.1 was used as a matrix for the determination of the potential distribution through Ecological Niche Modeling (ENM) [25].

Fig. 5.1 Plants of alternate species of *Ilex venulosa* collected from various regions of Meghalaya for the conducted investigations: *Ilex venulosa* **a**: Pomlum, **b**: Lad Mawphlang, **c**: Langkyrdem, **d**: Mawkajem, and **e**: Swer

Table 5.1 An enumeration of the sites and aspects of the study conducted on different species of *Ilex* in Meghalaya, India

State	District	Locality	Latitude	Longitude	Altitude (m)	Aspect of the study
Meghalaya	East Khasi Hills	Swer	25.405317	91.802167	1825	Distribution mapping and potential habitat
		Lad Mawphlang	25.37485	91.749567	1732	
		Pomlum	25.411467	91.874017	1642	
		Langkyrdem	25.357433	91.8925	1536	
		Mawkajem	25.346225	91.891931	1567	

5.2.3 Selection of Environmental Predictors and Prediction of Geographical Distribution

For Ecological Niche Modelling, ENVIREM variables (Environmental Rasters for Ecological Modeling, http://envirem.github.io/) were taken to predict potential habitats in Northeast region of India for the selected species. Data sets comprising of 2 topographic and 16 climatic variables, that are important to species ecological and physiological processe, were taken [26]. Maximum entropy modeling software MaxEnt version 3.4.1, [27] was used to reconstruct the ecological niche and predict the geographical distribution of *Ilex venulosa.*

5.2.4 Validation of Model Robustness

The quality of the model was evaluated based on AUC value following the conservative estimate as (i) poor (AUC < 0.8), (ii) fair (0.8 < AUC < 0.9), (iii) good (0.9 < AUC < 0.95) and (iv) very good (0.95 < AUC < 1.0) [28]. Field surveys were undertaken to validate the areas for independent and correct justification of the model.

5.2.5 Identification of Determinants for Environmental Niche

Major determinants of the environmental niche of the chosen species were recognized through the position of the predictor variables and jackknife analysis [29]. Percent contribution and permutation importance are the two metrics that determine the ranking of the predictor. Determination of the percent contribution in each training algorithm involved an addition to the contribution of the equivalent variable or substraction from it if the change to the absolute value of lambda is negative. The scores of each ecological unit on training existence and supporting details are randomly permuted to determine the relevance of each permutation.

5.2.6 DNA Extraction and PCR Amplification

DNA was isolated from leaves of 23 individuals of *Ilex venulosa* and amplification of ITS, *matK* and *rbcL* regions was carried out with primer pairs IlexIF-IlexIR for ITS, IMF-IMR for *matK* and IrF-IrR for *rbcL* as shown in Table 5.2. The reaction mixture was optimized with 10X reaction buffer (2.5 μL) (Merck, India), 25 mM$MgCl_2$ (1.0 μL) (Merck, India), 10 mM dNTPs mixture (1.0 μL) (Merck, India), 1 U Taq polymerase (0.9 μL) and 50 ng of genomic DNA template. Applied Biosystems thermal cycler (Gene Amp PCR system 9700) was used for the amplification with the

Table 5.2 A summary of primer names, primer sequences, melting temperature (Tm °C) and reference of ITS, mat*K*, rbc*L* primer pairs for amplification of target regions of the DNA of 23 individual of *Ilex venulosa*

Taxa	Primer	Sequence (5′-3′)	Tm (°C)	References
ITS	Ilex IF	AACAAGGTTTCCGTAGGTGA	53.4	[30, 31]
	Ilex IR	TCCTCCGCTTATTGATATGC	55.9	
matK	IMF	GCA CTT GCT CAT GAT CAT GGT	55.7	Present study
	IMR	CCC TTC CTA CCT GCT ACA TCA	55.9	
rbcL	IrF	CGA GTA TCT CCT CAA CCT GGA	55.5	Present study
	IrR	AGC AAG ATC ACG CCC CTC ATT	59.2	

reaction cycle comprising of one cycle of "hot start" at 94 °C for 5 min denaturation at 94 °C for 1 min (35 cycles), annealing, extension at 72 °C for 1 min and one cycle of chain elongation at 72 °C for 10 min. Gel electrophoresis was carried out on 1.2% agarose gel and the amplicons were then visualized under UV light in a gel imaging system (BIORAD).

5.2.7 *Nucleotide Sequencing and Analysis*

Amplicons were sequenced in an automated sequencer (ABI3130). Sequences were individually subjected to the BLAST tool of NCBI for sequence identification. The Muscle program produced the multiple sequence alignments [32] through MEGA6 software [33] to determine sequence statistics viz. conserved sites (C), variable sites (V), parsimony informative sites (PIS), singleton sites (S), transition/transversion ratio and nucleotide pair frequencies viz. identical pair (ii), transitional pair (Si) and transversional pair (Sv). The GC content and indels were coded using a seqstate tool [34].

5.3 Results and Discussion

5.3.1 *Distribution Mapping of* Ilex venulosa in Meghalaya

Ecological niche modeling was carried out to predict the distribution of *I. venulosa* across North East India. The model constructed from the set of environmental layers revealed potential evapotranspiration seasonality (67.7%), embergerQ (14.6%) and topographic wetness (11.9%) as the set of cumulative contributors for this species (Table 5.3). Jackknife test of variable importance further showed potential evapotranspiration seasonality as a dominant environmental contributor in determining the model fitness of *I. venulosa* (Fig. 5.2a). The distribution map of *I. venulosa*

Table 5.3 A summary of replicate runs based on average data of the weighted effect size of environmental factors to the Maxent Model for *Ilex venulosa*

Variable	Percent contribution	Permutation importance
PET seasonality	67.7	91.8
Emberger Q	14.6	0
Topographic wetness	11.9	5.2
Thermicity index	1.1	0
Topographic roughness index	1.5	0.7
Continentality	2	2.2
Thornthwaitre aridity index	0.9	0
PET wettest quarter	0.3	0.1

predicted East Khasi Hills of Meghalaya as the most potential area for its occurrence. The species is also predicted scarcely towards Myanmar bordering Tuensang district of Nagaland (Fig. 5.2b). Niche modeling for species of *I. venulosa* yielded a mean AUC value of 0.96, indicating a high accuracy rate of the true presence of the species (Fig. 5.2c).

Field surveys carried out at the specified locations of Meghalaya, as predicted by ENM, confirmed the occurrence of *I. venulosa* at the sites predicted by ENM. Collections were made from Lad Mawphlang (91.749567E, 25.37485N), Swer (91.802167E, 25.405317N), Pomlum (91.874017E, 25.411467N), Mawkajem (91.891931E, 25.346225), and Langkyrdem (91.8925E, 25.357433N) (Table 5.1, Fig. 5.1). The distribution map of *I. venulosa* predicted East Khasi Hills of Meghalaya as the area with the highest potential for the probability of occurrence of the species. While earlier reports recorded the presence of *I. venulosa* in Sohra Rim, East Khasi Hills of Meghalaya [22] and Jowai-Jarain, East Jaintia Hills and Shangpung, West Jaintia Hills of Meghalaya [35], validation of occurrence through field surveys revealed five new locations for the occurrence of *I. venulosa*. These included Lad Mawphlang (91.749567N, 25.37485E), Swer (91.802167N, 25.405317E), Pomlum (91.874017N, 25.411467E), Langkyrdem (91.8925N, 25.357433E) and Mawkajem (91.891931N, 25.346225E). However, it has not been located in any population/individual of *I. venulosa* from Jowai-jarain and Shangpung. It is possible that the absence of any individual/population of *I. venulosa* was a consequence of its depletion due to various anthropogenic pressures. The distribution maps for *I. venulosa* developed in the present study revealed a narrow tolerance range and habitat specificity. It is possible that the narrow ecological niche, habitat specificity and limited tolerance range could be the reasons for *Ilex venulosa* to be endemic to Meghalaya, particularly in East Khasi Hills. These findings are also in accordance with relevant prior art [36]. The predicted output from ENM was validated by our study and shows that the modeled ecological niche matches the environment of the actual habitat of *I. venulosa*.

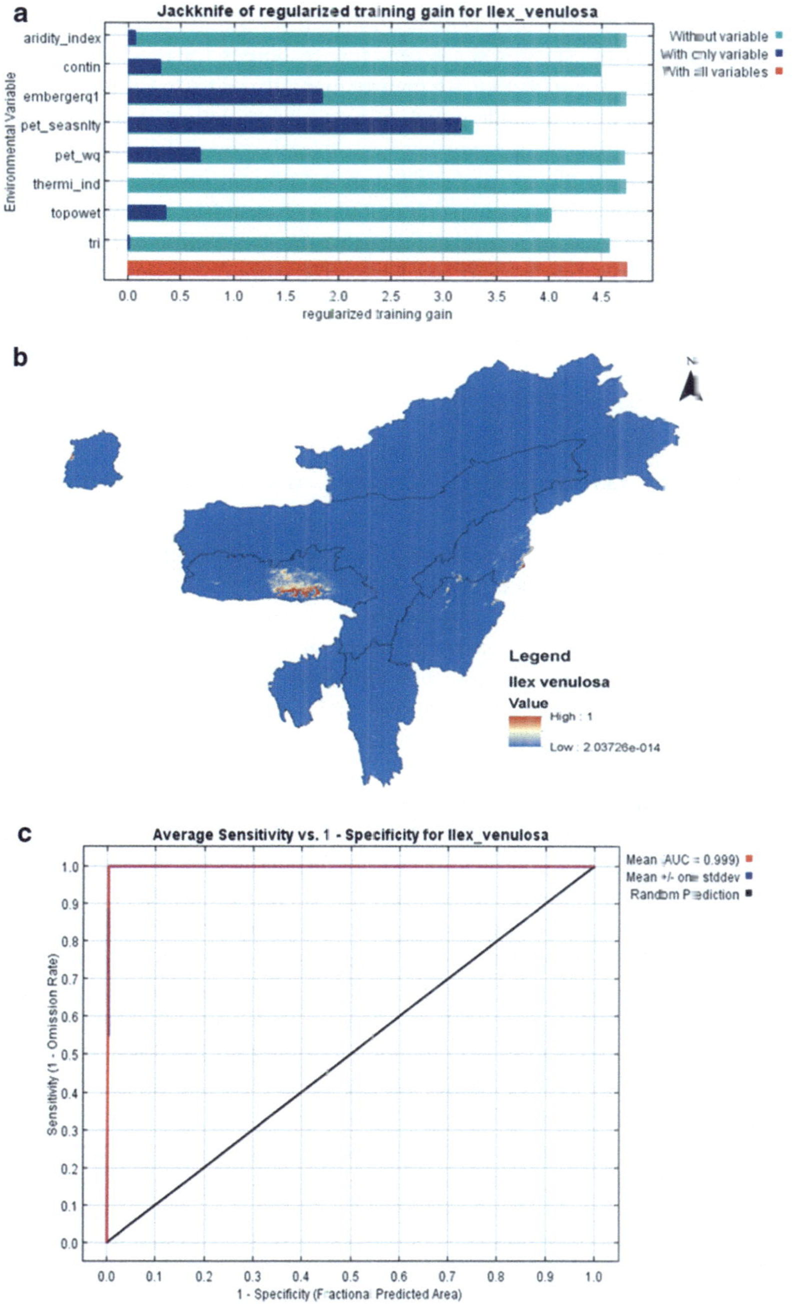

Fig. 5.2 **a** Jackknife of regularized training gain for *I. venulosa* elucidating on the importance of variables. Values refer to averages over replicate runs. **b** Distribution map of *I. venulosa* in the Northeast region of India. **c** Receiver operating characteristic (ROC) curve along with the area under the curve (AUC) to infer upon model performance

The ecological niches for *Ilex venulosa* range from limited to moderate overlap which indicates that the species share very few resources for their survival. ENM can effectively delineate areas that are environmentally suitable for a given species facilitating in maintaining a viable population and interpolation of potential species distribution in areas which have not been completely covered in the survey [37–39]. The distribution of the species studied in the present investigation could also be an outcome of over-harvesting, deforestation, forest fragmentation and other anthropogenic activities. Our results indicate that the distribution of *Ilex venulosa* is restricted to Meghalaya. Further, most such areas fall within areas impacted by a range of anthropogenic activities. Considering these, it is anticipated that the species may become extinct in near future from the current threat classification of 'Endangered' for *Ilex venulosa.*

5.3.2 Genetic Assessment of ITS, rbcL and matK Regions

Assessing the evolutionary connection, gene flow, genetic drift, and level of outbreeding in various taxa has been made easier by phylogenetic research. Therefore, prior to any conservation action, genetically diverse populations of a species need to be identified through a combination of morphological and molecular approaches. The present study describes the molecular similarity or dissimilarity in species.

The internal transcribed spacer (ITS), *rbc*L and *mat*K regions of the DNA of the 23 individuals of *Ilex venulosa,* were amplified by PCR with primer pairs IlexIF-IlexIR for ITS, IMF-IMR for *matK* and IrF-IrR for *rbcL.* While the primer pair IlexIF-IlexIR generated amplicons of an apparent molecular mass of 0.7 kb from each individual, PCR carried out with primer pair IMF-IMR and IrF-IrR generated 1.2 kb amplicons from each individual. The amplification profiles on agarose gel revealed only a single sharp band without any nonspecific amplification (Fig. 5.3). BLAST analysis of the nucleotide sequences of the amplicons generated using primer pair IlexIF-IlexIR revealed 92–93% homology with the sequences of nuclear ITS region from other species of *Ilex* with an e-value of zero. BLAST-based approach was used to annotate the ITS1, 5.8S and ITS2 portion of the nuclear internal transcribed spacer region of each individual. The annotation identified a length of 244–264 bases for ITS1, 165 bases for the 5.8S and 268–279 bases for the ITS2 region in all the twenty-three individuals of *Ilex* investigated in the present study.

Sequence analysis within the species *I. venulosa* revealed the highest conserved sites of 100% from Lad Mawphlang population with zero percent sequence divergence, that to sequence analysis from Langkyrdem population with 93.75% conserved sites, 6.25% variable sites and R value of 0.73 (Table 5.4). An important feature detected in the ITS1 region of the sequences was the presence of the 'CAAGGAA' motif at $P'_{165-171}$, which is a part of the conserved angiosperm motif GGCR-(4–7n)-GYGYCAAGGAA. In the same context, the 5.8S region showed the

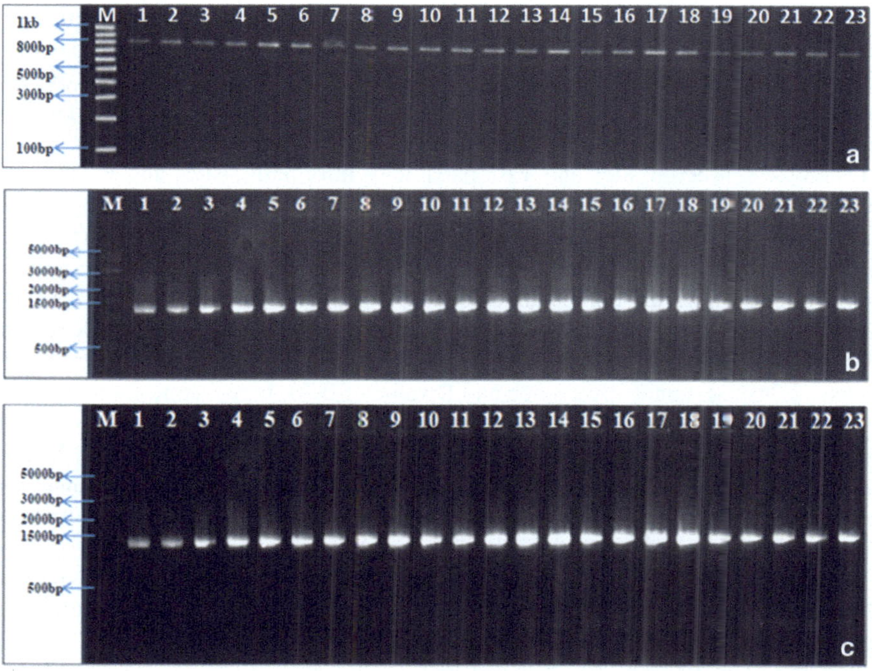

Fig. 5.3 Image depicting the findings of the amplification of 23 individuals of **a** nuclear ITS, **b** chloroplast*matK* and **c** *rbcL* region using primer pair IlexIF/IlexIR, IMF/IMR and IrF/IrR, respectively

10 bp motif represented by the sequence "TTTGAACGCA" at $P'_{348-357}$ and the 14 bp angiospermic motif "GAATTGCAGAATCC" at $P'_{318-331}$ which is 20 bases pairs downstream of the 16 bp motif CGATGAAGAACGTAGC that is located at $P'_{283-298}$. *I*ITS1 of *I. venulosa* revealed one indel one and one single base change, 94.83% of conserved sites, 5.16% of parsimony informative sites and 3.42% of sequence divergence. Further, ITS2 showed one indel one and seven single base change, 75.74% of conserved sites, 21.70% of parsimony informative sites and 8.25% of sequence divergence (Table 5.4). The two standard barcoding regions of the chloroplast genome, *mat*K and *rbc*L, showed the highest sequence homology of 93% and 100% respectively with the corresponding sequences of different species of *Ilex* available in the database. BLAST analysis of the nucleotide sequences of the amplicons generated with primer pair IMF-IMR revealed 99–100% homology with the *mat*K sequence from other species of *Ilex* with an e-value of zero. The analysis of sequences from individuals belonging to *I. venulosa* revealed 99.38% conserved sites, 0.46% parsimony informative sites, 0.88% sequence divergence and an R value of 3.0. While sequence analysis for the populations of *I. venulosa* from the Sohra region revealed 99.84% conserved sites, 0.46% parsimony informative sites and 0.88% sequence divergence, that from the Pynursla region revealed 99.84% conserved sites, 0.15%

parsimony informative sites and 0.07% sequence divergence (Table 5.4). BLAST analysis of the nucleotide sequences of the amplicons generated with primer pair ErF-ErR revealed 92–93% homology with the sequences of *rbc*L from other species of *Ilex* with an e-value of zero. Sequence analysis from individuals belonging to populations of *I. venulosa* from Sohra and Pynursla regions revealed similar values of 98.39% conserved sites and 1.62 parsimony informative sites and R value of 0.87. The sequence divergence was 0.77% for individuals belonging to populations from the Sohra region and 0.50% for individuals belonging to populations from the Pynursla region.

Table 5.5 shows the comparative sequence analysis of populations of *I. venulosa* from two regions namely Sohra and Pynursla. It revealed for the ITS region 93.32% conserved sites, 5.04% variable sites, 1.03% parsimony informative sites, 5 indels, 2.74% sequence divergence and R value of 0.73. In the same context, analysis of sequences of individuals from the Pynursla region revealed 90.80% conserved sites, 8.75% variable sites and 8.75% parsimony informative sites, 6 indels, 1.89% sequence divergence and R value of 0.66. While the comparative sequence analysis of *matK* and *rbcL* from these populations did not affirm any significant variation (Table 5.5).

Moderate to low genetic diversity and limited isolated population partitioning in the seeded plant species are due to various factors such as fragmentation of continuous genetic system and adaptation of genetic systems in small populations. In the present study, the plant *Ilex venulosa* showed low genetic diversity making it short-lived and approaching the threshold of extinction. Thus, immediate steps needed for the conservation of *Ilex venulosa* include enriching the improvised populations with genetically diverse germplasms and allowing the plants to propagate and multiply in number through natural regeneration and checking anthropogenic destruction.

Table 5.4 Data summary of the ITS region of 23 individual of *Ilex venulosa*

Sl. No.	Region	Sequence statistics								Nucleotide pair frequencies			
		Indels	% divergence	G + C (%)	CS (%)	VS (%)	PIS (%)	SS (%)	CpG (100 coverage)	ii	Si	Sv	R
1	Whole ITS	2	3.47	57.52	88.94	11.05	10.09	0.96	194	602	7.00	15.00	0.47
2	ITS1	1	3.42	57.52	94.83	5.16	5.16	0	70	211	0.00	1.00	0.22
3	ITS2	1	8.25	57.19	75.74	24.25	21.70	2.55	44	215	6.00	13.00	0.50
4	*matK*	0	0.15	33.80	99.38	0.61	0.46	0.15	170	645	1.00	0.00	3
5	*rbcL*	0	0.50	43.95	98.01	1.98	1.44	0.54	350	1087	2.00	3.00	0.75

Indels, percentage diversity, G + C content, conserved sites (CS), variable sites (VS), parsimony informative sites (PIS), singleton sites (SS), CpG, identical pairs (ii), transistional pair s(si), transversional pairs (sv) and R (si/sv)

Table 5.5 Data summary of the C sequence analysis of *I. venulosa* populations from two regions namely Sohra and Pynursla

Region	Parameters	*I. venulosa*	
ITS region		Sohra region	Pynursla region
	CS (%)	93.32	90.80
	VS (%)	5.04	8.75
	PIS (%)	1.03	1.89
	Indels	5	6
	G + C content (%)	56.71	56.99
	Sequence divergence (%)	2.74	4.26
	Transition/transversion (R)	0.73	0.66
matK	CS (%)	99.84	99.84
	VS (%)	0.15	0.15
	PIS (%)	0.46	0.15
	Indels	1	0
	G + C content (%)	33.79	33.91
	Sequence divergence (%)	0.88	0.07
	Transition/transversion (R)	0	0
rbcL	CS (%)	98.39	98.39
	VS (%)	1.62	1.62
	PIS (%)	1.62	1.62
	Indels	0	0
	G + C content (%)	43.97	43.95
	Sequence divergence (%)	0.77	0.5
	Transition/transversion (R)	0	0
	Transition/transversion (R)	0.87	0.87

Indels, percentage diversity, G + C content, conserved sites (CS), variable sites (VS), parsimony informative sites (PIS) and R (si/sv) estimated from the sequences representing nuclear ITS region of 23 individual of *Ilex venulosa* investigated in the present study

5.4 Conclusions

Genetic drift is one of the most vital intrinsic drivers of species extinction. It is due to the narrow genetic base within the population. The distribution mapping of *Ilex venulosa* showed a small population size which indicates that the extinction probability due to genetic drift is very high. The distribution map of *I. venulosa* predicted East Khasi Hills of Meghalaya as the area with the highest potential for the probability of occurrence of the species. While earlier reports recorded the presence of *I. venulosa* in Sohra Rim, East Khasi Hills of Meghalaya and Jowai-Jarain, East Jaintia Hills and Shangpung, West Jaintia Hills of Meghalaya, the validation of occurrence through field surveys revealed five new locations for the occurrence of *I. venulosa*. These included Lad Mawphlang (91.749567N, 25.37485E), Swer (91.802167N, 25.405317E), Pomlum (91.874017N, 25.411467E), Langkyrdem (91.8925N, 25.357433E) and Mawkajem (91.891931N, 25.346225E). However, any population/individual of *I. venulosa* from Jowai-jarain and Shangpung could not be located. It is possible that the absence of any individual/population of *I. venulosa* was a consequence of its depletion due to various anthropogenic pressures. Genetic variations do play a crucial role in species adaptation and the subsequent evolutionary processes. In this investigation, it has been affirmed that the populations of *I. venulosa* from Pynursla showed a relatively higher level of diversity than the Sohra population. The later can be used for expanding the extent population of a species. Thereby, the probability of genetic drift can be minimized to ensure its conservation. The low degree of genetic variation found in *Ilex* species could most likely be ascribed to the fact that all collected individuals originate from small populations. Thus, immediate steps needed for the conservation of Ilex venulosa include enriching the improvised populations with genetically diverse germplasms and allowing the plants to propagate and multiply in number through natural regeneration and checking anthropogenic destruction. The presented research methodology is generic in nature and can be applied for the bio-diversity conservation of any endangered or endemic wild plant species in the North-East India. Through such studies, greater human efforts can be devoted towards bio-diversity conservation and ecological sustainability.

Acknowledgements This work was carried out as part of the grant received by S.K.B. from the DBT, Government of India under the network project 'Preventing extinction and improving the conservation status of threatened plants through the application of biotechnological tools'.

References

1. Shiva, V.: Protecting Our Biological and Intellectual Heritage in the Age of Biopiracy: Paper Prepared for the Seminar on IPRs, Community Rights and Biodiversity: A New Partnership for National Sovereignty Held at New Delhi, February 20, 1996. Research Foundation for Science, Technology and Natural Resource Policy (1996)
2. Das, A.K., Dutta, B.K., Sharma, G.D.: Medicinal plants used by different tribes of Cachar district, Assam (2008)
3. Mir, B.A., Mir, S.A., Koul, S.: In vitro propagation and withaferin A production in *Withania ashwagandha*, a rare medicinal plant of India. Physiol. Mol. Biol. Plants **20**(3), 357–364 (2014)
4. Vijayalatha, K.R., Chezhiyan, N.: Correlation and path analysis studies in coriander (*Coriandrumsativum* L.). South Indian J. Hortic. **52**(1/6), 248 (2004)
5. Singh, D., Jain, U.K., Rajput, S.S., Khandelwal, V., Shiva, K.N.: Genetic variation for seed yield and its components and their association in coriander (*Coriandrumsiltivum* L.) germplasm. J. Spices Aromat. Crops **15**(1), 25–29 (2006)
6. Hongtrakul, V., Huestis, G.M., Knapp, S.J.: Amplified fragment length polymorphisms as a tool for DNA fingerprinting sunflower germplasm: genetic diversity among oilseed inbred lines. Theor. Appl. Genet. **95**(3), 400–407 (1997)
7. Cervera, M.T., Cabezas, J., Sancha, J.: Theor. Appl. Genet. **97**, 51 (1998)

8. Jain, S.K., Rao, R.R.: A Handbook of Field and Herbarium Methods. Today and Tomorro's Printers and Publishers, New Delhi (1997)
9. Menkir, A., Goldsbrough, P., Ejeta, G.: RAPD based assessment of genetic diversity in cultivated races of sorghum. Crop Sci. **37**(2), 564–569 (1997)
10. Barrett, B.A., Kidwell, K.K.: AFLP-based genetic diversity assessment among wheat cultivars from the Pacific Northwest. Crop Sci. **38**(5), 1261–1271 (1998)
11. Novy, R.G., Vorsa, N.: Identification of intracultivar genetic heterogeneity in cranberry using silver-stained RAPDs. Hortic. Sci. **30**(3), 600–604 (1995)
12. Virk, P.S., Zhu, J., Newbury, H.J., Bryan, G.J., Jackson, M.T., Ford-Lloyd, B.V.: Effectiveness of different classes of molecular marker for classifying and revealing variation in rice (*Oryza sativa*) germplasm. Euphytica **112**(3), 275–284 (2000)
13. Margale, E., Herve, Y., Hue, J., Quiros, C.F.: Determination of genetic variability by RAPD markers in cauliflower, cabbage and kale local cultivars from France. Genet. Resour. Crop Evol. **42**(3), 281–289 (1995)
14. Kelly, J.D., Miklas, P.N.: The role of RAPD markers in breeding for disease resistance in common bee. Mol. Breed. **4**(1), 1–11 (1998)
15. Tsuji, K., Ohnishi, O.: Origin of cultivated Tatary buckwheat (*FagopyrumtataricumGaertn.*) revealed by RAPD analyses. Genet. Resour. Crop Evol. **47**(4), 431–438 (2000)
16. Potokina, E., Vaughan, D.A., Eggi, E.E., Tomooka, N.: Population diversity of the *Vicia sativa*agg. (Fabaceae) in the flora of the former USSR deduced from RAPD and seed protein analyses. Genet. Resour. Crop Evol. **47**(2), 171–183 (2000)
17. Gupta, R.C., Folta-Stogniew, E., O'Malley, S., Takahashi, M., Radding, C.M.: Rapid exchange of A:T base pairs is essential for recognition of DNA homology by human Rad51 recombination protein. Mol. Cell **4**(5), 705–714
18. Sivolap, Y.M., Chebotar, S.V., Topchieva, E.A., Korzun, V.N., Totskiy, V.N.: RAPD and SSRP analyses of molecular-genetic polymorphism in *Triticumaestivum* L. cultivars. Russ. J. Genet. **35**(12), 1433–1440 (1999)
19. Strelchenko, P., Kovalyova, O., Okuno, K.: Genetic differentiation and geographical distribution of barley germplasm based on RAPD markers. Genet. Resour. Crop Evol. **46**(2), 193–205 (1999)
20. Hu, C.Y.: Holly (Ilex spp). In: Bajaj, Y.P.S. (ed.) Biotechnology in Agriculture and Forestry, vol. 5, pp. 412–427. Springer-Verlag, Berlin (1989)
21. Kanjilal, U.N., Kanjilal, P.C., Das, A., De, R.N., Bor, N.L.: Flora of Assam, vols. 1–5 (1934–1940)
22. Haridasan, K., Rao, R.R.: Forest Flora of Meghalaya, vol. 2 (1985–1987)
23. Balakrishnan, N.P.: Flora of Jowai and Vicinity, Meghalaya (1981)
24. Rao, R.R., Haridasan, K.: Threatened plants of Meghalaya—a plea for conservation. In: Jain, S.K., Rao, R.R. (eds.) Assessment of Threatened Plants of India: Proceedings of the Seminar Held at Dehra Dun, 14–17 September 1981. Botanical Survey of India, Department of Environment, Howrath (1983)
25. Warren, D.L., Glor, R.E., Turelli, M.: ENMTools: a toolbox for comparative studies of environmental niche models. Ecography **33**(3), 607–611 (2010)
26. Title, P.O., Bemmels, J.B.: ENVIREM: an expanded set of bioclimatic and topographic variables increases flexibility and improves performance of ecological niche modeling. Ecography **41**(2), 291–307 (2018)
27. Phillips, S.J., Anderson, R.P., Dudík, M., Schapire, R.E., Blair, M.E.: Opening the black box: an open-source release of Maxent. Ecography **40**(7), 887–893 (2017)
28. Thuiller, W., Richardson, D.M., Pyšek, P., Midgley, G.F., Hughes, G.O., Rouget, M.: Niche-based modelling as a tool for predicting the risk of alien plant invasions at a global scale. Glob. Change Biol. **11**(12), 2234–2250 (2005)
29. Phillips, S.J., Anderson, R.P., Schapire, R.E.: Maximum entropy modeling of species geographic distributions. Ecol. Modell. **190**(3–4), 231–259 (2006)
30. Powell, M., Savolainen, V., Cuénoud, P., Manen, J.F., Andrews, S.: The mountain holly (*Nemopanthus mucronatus*: Aquifoliaceae) revisited with molecular data. Kew Bull. 341–347 (2000)

31. White, T.J., Bruns, T., Lee, S.J.W.T., Taylor, J.: Amplification and direct sequencing of fungal ribosomal RNA genes for phylogenetics. PCR Protoc.: Guid. Methods Appl. **18**(1), 315–322 (1990)
32. Edgar, R.C.: MUSCLE: multiple sequence alignment with high accuracy and high throughput. Nucl. Acids Res. **32**(5), 1792–1797 (2004)
33. Tamura, K., Stecher, G., Peterson, D., Filipski, A., Kumar, S.: MEGA6: molecular evolutionary genetics analysis version 6.0. Mol. Biol. Evol. **30**(12), 2725–2729 (2013)
34. Müller, K.: SeqState. Appl. Bioinform. **4**(1), 65–69 (2005)
35. Balakrishnan, N.P.: Flora of jowai and vicinity united Khasi Jaintia hills district Meghalaya Assam, vol. 2 (1971)
36. Adhikari, D., Barik, S.K., Upadhaya, K.: Habitat distribution modelling for reintroduction of Ilex khasiana Purk., a critically endangered tree species of northeastern India. Ecol. Eng. **40**, 37–43 (2012)
37. Borthakur, S.K., Baruah, P.S., Deka, K., Das, P., Sarma, B., Adhikari, D., Tanti, B.: Habitat distribution modelling for improving conservation status of *Bruceamollis* Wall. exKurz.—an endangered potential medicinal plant of Northeast India. J. Nat. Conserv. **43**, 104–110 (2018)
38. Adhikari, P.P., Talukdar, S., Borah, A.: Ethnomedgicobotanical study of indigenous knowledge on medicinal plants used for the treatment of reproductive problems in Nalbari district, Assam, India. J. Ethnopharmacol. **210**, 386–407 (2018)
39. Adhikari, J.N., Bhattarai, B.P., Thapa, T.B.: Factors affecting diversity and distribution of threatened birds in Chitwan National Park, Nepal. JoTT **11**(5), 13511–13522 (2019)

Chapter 6
Application of Allometric Equations to Estimate Carbon Stock and Above-Ground Biomass in Narpuh Wildlife Sanctuary, Meghalaya

Pynshailang Syiemiong and Shiva Shankar Chaturvedi

6.1 Introduction

In nature, carbon is found in the challenging form of atmospheric CO_2. It constitutes only a negligible amount of the atmosphere and about 0.04%. Even with such a little concentration, it contributes significantly to the biogeochemical cycle (carbon cycle) that supports life on earth. It was also reported that up to the year 2020, CO_2 level had increased significantly from 280 to 412 ppm which is a 46% increase [1]. Concerning the Indian scenario, the greenhouse gas emission in the year 2016 has been reported to be 2,531.07 million tonnes equivalent (MT eq) with the inclusion of "Land Use Land Use Change and Forestry" (LULUCF). In these, emissions from carbon dioxide accounted for 2,231 MT (78.59%) [2]. India also committed to adding a carbon sink of 2.5–3.0 billion tonnes of CO_2 capacity by 2030 through forest and tree cover [3]. In terms of total carbon stock, the global forest resource assessment reported that the total forest carbon stock from all its pools was estimated to be 662 Gt (163 tonnes ha^{-1}) and it was mentioned that the global carbon stock has decreased from 668 to 662 Gt between the year 1990 and 2020 due to the decline in forest area [4]. At the national level, carbon stock (from all carbon pools) from the Indian forest, including forest, scrubs and plantation, was estimated in the year 2021. This carbon stock was 7204.0 MT, with an increase of 79.4 MT in comparison with the previous year's emission of 7124.6 MT. The estimate of carbon stock from different components for the year 2021 and for the above-ground biomass was 2319.9 MT. For below ground biomass case, it was 718.9 MT. Similarly, for deadwood it was 47.7 MT, for Litter, it was 107.3 MT, and for soil, it was 4010.2 MT. In the North-East state of Meghalaya, the total Carbon stock for the year 2021 is estimated at around 107.59 MT. The primary contributor of carbon stock for the State comes from the Soil Organic Carbon at 63.37 MT and above-ground biomass at 32.41 MT. In India,

---

P. Syiemiong (✉) · S. S. Chaturvedi
Department of Environmental Studies, North-Eastern Hill University, Shillong 793022, India
e-mail: psyiemiong@gmail.com

R. V. S. Uppaluri and L. Rangan (eds.), *Conservation of Biodiversity in the North Eastern States of India*, https://doi.org/10.1007/978-981-99-0945-2_6

the overall carbon stock increased by 541 MT from 6663 MT between the year 2011 to 2021 [3].

Indian forests have been divided into three classes based on the canopy density and as per the assessment of the Indian State Forest Report (ISFR) 2021. The forest has been classified as very dense forest (VDF), moderately dense forest (MDF), and open forest (OF). Together, these three classes of canopy density account only for 713,789 km^2 (21.71%) of the entire geographical area of the country. However, the forest cover of the North-East region is 64.66% of its total geographical area (169,521 km^2) [3]. Approximately 80% of India's terrestrial biodiversity is found in the forest and the forest also caters to around 300 million people to support their livelihood [3].

As per the ISFR 2021 assessment, the forest cover in Meghalaya was 17,046 km^2 (75.99%) of the total geographical area. It was found that there is a decline of −0.43% in forest cover with respect to ISFR 2019 assessment [3]. Meghalaya is home to a variety of plants and animals with high species composition. The State is a part of the Indo-Malayan biodiversity hotspot. The region is blessed with a rich biodiversity of flora and fauna, the majority of which is found in the state's southern tropical belt and upland subtropical forest. The forests of Meghalaya can broadly be categorised into tropical, subtropical and temperate forests. The climax vegetation of the area, according to Champion and Seth [5] can be categorised as a subtropical wet hill forest. The climatic condition, topography, rainfall patterns, flora, land use, culture, diversity and ethnicity of the state varies greatly. Hence, it is certain that the carbon stock, rate of carbon sequestration and carbon dynamics will also vary across the state.

One of the most important environmental services that a forest provides is the sequestration and storage of carbon. Tropical forests are considered to be accountable for about 25% of the world's biomass and carbon. It provides up to 40% of the terrestrial net primary output, although these type of forests covers only 12% of the total geographical area of the earth [6, 7]. The tropical forest is known to be the last standing forest that can give us hope to tackle climate change in addition to being a major carbon sink and home to the rich diversity of flora and fauna. They play a vital role in the proper functioning of the biogeochemical cycle, especially the carbon cycle. In order to be able to predict climate change and to implement efficient mitigation methods to reduce global warming, a thorough understanding of natural forests is extremely important as they contribute significantly to the carbon pool and carbon flux relating to net primary production, carbon assimilation and decomposition in biome [8]. On the other hand, governments from different nations have raised concerns about global warming and climate change for the past many decades. They have come up with laws and protocols both at the national and international levels to mitigate various problems being faced today and those that will be encountered in the near future. The key solution to all these environmental problems can be achieved from the forest as they are essential for limiting greenhouse gas emission. If one can estimate how much carbon can be stored and sequest by a particular type of forest and if one can also estimate how much carbon has been stored, the data can then be used to update and improve forest management practices. This will assist in the adoption of sustainable methods to maximize the quantity of carbon dioxide sequestration by the forest. Forest management is also known as the land-use management system. It can help to mitigate and control environmental issues such as climate change, global warming, and biodiversity conservation [9].

As a result, the estimation of carbon stock and above-ground biomass is the focus of this research. Such efforts deploy a variety of allometric equations. Thereby, the type of allometric equation and its suitability and relevance to represent a specific forest type can be assessed. To have an in-depth understanding of carbon stock and biomass of a particular forest type, two methods have been adopted: the direct and the indirect method. The direct method measures plant biomass by felling down trees within the quadrat (which is a frame along a transect to estimate plants abundance). The components such as the roots, stem, leaves, branches and stumps are segregated and measurements for each of the components were carried out. The overall biomass of a tree is then calculated by adding the oven-dried masses of these components. The direct method is considered to be the most precise method in determining the dry biomass of the plant and this enables the development of allometric equations. Hence, the allometric mathematical model may be used to explain the connection between a tree's dry weight and its wood density, breast height diameter, or crown area. The equations developed by the direct method can be applied in the field of forestry which can also be used to measure biomass using a non-destructive method called the indirect method [10, 11].

However, the drawbacks of using the direct method are that only a limited area can be studied and the falling down of trees is carried out in large numbers. This method cannot be carried out if the forests are Government protected forests or if it is a community-protected area like sacred groves. Endangered and threatened species are affected by this method as the felling of trees is required [12]. The indirect method of biomass estimation does not involve cutting down trees or their components. Instead, a mathematical equation developed from the direct method can be deployed in the biomass estimation of trees and their components. The advantages of using the non-destructive or the indirect method are that it is less time-consuming, less labour is intensive, and low capital investment. Further, biomass study in protected areas can be carried out without any compliances. Targeting such needs research, the book chapter delineates upon the relevance of allometric equations to fairly represent carbon stock in Narpuh Wildlife Sanctuary.

6.2 Materials and Methods

6.2.1 Study Area

The selected research site is the Narpuh Wildlife Sanctuary (Fig. 6.1), located in the East Jaintia Hill district of Meghalaya. It lies between 25°05′19″ to 25°06′14″ North Latitudes and 92°21′04″ to 92°31′38″ East latitudes (Fig. 6.2). Narpuh Wildlife

Sanctuary covers an area of 59.9 km^2, and it is the only pristine forest in the eastern part of Meghalaya. The sanctuary has one of the tallest, largest evergreen and the only semi-evergreen forests left in Meghalaya. The sanctuary hosts a wide variety of fauna, having around 92 species of birds, approximately 91 species of butterflies, and Schedule I species such as Leopard, Hollock gibbons, Slow Loris, etc. [13]. By incorporating parameters such as tree height and Breast Height Diameter (DBH), an allometric equation was proposed which is also known as the regression model and the biomass or volume of the above-ground trees component can then be calculated using these models. These equations have been generated from sample trees and measured weights in relation to their DBH and height. As a result, a common and economical approach for the estimation of tree species biomass in a forest or plantation is much required in order to cover more area and be able to carry out the work rigorously and efficiently [14].

Fig. 6.1 Panoramic view of Narpuh Wildlife Sanctuary

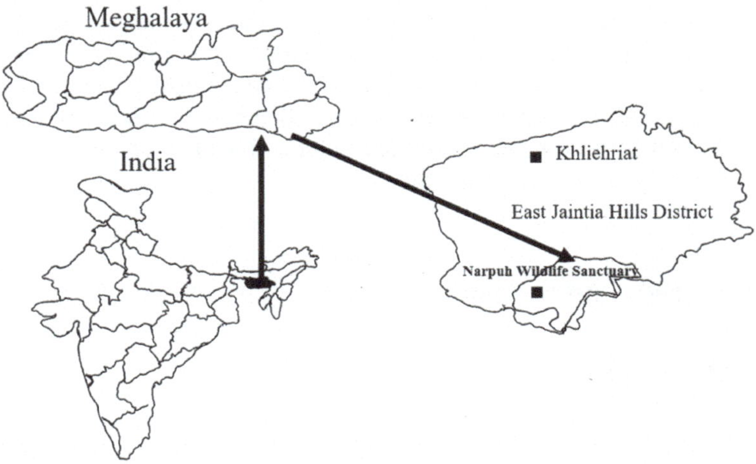

Fig. 6.2 Map of the Narpuh Wildlife Sanctuary

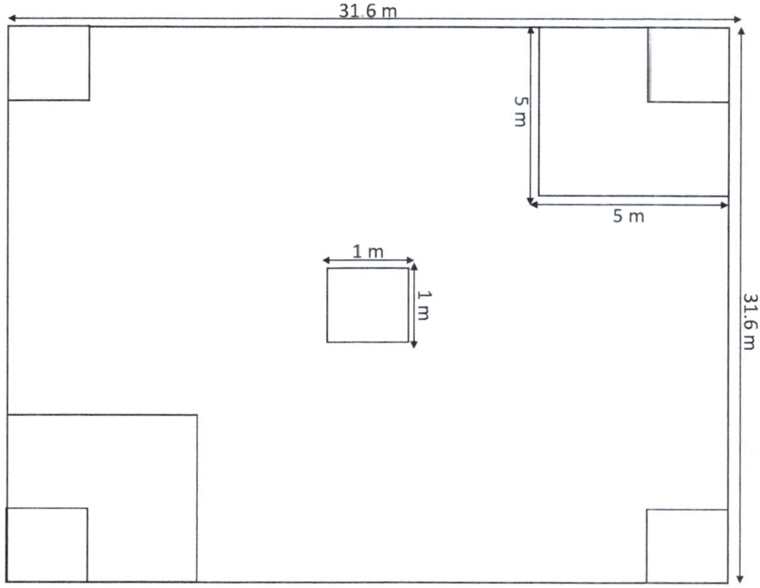

Fig. 6.3 Layout of permanent plot

The study involved the setting of eleven permanent plots within the Narpuh Wildlife Sanctuary during the pre-monsoon season of 2019. Each permanent plot was of the size of 31.6 × 31.6 m and within the perimeter of the permanent plot, a well-defined dimension of 5 × 5 m for shrubs and 1 × 1 m for herbs were marked. The relevant plan has been laid out in Fig. 6.3. In addition to that, a field survey, the elevation of around 50–300 m above sea level, a distance of 0.3–5 km between the plots was maintained. Thereby, field data were collected through physical measurement of tree samples above 15 cm in girth size from all eleven plots. The regression analysis was conducted on IBM SPSS statistic 2.0.

6.2.2 Allometric Equations

The traditional method for calculating the above-ground biomass of trees in a plot and region is to develop or apply allometric equations [10]. Allometric equations are required for employing a non-destructive (indirect) technique to calculate the biomass and carbon storage in the terrestrial ecosystem. It has also been observed that many regions lack accurate estimation of carbon sequestration in tropical forests due to a lack of suitable allometric models for estimating the biomass estimation in a tropical forest with a diverse range of flora [12]. As a result, the development of generalised allometric models for tropical trees has drawn the interest of many researchers. This is due to the vast species diversity being hosted by the tropical forest [10, 15–17].

In this study, the number of stems from all woody species with a diameter of more than 15 cm was measured and counted for each plot in all eleven plots. The girth of a tree was measured by using a measuring tape, and the tree height was measured with a Nikon forestry Pro. The herbarium and other published literature were used to identify all woody species. The allometric equation or regression equation used for the estimation of biomass in this research was developed by Chambers et al. [18], FAO [19], Brown et al. [20], Nath et al. [21], Chave et al. [22], (Table 6.1). While the total above-ground carbon was calculated by approximately considering 55% of the total above-ground biomass [23], the below-ground biomass estimation was calculated by using the regression model of Cairns et al., regression equation [24] (Table 6.2). The allometric used in this study were then selected based on the model's input components, with DBH as the only component required for the allometric equations. Furthermore, other factors that affect the selection of allometric equations also include the type of forest to which they are applied. A statistical test was done on all equations to assess the R^2 value and the adjusted R^2. Thereby, the best-fitted model with the highest R^2 value has been selected for the estimation of biomass and carbon stock. A higher R^2 value implies that the allometric equation is more suitable for its utility to a particular forest in terms of biomass and carbon estimates.

6.3 Results and Discussion

From eleven plots, 155 woody species were identified. Among these, some plots had the lowest count of 49 trees and the greatest count of 87. In total, 746 trees exist in all eleven plots. These eleven plots together have an area equivalent to one hectare. Common trees species found in the forest include *Ficus sp., Michelia champa, M. globosa, Castonopsis sp., Eleocarpus sp., Albezia lebbeck, Prunus undulate, Arvaena coronania, Docynia indica.* It was found that out of 746 trees, 542 had girth sizes of 15–70 cm, accounting for 32.759 tonnes ha^{-1} of total aboveground biomass and 18.017 tonnes ha^{-1} of total aboveground carbon. The balance 204 trees have a girth size of 70 cm and above, accounting for 201.669 tonnes ha^{-1} of total aboveground biomass and 110.918 tonnes ha^{-1} total above-ground carbon.

Eight models were used and compared for above-ground biomass estimation in the Narpuh Wildlife Sanctuary before deciding upon the best model fit by considering all dependent variables (above-ground biomass) and (DBH) from all the woody species within the 31.6 × 31.6 m plot. The dependent variables have been calculated by using regression analysis in all eight equations.

The findings from the regression analysis in terms of above-ground carbon for the altered DBH have been presented in Fig. 6.4. It was found that Chambers et al. [18] developed the allometric equation with the greatest R^2 value of 0.804, whereas FAO [14] developed the equation with the lowest R^2 value of 0.5902 (Fig. 6.4). Figure 6.4

Table 6.1 A summary of alternate allometric equations deployed for the estimation of biomass and carbon stock

Authors	Publication year	Allometric equation/model	Vegetation type	Region	Country	Notes
Chambers et al.	2001	$AGB = \exp[0.37 + 0.33 \times \ln(D) + 0.933 \times \ln(D)^2$ $0.122 \times \ln(D)]^3$	Tropical rainforest	Latin America	Brazil	The equation was developed by harvesting 315 trees from 0.04 ha (20 × 20 m) plot
FAO 3.2.3	1997	$AGB = 42.69 - 12.800 \times (D) + 1.242 \times (D)^2$	Tropical trees in moist climatic zone	Pan-tropical	N/A	The model was developed by harvesting 170 trees from three sites on the montane forest in New Guinea and from the unpublished data near La Selva, Costa Rica
FAO 3.2.4		$AGB = \exp\{-2.134 + 2.530 \times \ln(D)\}$				
FAO 3.2.5		$AGB = 21.297 - 6.953 \times (D) + 0.740 (D)^2$				
Brown et al.	1989	$AGB = 38.4908 - 11.7883 \times (D) + 1.1926 \times (D)$	N/A	N/A	N/A	N/A
		$AGB = 13.2579 - 4.8945 \times (D) + 0.6713 \times (D)$	N/A	N/A	N/A	N/A
Nath et al.	2019	$AGB_{est} = 0.18D^{2.16} \times 1.32$	Tropical semi-evergreen and wet evergreen, sub-tropical broad leaved and pine forest of North East India	North-East India	India	The model was developed by sampling 303 trees from four major forest types ranging from tropical to sub-tropical

(continued)

Table 6.1 (continued)

Authors	Publication year	Allometric equation/model	Vegetation type	Region	Country	Notes
Chave et al.	2001	$AGB = \exp(-2.00 + 2.42) \times \ln(D)$	Wet rainforest in Lowland	Latin America	French Guiana	The Regression model was inferred from previously published datasets of Araujo et al. (1999) [25], Ovington and Olson (1970) [26], Edward and Grubb (1977) [27], and Brown (1997) [28]—equivalent to 378 trees with DBH greater than 10 cm

Table 6.2 Regression model for estimation of the belowground biomass

Model	Regression equation	Parameters
Cairns et al. (1997)	Exp (−1.0589 + 0.8836 × ln (AGB))	AGB

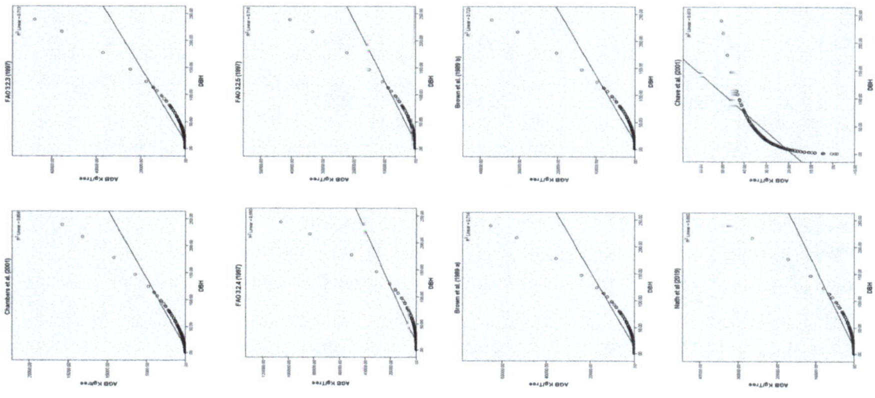

Fig. 6.4 Fitness plots for alternate allometric equation models to estimate biomass in Narpuh Wildlife Sanctuary

shows the pictorial representation of various regression analysis of various allometric equations developed by various authors for a specific vegetation type, region and country. The result indicates that using the allometric model developed by Chambers et al. was suitable and more acceptable for the Narpuh Wildlife Sanctuary. The equation developed by Chambers et al. can be used for the forest types similar to the Narpuh Wildlife Sanctuary i.e., that is tropical evergreen and semi-evergreen forest of North-East India.

The analysis affirmed that the Chambers et al. [18] model possessed the best R^2 value of 0.804. The corresponding total above-ground biomass value was 234.43 tonnes ha^{-1}, and the total above-ground carbon was 128.94 tonnes ha^{-1}. The Chambers model did not overestimate or underestimate the total above-ground biomass and carbon. Depending on the deployed model, the above-ground biomass estimates altered from 17.89 to 736.14 tonnes ha^{-1}, (Table 6.3). In comparison to the model created by Chambers et al. [18], the FAO [14] model had the lowest R^2 value of 0.590, with 736.14 tonnes ha^{-1} of total above-ground biomass and 404.87 tonnes ha^{-1} of total above-ground carbon. Thereby, the model overestimated the total above-ground biomass and carbon by 3.14 times that indicated by Chambers et al. model [18]. The model created by Chave [22] with an R^2 value of 0.673 produced the lowest total above-ground biomass and carbon content of 17.89 and 9.84 tonnes ha^{-1}. Correspondingly, the biomass and Carbon were underestimated by 13.1 times by the model in comparison to the Chambers et al. model [18] (Table 6.3). The other allometric models which include FAO 3.2.3, FAO 3.2.5, Brown et al. (1989) and Nath et al. (2019) models enabled an overestimate of the total above-ground biomass and

Table 6.3 Data summary of regression parameters and predicted values of total carbon, biomass below ground and biomass above ground

Allometric models	R^2	Adjusted R^2	TAGB (tonne/ha)	TAGC (tonne/ha)	TBGB (tonne/ha)	TBGC (tonne/ha)
Chambers et al. (2001)	0.804	0.80	234.43	128.94	–	–
FAO 3.2.3 (1997)	0.713	0.71	692.56	380.91	–	–
FAO 3.2.4 (1997)	0.590	0.59	736.14	404.87	–	–
FAO 3.2.5 (1997)	0.716	0.72	419.67	230.82	–	–
Brown et al. (1989)	0.714	0.71	670.70	368.89	–	–
Brown et al. (1989)	0.723	0.72	397.43	218.59	–	–
Nath et al. (2019)	0.692	0.69	322.99	177.65	–	–
Chave et al. (2001)	0.673	0.67	17.89	9.84	–	–
Cairns et al. (1997)		–	–	–	36.57	20.11

carbon by 1.37–2.96 times with respect to the corresponding values obtained from the Chambers model.

A model established by Cairns et al. [24] was used for the estimation of total carbon and biomass below ground. The values were 36.57 tonnes ha^{-1} for total belowground biomass and 20.11 tonnes ha^{-1} for total below-ground carbon (Table 6.3). Thereby, the carbon stock and biomass asset data can be estimated as one of the important variables to prepare an assessment report and thereby submit to the State Forest Department and Ministry of Environment, Forest and Climate Change (MOEF).

6.4 Conclusions

The findings reported in this chapter indicate that the forest of Narpuh Wildlife Sanctuary has a high biomass and carbon stock capacity, only when the forest continues to be undisturbed and maintained with a sustainable management policy. The study shows that a wide variety of tree species and valuable live timber can effectively store the highest content of biomass and carbon. Therefore, a non-destructive (indirect method) method is preferred for the estimation of biomass and carbon asset.

The benefit of using an indirect method is that it consumes less time and work, and it can be carried out on all types of species including endangered and threatened species without disturbing the forest ecosystem. It is also important to consider trees with a girth size less than 70 cm because they are higher in numbers in comparison to those with of girth size greater than 70 cm. This confirmed natural regeneration and reforestation. These young trees with small girth sizes will ultimately replace the old aged trees in the forest to sequest atmospheric carbon and store large volumes of carbon and biomass. Mitigation of climate change by carbon sequestration through forests will develop initiatives to acquire more funding from the public and private sectors to safeguard the environment and address problems of global warming and promote biodiversity conservation for the locals living on the fringe of Narpuh Wildlife Sanctuary.

The above study concludes that all the allometric equations should be tested prior to the selection of a suitable allometric equation or model. Thereby, depending on the multiple regression findings and the type of forest best allometric equation can be identified. Such an approach will allow accurate estimation of biomass and carbon using only tree diameter as an explanatory variable. The equation also compliments the topography and floral biodiversity of the forest. Allometric equations derived for natural tropical evergreen and semi-evergreen are few and unreliable. Hence, allometric equation testing will allow the researcher to offer unbiased and consistent data evaluation of the immaculate forest's carbon stock and biomass of Narpuh Wildlife Sanctuary. In summary, allometric equations are critical in estimating carbon stocks and biomass. This is due to the variation in the tree species, sizes and age within forests and stand density. It can be concluded that the distribution of biomass among different girth sizes between woody species may also be affected by the topography, climate, edaphic properties and developmental activities outside the wildlife sanctuary. The Chambers model has been the best fit to represent the biomass and carbon asset of the sanctuary. Through such research emphasis, plant bio-diversity conservation in North-East India can be most practically achieved through the dissemination of knowledge that fosters renewed interest in the desired subject of interest.

Acknowledgements The authors thank the HOD of Environmental Studies, NEHU, Shillong for providing the infrastructure, Supervisor Dr. S. S. Chaturvedi for his support and guidance, the ICSSR Ph.D. doctoral fellowship, the Forest and Environment Department Government of Meghalaya for the Permit to work inside the Wildlife Sanctuary and all my friends who have helped with my fieldwork.

Conflict of Interest This is original research, and it has not been published in any journals, newspaper and no conflict of interest with anyone.

References

1. Website: http://www.epa.gov. Accessed 19 May 2022
2. India: Third Biennial Update Report to the United Nations Framework Convention on Climate Change. Ministry of Environmental, Forest and Climate Change (MoEFCC), Government of India, (2021)
3. Indian State Forest Report 2021, vol. 17. Forest Survey of India, Ministry of Environmental, Forest and Climate Change (MoEFCC) (2019–2020)
4. FAO: Global Forest Resource Assessment 2020: Main Report. FAO, Rome (2020)
5. Champion, H.G., Seth, S.K.: A Revised Survey of the Forest Ecosystems of India. Govt. of India Manager Publication, New Delhi, India (1968)
6. Cleveland, C.C., Townsend, A.R., Taylor, P., Alvarez-Clare, S., Bustamante, M.M.C., Chuyong, G., Dobrowski, S.Z., Grierson, P., Harms, K.E., Houlton, B.Z., Marklein, A., Parton, W., Porder, S., Reed, S.C., Sierra, C.A., Silver, W.L., Tanner, E.V.J., Wieder, W.R.: Relationships among net primary productivity, nutrients and climate in tropical rain forest: a pan-tropical analysis. Ecol. Lett. **14**, 939–947 (2011)
7. Townsend, A.R., Cleveland, C.C., Houlton, B.Z., Alden, C.B., White, J.W.: Multi-element regulation of the tropical forest carbon cycle. Front. Ecol. Environ. **9**(1), 9–17 (2011)
8. Grace, J., Meir, P.: Tropical rain forests as old-growth forests. In: Wirth, C., Gleixner, G., Heimann, M. (eds.) Old-Growth Forests. Function, Fate and Value. Ecological Studies, vol. 207, pp. 391–408 (2009)
9. Morote, F.G., Serrano, F.L., Andrés, M., Rubio, E., Jiménez, J.G., De Las Heras, J.: Allometries, biomass stocks and biomass allocation in the thermophilic Spanish juniper woodlands of Southern Spain. For. Ecol. Manag. **270**, 85–93 (2012)
10. Brown, S.: Estimating Biomass and Biomass Change of Tropical Forests: A Primer. FAO Forestry Paper, vol. 134. FAO, Rome (1997)
11. Gibbs, H.K., Brown, S., Niles, J.O., Foley, J.A.: Monitoring and estimating tropical forest carbon stocks: making REDD a reality. Environ. Res. Lett. **2**(4) (2007)
12. Petrokofsky, G., Kanamaru, H., Achard, F., Scott, J.S., Joosten, H., Holmgren, P., Lehtonen, A., Menton, C.S.M., Pullin, S.A., Wattenbach, M.: Comparison of methods for measuring and assessing carbon stocks and carbon stocks changes in terrestrial carbon pools. How do the accuracy and precision of current methods compare? A systematic review protocol. Environ. Evid. **1**(6) (2012)
13. Website: http://www.megforest.gov.in/. Accessed 10 May 2022
14. Ravindranath, N.H., Ostwald, M.: Carbon Inventory Methods: Handbook for Greenhouse Gas Inventory, Carbon Mitigation and Round Wood Production Projects. Advances in Global Change Research. Springer, Delft (2008)
15. Chave, J., Andalo, C., Brown, S., Cairns, M.A., Chambers, J.Q., Eamus, D., Fölster, H., Fromard, F., Higuchi, N., Kira, T., Lescure, J.P.: Tree allometry and improved estimation of carbon stocks and balance in tropical forests. Oecologia **145**, 87–99 (2005)
16. Zianis, D., Mencuccini, M.: On simplifying allometric analyses of forest biomass. For. Ecol. Manag. **187**, 311–332 (2004)
17. Pilli, R., Anfodillo, T., Carrer, M.: Towards a functional and simplified allometry for estimating forest biomass. For. Ecol. Manag. **237**(1–3), 583–593 (2006)
18. Chambers, J.Q., dos Santos, J., Ribeiro, R.F., Higuchi, N.: Tree damage, allometric relationships, and above-ground net primary production in central Amazon forest. For. Ecol. Manag. **152**, 73–84 (2001)
19. FAO: Estimating Biomass and Biomass Change of Tropical Forests: A Primer. FAO Forestry Paper No. 134. FAO, Rome, Italy (1997)
20. Brown, S., Gillespie, A.J., Lugo, A.E.: Biomass estimation methods for tropical forests with applications to forest inventory data. For. Sci. **35**(4), 881–902 (1989)
21. Nath, A.J., Tiwari, B.K., Sileshi, G.W., Sahoo, U.K., Brahma, B., Deb, S., Devi, N.B., Das, A.K., Reang, D., Chaturvedi, S.S., Tripathi, O.P.: Allometric models for estimation of forest biomass in North East India. Forests **10**(2), 103 (2019)

22. Chave, J., Riera, B., Dubois, M.A.: Estimation of biomass in a neotropical forest of French Guiana spatial and temporal variability. J. Trop. Ecol. **1**, 79–96 (2001)
23. Winrock International Institute for Agricultural Department: A guide to monitoring carbon storage in forestry projects (1997)
24. Cairns, M.A., Brown, S., Helmer, E.H., Baumgardner, G.A.: Root biomass allocation in the world's upland forests. Oecologia **111**, 1–11 (1997)
25. de Araujo, L.S., dos Santos, J.R., da Costa Freitas, C., Xaud, H.A.M.: The use of microwave and optical data for estimating aerial biomass of the savanna and forest formations at Roraima State, Brazil. In: IEEE 1999 International Geoscience and Remote Sensing Symposium. IGARSS'99 (Cat. No. 99CH36293), vol. 5, pp. 2762–2764 (1999, June)
26. Ovington, F.N., Olson, J.S.: Biomass and chemical content of El Verde lower montane rain forest plants. In: Odum, H.T., Pigeon, R.F. (eds.) A Tropica; Rain Forest: A Study of Irradiation and Ecology at El Verde, Puerro Rico, pp. H53–H77. National Technical Information Service, Springfield, Virginia (1970)
27. Edwards, P.J., Grubb, P.J.: Studies of mineral cycling in a montane rain forest in New Guinea: I. The distribution of organic matter in the vegetation and soil. J. Ecol. 943–969 (1977)
28. Brown, S.: Estimating Biomass and Biomass Change of Tropical Forests: A Primer, vol. 134. Food & Agriculture Organisation (1997)

Chapter 7
Gendered Knowledge, Conservation Priorities and Actions: A Case Study of On-Farm Conservation of Small Millets Among *Malayalar* of Kolli Hills, South India

R. Rengalakshmi and L. Vedavalli

7.1 Introduction

The recent FAO's voluntary guidelines for the Conservation and Sustainable Use of Farmers Varieties/Landraces reiterate that crop genetic diversity provides resilience to climate change, environmental and other shocks [1]. The threat assessment of on-farm crop genetic resources based on the pre-2001 IUCN Red List Categories [2] has shown eight different levels and such a detailed assessment is necessary to fix the conservation priorities. The FAO's voluntary guideline further restates the national action plans for the sustainable use and conservation of the crop biological resources. To prepare a national plan, it is essential to understand the knowledge base of the on-farm diversity and knowledge of farmers' varieties/landraces in the country. Farmers' knowledge about the varieties and landraces is crucial information in the database. The current global view among developmental practitioners and promoters is that rural men and women are the managers of agro-biodiversity and hold in-depth knowledge of crop species. However, no detailed methodology is available to chronicle the gendered farmers' knowledge. In Asian and African farm households, though men and women contribute to agricultural production, gender roles vary by agro-ecosystem, types of crops grown, social groups, and farming system. Choice of crops and cropping system, selection of crops and varieties, preparation for sowing, seed selection, pre-treatment for sowing and storage of subsistence crops are often the primary responsibility of women in many regions of the world. Several studies pointed out that the women's contribution to household food security and contribution

R. Rengalakshmi (✉) · L. Vedavalli
M S Swaminathan Research Foundation, Chennai, India
e-mail: rengalakshmi@mssrf.res.in

L. Vedavalli
e-mail: veda@mssrf.res.in

R. V. S. Uppaluri and L. Rangan (eds.), *Conservation of Biodiversity in the North Eastern States of India*, https://doi.org/10.1007/978-981-99-0945-2_7

of active labor in both production and post-harvest operations are substantial in the food crops [3]. For the improvement and protection of biological resources, women's participation in agro-biodiversity management and the knowledge that goes along with it, notably in seed production, supply, and interchange, has been essential. In Thailand, women play a major role in managing less-common species and varieties in home gardens [4] and in South Asia, women have traditionally chosen and preserved seeds [5, 6]. But in most of Sub-Saharan Africa both men and women customarily farm separate fields, men tend to grow cash crops while women till the land for subsistence crops to feed their families [7]. Thus, women are recognized as the custodians of plant genetic resources, which shows how important they are to systems for food generation focused on ecosystems and biodiversity [8].

Many of the crop species, largely small millets, which are neglected or underutilized like *Panicum, Paspalum, Setaria, Echinochloa,* etc. have been cultivated by women and they are responsible for managing at subsistence scale considering their agronomical and nutritional value to meet the household food requirements, besides its other functional values. Studies have revealed that traditional knowledge lays the groundwork for decisions made at the community level regarding issues such as food and nutrition security, veterinary and public health, ecological resource monitoring, and other topics [9, 10]. However, the knowledge and information vary with social and economic characteristics like age, gender, occupation, ethnicity, caste, and class. As a result of the possibility that various groups may have distinct interpretative frameworks depending on their circumstances, various socioeconomic gain prominence in the social production of knowledge. However, until recently gender-specific knowledge was ignored in planning and execution phases of many natural resource management initiatives [11, 12]. Understanding the perspectives of those who use, manage and benefit is essential for equitable and sustainable management of the resources. The participation of women in asset management and control has not been extensively exploited, and availability of such information databases and systems has been restricted due to a shortage of paperwork or records, which resulted in inequality in sharing the benefits that arise from legal tools and formulating policies.

Thus, interventions promoting the utilization of underutilized crops need to focus on gendered traditional knowledge, differential role, wisdom, and access to resources. Women's knowledge base, abilities, and authority well over change process must be recognized and built upon by the gender equality-supportive nature conservation management approach in order to be successful [12]. This goes beyond just "hearing them out".

7.2 Research Context and Methodology

The research was conducted in the Kolli Hills, which are part of the South India's Tamil Nadu province's Nammakkal district, and are situated at the extremity of the Eastern Ghats. The majority of the population belongs to Malayali or Malayalar tribal community, and farming is their main source of income. The other sources of income include pig-rearing, working as a laborer, migrating to other states or regions, and raising livestock. Among the most prevalent tribal communities, it is centered in Tamil Nadu's northern regions and accounts for roughly 55% of the state's overall tribal communities. The name Malayali derives from the words "malai" and "yali," which are both Sanskrit terms for dwellers or lords of hills. The Ve lala group of farmers initially made up the Malayalis, who left the plains at the start of the sixteenth century [13]. Women has a secondary place in the community since the system is paternal lineage and residency is patrilocal. The Malayali population is divided into many exogamous groups. Despite making a significant contribution to the domestic economic system, women do not acquire/own lands or other types of assets. However, if there are no male offspring in the home, the girls inherit the assets. The favored weddings are those among cross-cousins and those between the child of the sister and the brother of his own mother. Monogamy and serial monogamy are the particular rules of the marriage. A male may have a number of wives at once under certain circumstances. When it comes to marriage, the father is accountable for the offspring of the separated mother. The proposal for separation may come from any of them. Men or women may file for separation or divorce by bringing their petition prior to the traditional panchayat. Legal divorces are sometimes, but not often, requested via judicial processes in the contemporary period. Widowhood is not seen negatively, and both men and women often remarry. Men entirely control the traditional panchayat administrations, while women are not permitted to attend sessions. Their original basic foods were grain legumes, small millets, wild yams, wheat, and rice as supplements. 90% are marginal and small farmers, making up 88% of the overall workforce. Farm laborers make up 7% of the workforce, while other employees make up 5%.

The agricultural landscapes spread across the hill slopes and valleys that form diverse agro-ecosystems. *Malayali's* have classified their farming field according to the type of soil and its characteristics *as vayal* (Valley), *mettankadu* (terraces in slopes), *kollakadu/kothukadu* (rocky terrain). The cropping system varies with the landscape (Table 7.1) and the cultivation of small millets as dry land crops may take place in *kollakadu/kothukadu* or in *mettankadu*, where a slash-and-burn rotating system is in place. In *kollakadu*, the slash-and-burn agriculture was the most common technique for growing small millet. *Panicum miliaceum* L. (proso millet), *Panicum sumatrense* Roth ex Roem. & Schult. (little millet), *Paspalum scrobiculatum* L (Kodo millet), *Elusine coracana* (finger millet) and *Setaria italica* (L.)Beav (Italian millet) are the commonly grown small millet species with more diversity within their own genus as subsistence food crops. Besides, sorghum and pearl millet have been cultivated as inter or border crops. Recently these crops are on the verge of disappearance and consequently households are facing transitory food and nutritional insecurity problems coupled with resource (especially soil fertility) degradation. The decline in cultivation was primarily due to the introduction of market-oriented commercial crops, rice in the public distribution system, and less supportive government policies. Subsequently, their food systems have altered, with rice now serving as the major diet. Although the area under different landraces of small millets has been drastically

Table 7.1 A summary of various types of millets and associated landscape, cropping systems, and cropping pattern

Millet species	Land types	Cropping system	Cropping pattern
Little millet	*Mettankadu*—early maturing landraces	Pure	Millet—field beans
	Mettankadu—late maturing landraces	Combined	Little millet + lab lab + finger millet + Italian millet + mustard (*Brassica nigra*) + cucubits spp. + maize + amaranthus spp
	Kollakadu—late maturing landraces	Pure	Little millet—under slash and burn system
Italian millet	*Mettankadu*—early maturing landraces	Pure	Italian millet—field beans—coriander
	Mettankadu—late maturing landraces	Combined	Italian millet + finger millet + mustard + lab lab
	Kollakadu—late maturing landraces	Combined	Italian millet and little millet
Proso millet	*Mettankadu*	Pure	Field bean—coriander-prosomillet
Kodo millet	*Mettankadu*	Pure	Kodo millet—field bean
Finger millet	*Mettankadu*	Pure	Finger millet
	Mettankadu	Combined	Finger millet + lab lab + mustard + maize + redgram + amaranthus etc
	Kollakadu	Combined	Finger millet + Italian millet
Sorghum and pearl millet	*Mettankadu*	Combined	Finger millet + Italian millet + redgram + amaranthus + mustard

reduced across the hills over a decade, still they are under cultivation in a limited way.

Farmer's involvement in the small millet production is gendered, as shown by gender-specific tasks and responsibilities in the production and use of small millets [14].

The empirical information was gathered from 25 different villages across Kolli Hills since November 2015, although it has been updated during subsequent visits between then and December 2018. A mix of different participatory qualitative tools was used to elicit the information and triangulated the data for analysis (Fig. 7.1). A seasonal calendar for the millet-based farming systems was constructed by a group of men and women farmers to generate information about different crops and identify periods of different agronomic and cultural practices associated with small millets cultivation. The distribution of labor among men and women in the management and cultivation of small millet species and varieties was analysed using gender division of labor[1] (Table 7.2). Pairwise comparison[2] of landraces was conducted to discover the parameters utilized by men and women cultivators to distinguish their crops. Triangulation[3] (Cross-verification) at the conclusion assisted to authenticate the information. Details on post-harvest activities such as cooking, taste, and storability were gathered through focused group discussions with knowledgeable men and women farmers. In-depth interviews were carried out with selected informants on work allocation, decision-making procedures, and the passing down of expertise by gender to the next generations. Additionally, participant observation was done through field visits at various stages of grain development, along with harvest and post-harvesting procedures.

Based on the understanding of the deeper knowledge and associated practices in small millet cultivation, a series of on-farm management measures have been

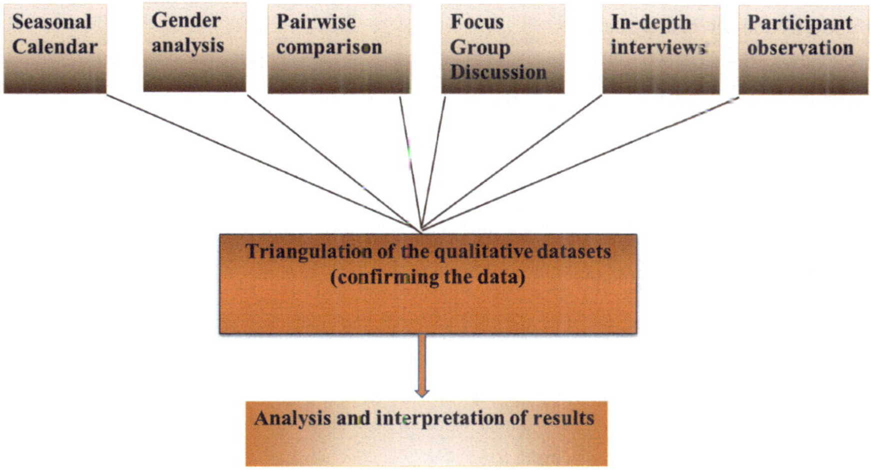

Fig. 7.1 A schematic of data collection and triangulation process

Table 7.2 A qualitative representation of gender roles in small millet cultivation in Kolli Hills

Roles	Women		Men		Both	
	R	DM	R	DM	R	DM
Crop/varietal selection						
1. Crop/landrace selection	–	–	–	–	√	√
2. Seed for sowing from own source/seed borrowing	√	√	–	–	–	–
Production						
3. Land preparation—tilling the soil	–	–	√	√	–	–
4. Bush clearance (cutting)	–	–	–	–	√	√
5. Collecting the bushes, firing, removing the smaller stones and arranging on the bunds	–	–	–	–	√	√
6. Manuring—cattle penning	–	–	–	–	√	√
7. Preparation of seed before sowing—drying and sometimes testing its germination percentage and vigor	√	√	–	–	–	–
8. Sowing	–	–	–	–	√	√
9. Maintaining optimum population—manual thinning	√	√	–	–	–	–
10. Weeding—hand pulling	√	√	–	–	–	–
11. Cross-plowing	–	√	√	–	–	–
12. Monitoring pest and diseases and general growth	√	√	–	–	–	–
13. Pruning (over growth)	√	√	–	–	–	–
14. Harvesting	–	–	–	–	√	√
15. Collection and making bundles	√	√	–	–	–	–
16. Transporting	–	–	–	–	√	√
17. Preparing the threshing ground—cleaning, smoothening the floor using cow dung slurry	√	√	–	–	–	–
Threshing						
18. Spreading the harvested material on the threshing floor	√	√	–	–	–	–
19. Tending the cattle around the floor	–	–	√	√	–	–
20. Turning the threshed material—twice	√	√	–	–	–	–
21. Sorting out straw and separating the grains	–	–	–	–	√	√
22. Pooling the grains for winnowing	√	√	–	–	–	–

(continued)

Table 7.2 (continued)

Roles	Women		Men		Both	
	R	DM	R	DM	R	DM
23. Transporting the straw either for fodder or other uses	–	–	–	–	√	√
Winnowing						
24. Winnowing the grains in the air	–	–	–	–	√	√
25. Removing chaff by hand winnowing	√	√	–	–	–	–
26. Drying the grains in the floor	√	√	–	–	–	–
27. Packing the materials	–	–	–	–	√	√
28. Preparing the storage materials or granary	√	√	–	–	–	–
29. Transporting the materials to the house	–	–	–	–	√	√
Seed management						
30. Identifying good quality grains (selection)	√	√	–	–	–	–
31. Separating the good quality grains for seeds (threshing)	√	√	–	–	–	–
32. Drying with care	√	√	–	–	–	–
33. Seed treatment (using plant leaves and others)	√	√	–	–	–	–
34. Arranging storage containers and cleaning, making it ready	√	√	–	–	–	–
35. Monitoring for storage pest and periodical drying	√	√	–	–	–	–
36. Checking the seed quality	√	√	–	–	–	–
Seed exchange						
37. Managing the lending process	√	√	–	–	–	–
38. Checking the seed quality before exchange with others	√	√	–	–	–	–
39. Getting seeds through exchange (if it is not available)	√	√	–	–	–	–
Utilization						
40. Monitoring for the pests in the grain	√	√	–	–	–	–
41. Winnowing	√	√	–	–	–	–
42. Drying	√	√	–	–	–	–
43. Pounding and polishing	√	√	–	–	–	–

(continued)

Table 7.2 (continued)

Roles	Women		Men		Both	
	R	DM	R	DM	R	DM
44. Removing the stones and unwanted materials	√	√	–	–	–	–
45. Preparation of different food items	√	√	–	–	–	–
46. Storing and using the husk as a pig feed	√	√	–	–	–	–

R—role, DM—decision making

promoted to revive the farmers' varieties and landraces. In the revival process, due importance is given to women as conservation champions engaged in conservation and sustainable use through post-harvest value addition and markets.

7.3 Results and Discussion

The involvement of women in the production of food for sustenance is shown through a detailed investigation of gender-specific responsibilities and engagement in the production and use of small millets. The research demonstrates that men and women mutually decide on the crops, the landrace selection, and crop choices on their fields. Men only carry out tasks like initial ploughing, seeding, and cross-plowing (a method of operation at 20–25 days after sowing for thinning and weeding purposes using a small wooden plough), which is common throughout South Asia for rice, millets, slash and burn farming and other dryland crops [15]. While both men and women work together to determine the best period to sow seeds, women choose the best period to do cross ploughing. Malayali women make the majority of the decisions and carry out the majority of the work for important tasks like weeding, continual surveillance, and seed handling, which includes seed exchange, preservation, and choice of seeds. However, often men are also involved in weeding if there is no adequate hand, a task exclusively done by women. Both women and men work together to do threshing, harvesting, packing, and winnowing which are widely prevailing in the rice farming system of South Asia and other millet-based farming systems of the Indian subcontinent [15].

It is necessary to understand not only who is doing what, but also it is important to know who is contributing to take and make the decisions. The direct involvement in an activity empowers a person with decision-making power related to that activity. Table 7.2 clearly discloses the prevailing role as well as decision-making of men and women of different households across Kolli Hills in small millet cultivation, management, and utilization. Since it is a more or less economically homogeneous group the same pattern of division of labor and decision-making prevails across households/hamlets.

When it comes to work, women play a disproportionately large contribution in the production of small millet. The categorization of many tasks indicates the tendency to distribute monotonous activities among women. Small millets are being cultivated on a subsistence scale, even though men and women both have to take care of things. Thus, the contribution of women is more. It is unlike the situation in African countries like Gambia, Mali, and Liberia, where the women are mostly in charge of cultivating food crops for livelihood and men take a lead in commercial or cash crops [16, 17]. The research reiterates that socioeconomic, and culturally important aspects decide the gendered responsibilities, that eventually impact their roles and ability for making decisions related to various tasks in small millet production.

7.3.1 Malayali Women and Seed Management

Though men share the responsibility in other agronomical practices, women alone are responsible for carrying out the crop improvement and development activities like seed selection, management, and utilization associated with small millets. It agrees with the view of Shiva [6] that 'Indian women are seed collectors and sustainers, having more agro-ecological expertise than men'. The seed management practices encompass the activities viz., selection of panicle, harvesting, drying, cleaning, storing, monitoring, seed exchange, and quality testing at the time of sowing.

The selection of elite seed material for the next sowing has an implication on the population genetic structure and degree of genetic diversity. The *Malayali* women farmers follow crop-specific selection strategies that help in the maintenance of intra-specific diversity. It is mostly based on crop architecture and biology either at pre or at post-harvest stages. Similar observation was recorded among the women of Malawi, Africa in managing local bean diversity by Ferguson and Mkandawire [18]. In the little millet and proso millet cases, due to the loose and drooping panicle structure, post-harvest selection is commonly practiced and seeds are collected from the threshing floor. The crop architecture at the time of maturity is not suitable for individual plant or panicle selection based on the phenotypic character of the crop. Women collect the matured, well-formed grains (as they fall before the first threshing operation) and manage them separately for seed. Sometimes superior plants from a wide-spaced area are identified and harvested separately by them. Alternatively, from the bulk of the threshed grain, women take nearly 20–30% for seed purposes and manage separately.

In italian millet and kodo millet farming, pre-harvest selection has been practiced. This needs special identification and selection skills. The criteria for selection are based on agronomic and genetic characteristics coupled with disease free, true-to-type for head shape and grain size. It is carried out either by men or women but women make the decision on when to carry out the task by closely monitoring the crop growth and maturity. While selecting for seeds, men choose quality panicles as their primary criterion whereas women consider the vigor of whole plant along with well-filled grains in panicles. Best panicles are selected based on the size, fully matured grains on both ends of panicle and absence of chaff, pest and disease infection and ill-filled grains. Best true-to-type panicles are selected from the field and stored as such or threshed and stored separately.

The important activities in seed storage are cleaning, winnowing, drying, and storing. The material taken for seed require special care, which is carried out exclusively by women: Sun-dried seed was kept for two to three days to the required moisture level of around 10–12% and winnowed properly to remove chaffy, immature seeds, and dust materials. Women use specific indigenous methods to assess the drying process. Women assess the hardiness by biting a grain (stone hard) or by threshing whole debraned grain between fingers or on any surface. The seed materials are stored either in mud containers or bag (gunny/polyethylene) tied and stored separately. The gaps in the bag allow moisture and the seeds deteriorate much faster. To avoid that, nowadays, the gunny bags are kept near or above the hearth. According to them, the seed holds its viability for 2–3 years when properly dried and stored in dry containers. They prefer mud pots for long-term storage as it protects the seed material from rats and at the same time, it is moisture proof. The traditional cultural practices encouraged the indigenous seed storage practice using mud pots and the practice of giving a mud container to the bridegroom on marriage as a gift supports this view. It is again dried under sun before sowing to break the dormancy in storage. Some households have a separate seed storage mud structure outside their houses.

Women play a unique role in monitoring the quality of stored seeds by using process indicators. The grain and seed storage activity is generally the responsibility of women. As Albu [19] pointed out, the knowledge generation needs a process of learning in Kolli Hills due to the continuous involvement of women in the millet cultivation and utilization process. Hence, the elder women are more knowledgeable than the middle and young aged ones. The skills and information on seed quality are passed to the young women by involving them in the process, which provides an opportunity for them to learn and gain knowledge. Their involvement in the process helps them to gain knowledge and skills to identify, define and practice the process indicators for monitoring. The specific process indicators are color change to greyish black, crumby structures in the grains, insect webbings, and powdery structure of the seeds. The quality of the seed is assessed using seed color (greyish) and odor. In case of doubt on the quality and rate of germination, women do simple in-house germination tests to determine the quantity of seed required for sowing. Since it is a rainfed system, ensuring pre-sowing seed quality i.e., germination rate assumes greater importance.

7.3.2 Seed Network and Exchange

Farmers rely on their own source of seed and women of households to ensure upon the availability of seed for the next sowing. Women have the practice of storing five times higher seed requirement of forth-coming season. The women ensure that the excess seeds after meeting their requirement in the current season would be stored as buffer stock for the next season in order to cope up with the crop failures in the rainfed system. Also, they provide seeds to needy farmers on their request by following the locally evolved seed exchange policies and methods.

Based on neighborhood and kinship, the seed network (horizontal) among farmers is a traditional practice, which ease and ensure the accessibility of seed to them. Exchange takes place without cash transaction by local ethics through an informal network. Women play a dominant role in exchange network by sharing information, which is the base for network operation. Women ensure upon the seed quality while lending and they take the responsibility of returning 1½–2 times the borrowed amount after harvest. Farmers trust neighborhood certification regarding the seed quality in terms of vigor and germination percentage. If harvest is affected in a particular season, the borrowed amount can be returned after subsequent harvest and in these situations, the borrower has to return three to four times of the seed borrowed.

Women decide the norms in exchange process and normally exchange takes place for the products having equal value. For example, *samai* or *thina* is not transacted for paddy because paddy gives nearly 60% of edible part after threshing while *samai* and *thinai* gives around 75% of the edible portion, after dehusking. Similarly, they never exchange the products, which need energy and labour for processing before converting them into useful product. For instance, even among thinai landraes, killanthinai is not transacted for koranthinai, as the former needs less energy for pounding than the latter. It shows the intrinsic knowledge of the women of the different landraces of small millet species. Usually, farmers exchange seeds due to the seed scarcity that occurs due to the poor harvest or lack of rainfall in the right season. Also, they exchange the germplasm with other farmers if the seeds are not true to type or mixed with other landraces of the same crop. Thus, in Kolli Hills, in addition to farmers own source, the local seed exchange network ensures the availability of seed within the community.

7.3.3 Genetic Enhancement

For broadening the crop genetic base and adopting agronomic procedures, women often take a lead. The deliberate mix of different landraces, which differ in their quantitative and qualitative features in a single field, as a risk aversion approach is one of the means for genetic improvement. They do not mix the landraces with the same duration and only mix landraces that differ especially in duration and plant architecture in the mixed cropping system. The introgression of genes was introduced to produce novel and valuable genotypes to widen diversity and variation. The developed variability is the base for the selection of ideal plants. Women used to say 'the crop mix is always seen as a risk aversion strategy and staggered harvest of different crops in a year results in the supply of produces at different times which is vital to meet the household food requirement'. The landrace mixtures or wild relatives/volunteers that are found in the field are not rouged out deliberately which may facilitate the introgression.

7.3.4 Cultivation and Management Practices

The socio-economic and agro-ecological environments are directly linked with the knowledge system through adaptation, conservation, and experimentation. Social norms and customs dictate underlying principles for gender differential patterns of access, role, responsibilities, and rights related to various resources. The gender differential skill and knowledge existing in the society is caused by the consequence of cultural and social work of people along with their power disparity. Efficient resource utility is necessary which is provided by the knowledge that includes a set of empirical observations for local ecosystems along with self-management and their classifications.

A close scrutiny of the work pattern in small millet cultivation, seed management, exchange, and utilization in Kolli Hills reveals that all the work done exclusively by women involves intensive knowledge, more drudgery (in terms of monotonous activity and energy), and greater physical strain. The processing of millet farming and cultivation system requires women to stay in the fields for a longer duration of time to gain better knowledge and help them observe the entire process. Almost 80% of the cultivation inputs for millets are contributed by women for their dedicated tasks in utilization, production, and processing chain. The difference in the workload is brought out by the fact as women say, "though we both are involved in planning and other activities, women alone are left with the responsibility to keep a track of the production process". The differential gender roles in specific resource management and production are creating access to experimentation and innovation and finally knowledge generation and management. Thus, the knowledge acquisition differs with roles and responsibilities, access and utilization of the resources. In this regard, a critical insight for soil environment and rainfall are needful and hence, a deeper interaction and obligation will help them. Nevertheless, men are often engaged in the plowing activity and women provide better information about soil fertility and its texture and its fitness for millet cultivation. Usually, the black clay soil after first rain can hold moisture for upto 7–10 days. However, red gravely soil after receiving first rain do not have any moisture-holding capacity. But the third category of soil 'kassanku kadu' (the red laterite soil) after first rain forms crust on the surface and becomes hard. Irrespective of various soil types as described above, women makes the right claim over the soil moisture suitable for plowing and sowing, to thereby make a knowledgeable lead decision regarding plowing, sowing, and weeding on the account of soil category and amount of rainfall. This enable them to understand and correlate rainfall intensity on soil profile characteristics and field conditions. In the same way, their understanding on biology of the species helps them to harvest a good yield. Many landraces of small millet species are highly sensitive to light and soil fertility which influence the growth physiology and especially crop duration and flowering. For instance, if the soil is highly fertile, landraces like kattavettisami and perumsamai are preferred as they grow for a longer vegetative phase and need more days to flower. It needs topping activity i.e. pruning the top foliage growth to the desired level to induce reproductive growth; otherwise, crop never flowers. Through a close observation on the growth of the crop, women easily identify and prune the foliage at appropriate time to activate flower initiation.

The active role of women in the selection of quality seed material aid them to have a deeper understanding for good breeding information. The taxonomic viewpoints at intra-specific level and an example for little millet landraces are shown in Table 7.2. They gain knowledge through meticulous involvement and experience on reproductive, physiological, and morphological aspects of the cultivars in each crop cycle. Thus, knowledge development is continuous and dynamic in response to the evolutionary adaptions and their changes. The women farmers adopt diverse crop-specific selection strategies that they acquired through their own experiences and knowledge received from the elders in the family. The roles and responsibilities of Malayali women were based on seed selection. Paris et al. [20] and Sahu et al. [21] observed that women mentioned straw quantity and quality as significant factors in eastern India, whereas men did not identify these factors as important features for varietal selection.

The process of differentiation, categorization, and nomenclature (folk biology) reflects the differential knowledge of men and women. The differences in the traditional separation of labor were characterized by women and men with respect to different folk varietals and folk-specific attributes. They also use plant vigour and seed color. Based on climatic and edaphic needs, plant height, grain shape and its arrangement in the panicle, meal quality and processing, geographical distribution as well as health impact, women could distinguish Italian and small millet landraces. On the other hand, Malayali men could distinguish them on the basis of its productivity and agronomic approach (like method of harvesting, cropping system, and maturity periods). Based on susceptible damage from birds, pounding, threshing requirements, and early ripening, women are able to discriminate between different landraces. The deeper involvement of women in food preparation, seed processing, and management will aid them to attain useful biological and gastronomic insights reflected in a better classification utilizing elusive features.

With respect to the landraces value, women differ from their counterparts in expertise and in their opinions. This includes the women's crop-choosing criterion that is composed of numerous concerns or preferences surrounding challenging problems in comparison with their counterpart. For women, resource availability, multiple use of the crop, the meal quality, and processing ease are the primary concerns, whereas men consider productivity and stability as major factors in landrace selection. Women favors vellaperumsamai/perumsamai for their meal quality, taste, and consistency and are responsible for cooking while men prefer the early maturing variety namely malliasamai. Women suggested kottapattisamai is the most wonderful and beautiful among all the small millet landraces. In Tamil it is stated as 'samaikku azhaku kottapattisamai' and is preferred due to its ease in productivity, processing meal quality, and adaptation to a diverse agro-ecosystem (Table 7.3).

The landraces and crop species were categorized by women according to the meal quality and ease in processing. Pounding is most difficult in little millet followed by common millet, kodo millet, and italian millet. The landraces could be rated by the meal quality and taste. The meal quality is in the order of perumthinai followed

Table 7.3 List of folk names and keys used for the differentiation of different populations

Farmer landrace names	Keys used for identification
Malliasama/vellamallisama/chinnasama	Meal colour, texture, caryopsis colour
Sadangsama/kullangsama/chadasama	Inflorescence shape/crop duration
Thrikulasama/kurungsama/karumsama	Inflorescence shape, crop duration caryopsis colour
Pillusama/elangsama	Plant structure and crop duration
Koottasama/kottappattisama	Grain yield
Kattavettisama/kattasama/kuthusama	Grain yield
Karumperumsama/kathasama/periyasama	Inflorescence shape, caryopsis colour
Vellaperumsamai/kothusama/vellasama	Inflorescence density and plant structure

by palanthiani, senthinai, mukkanthinai, koranthinai for Italian millet. Also, women categorise the small millet landraces according to the proportion of bran to edible parts, and they favor landraces with less bran. They evaluate the bran proportion and regulate the production of landraces based on pounding intensity and rate them as vellaperumsamai > kattavettisamai > karumperumsaamai > kottapattisamai. The knowledge about the grain characteristics in milling will help them for their responsibility and involvement in pounding process. However, despite men contributing to work with women in pounding, the responsibility lies with women only. Women carry out drying and destoning before pounding and winnowing and de stoning after pounding. Due to the continuous involvement of women in processing, they are able to decide the dryness (texture and period) for quick dehusking.

Despite small millets cultivation is less beneficial on an economic basis with respect to other cash crops, due to their subsistence importance, women take a lead role in its management. The grain and seed storage approaches and their techniques had progressed in an innovative manner. Their knowledge on nutritional characteristics on agronomy and of different landraces enable them to gain useful insights knowledge in plant breeding and seed selection. This allows them to preserve genetic diversity essential to adapt in a changing weather and biotic pressures. This will also safeguard the survival of traditional crops to local preferences and conditions. Thus, the evolved and accrued knowledge is based on collective understanding and interpretation of biological and social spheres on experiences and perceptions about millet species. Women have better understanding in the process of knowledge transmission to younger generation by purposefully involving them in the job; it is a kind of hands-on training. Thus, the degree of gender specific knowledge and skill depends on the pattern in which roles and responsibilities are assigned to men and women as well as on the degree of flexibility they have to carry out other tasks.

7.3.5 *Women and Property Rights*

The cultural norms of Malayali tribe deny women's rights to own and inherit the land and other resources, which are vital for cultivation. This is mainly due to the practice of patrilocal residence, which makes women move out of her natal home after marriage and leave behind all permanent assets to the male heirs of the family. The traditional role of men and women and associated values also reinforce the inheritance of parental property by men. Such differences in male and female rights to land are evident especially in the Middle East and most of the African and Latin American countries [22]. This social pressure has a direct impact on women's knowledge, power, position and status within home with regard to negotiating capacity, decision-making, and implementing the decisions. All these could be important factors in sustainable and conservation use of millet genetic resources. Lack of possession rights over cultivable land for Malayali women negatively impacts their freedom to take decisions and implement the activities related to millet cultivation.

7.3.6 *On-Farm Conservation for Revival of Small Millet Genetic Diversity*

According to IUCN Red list categories, most of the small millet species as well as its intra-specific diversity are either under Extinct On-Farm (ExF) or Endangered On-Farm (EF). Not only the crop species, but also the rich gendered traditional knowledge associated with its cultivation are also under threat. In order to revive the cultivation of small millets species, set of on-farm conservation strategies are rolled out integrating women's knowledge systems. These crops are under on-farm conservation prioritization categories to improve the availability of seed resources, ensuring availability of information about such landraces/farmer varieties, improving its cultivation methods for higher productivity under intensification and market creation and value addition; the specific measures are (i) reintroduction of seeds collected from different parts of the Kolli Hills across villages (ii) established community seed banks and seed producing farmers, (iii) diversity fairs, (iv) demonstration of different landraces/varieties from similar ecosystems and promoted participatory varietal systems (v) field days and cross-field visits (vi) micro-credit for its cultivation and processing (vii) introducing alternate agronomic measures for higher productivity (viii) cultivation under different production systems (ix) introducing decentralized processing units (x) building the capacity of women in preparing value added products (xi) expanded institutional links and markets etc. In such interventions, considering women's exclusive engagement and knowledge systems in millet production to sustainable use, women were promoted as seed bank managers, champion farmers, entrepreneurs producing value-added foods, etc.

7.4 Conclusions

In Kolli Hills, small millet genetic resources have been on the verge of disappearance in the past two decades due to a combination of pressures. The introduction of market-oriented cash crops, drudgery in processing, lack of market channels, low productivity, and rice availability in state-sponsored public distribution system for a subsidized price are some of the significant triggering factors for declining interest among small millets growers. Although agriculture is a family enterprise among Malayali tribe, women have a significant role in small millet cultivation and management. Analysis of gender roles and responsibilities, clearly reveal women's dominant role and knowledge in millet production, selection, post-harvest storage, seed management and genetic enhancement as well. Also, it indicates how women are related to the cultivation of these traditional subsistence food crops having high nutritional profile.

Gendered roles and responsibilities in decision-making for various activities in small millet cultivation show women's crucial role in different aspects of its cultivation and utilization. Even though the division of labor assigns exclusive activities for men and women, both share certain activities, the overall management however keeps women in a way to make almost all the decisions regarding millet management and cultivation. The crop selection criteria of men and women differ; women use multiple criteria for planting when compared to men who usually look for yield and stability. The intimate interaction of women with the millet cropping system especially in seed selection, storage, exchange, processing, and utilization resulted in specialized knowledge, understanding, and skills in crop production, which indirectly facilitates millet diversity and optimal utilization of natural resources.

Though the area under small millets has reduced during the past, it is continuing to be cultivated by a few households that ensure upon the continuous availability of millet genetic resources. This is made possible primarily due to the women who take care to store the seeds for future use. While women have access to land, the complete ownership lies with the men, which have an implication on conservation such under-utilized crop species. The nuanced understanding of women's knowledge systems in small millet diversity, helped to prioritize the conservation actions as well as on-farm management strategies for its promotion.

The demonstrated research pedagogy in this chapter can be applied for the resolution and empowerment of gendered knowledge, conserving priorities and actions in the North-East India. Thereby, similar know-how could be generated through survey-based case studies to infer upon the greater role of women in the conservation of bio-diversity in agriculture and farming in the North-eastern region of India.

Notes

1. Gender activity profile analysis helps to know the gender division of labour, which defines men's, and women's socio economic opportunities and constraints. The analysis focuses on

who does what, where do men and women work, when do men and women work and for how long.

2. Pair wise comparison was done by matrix method in which landraces and salient characteristics of species were listed out. Each one of them was relatively scored separately for men and women, which helped to get the details on criteria for selection as well as their prioritization.
3. Triangulation is one of the essential principles of PRA that tries to ensure that the results of the analysis are as accurate and unbiased as possible. It helps to cross check the different information and eliminate the unnecessary details.

References

1. FAO: Voluntary Guidelines for the Conservation and Sustainable Use of Farmers' Varieties/Landraces. Rome (2019)
2. Antofie, M., Sand, M., Ciotea, G., Iagraru, P.: Data sheet model for developing a red list regarding crop landraces in Romania. Ann.: Food Sci. Technol. **11**(1), 45–49 (2010)
3. Aragon, J., Miller, M.: Global Women's Issues: Women in the World Today, Extended Version. United States Department of State, Bureau of International Information Programs (2017)
4. Moreno-Black, G., Prapimporn, S., Sampong, T.: Women in northeastern Thailand: preservers of botanical diversity. Indig. Knowl. Dev. Monit. (2), 24 (1994)
5. Seshu, D., Malavika, D.: The role of women in seed management with special reference to rice. IRTP Bull. (5), 1–7 (1989)
6. Shiva, V.: Women's indigenous knowledge and biodiversity conservation. In: Sen, G. (ed.) Indigenous Vision: Peoples of India, Attitudes to the Environment, pp. 205–214. Sage Publications, New Delhi, India (1992)
7. FAO: Gender and food security: division of labour. www.fao.org/Gender/en/lab-e.htm. Last accessed 20 Aug 2022
8. Berlin, B.: Ethnobiological Classification: Principles of Categorization of Plants and Animals in Traditional Societies. Princeton Univ., Press, Princeton, USA (1992)
9. Juana, K., Francisco, S.: Integrating local and global knowledge, technology and production systems: challenges for technical cooperation. Sci., Technol. Soc. **7**(2), 215–247 (2002)
10. MOST and CIRAN: Best practices on indigenous knowledge. Management of Social Transformations Programme (MOST) and Centre for International Research and Advisory networks (CIRAN), Paris and The Hague. www.unesco.org/most/bpikreg.htm. Accessed 20 Aug 2022
11. Datar, C.: Nurturing Nature: Women at the Center of Natural and Social Regeneration. Earthcare, Bombay, India (1998)
12. Kapoor, A.: Gender: the missing link in environment conservation. http://www.bcpp-india.org/chapter31.htm. Accessed 24 Aug 2022
13. Thurston, E., Rangachari, K.: Castes and tribes of Southern India. **7**. New-York Johnson Reprint Corp. (1st ed 1909, Madras Government Press) (1975)
14. Rengalakshmi, R., et.al.: Rural and tribal women in agrobiodiversity conservation: An Indian Case Study (29 pp). M. S. Swaminathan Research Foundation, Chennai, India & Food and Agriculture Organization of United Nations, RAP 2002/08. ISBN: 971-614-019-3 (2002)
15. Jewit, S.: Unequal knowledge's in Jharkhand, India: de-romanticizing women's agroecological expertise. Dev. Chang. **31**, 961–985 (2000)
16. FAO: Women in Rice Farming Systems: Focus: Sub-Saharan Africa. Women in Agricultural Production and Rural Development Service. Dey, Rome (1984)
17. Synneveg, G.: Gender differentiated management of local crop genetic resources in Bafoulabe Cercle, Kayes Region of Mali—a case study. In: Actes du Collouque, Gestion des Resources Genetiques de plantes en afrique des Savanes, Bamako, Mali, 24–28 Feb 1997
18. Ferguson, A., Mkandawire, R.M.: A crop diversity improvement strategy. Indig. Knowl. Dev. Monit. **4**(1), 10–13 (1996)

19. Albu, M.: Technological learning and innovation in industrial clusters in the south. Electronic Working Paper Series, 1–49 (1997)
20. Paris, T.R., Singh, A., Luis, J.: Listening to male and female farmers in rice varietal selection: a case in eastern India. In: CIAT (ed.) The Proceedings of the International Symposium on Participatory Plant Breeding and Participatory Plant Genetic Resources Enhancement: An Exchange of Experiences from South and South East Asia, Pokhara, Nepal, 1–5 May 2000, pp. 90–95 (2001)
21. Sahu, R.L., Sahu, V.N., Sharma, M.L., Paris, T.R., Mc Allister, K., Singh, R.K., Sarkarung, S.: Understanding farmers selection criteria for rice varieties: a case in Madya Pradesh and Uttra Pradesh, Eastern India. In: CIAT (ed.) The Proceedings of the International Symposium on Participatory Plant Breeding and Participatory Plant Genetic Resources Enhancement: An Exchange of Experiences from South and South East Asia, Pokhara, Nepal, 1–5 May 2000, pp. 90–95 (2001)
22. FAO: Gender, land and fertility—women's access to land and security of tenure by Jacques de Guerny and Daphne Topouzis. In: Gender, Population and Rural Development with a Focus on Land Tenure and Farming Systems (1996)

Part II
Conservation of Diversified Microbial and Animal Bio-Diversity for Its Restoration

Chapter 8
Phenotypic and Genotypic Characterization of Predominant Lactic Acid Bacteria Isolated from Traditionally Fermented Brassica Leaf (Gundruk)

Arindam Barman, Nillo Yedi, and Neha M. Sangma

8.1 Introduction

Gundruk (*Brassica spp.* belongs to Brassicaceae family) and locally known as mula pata gundruk is a fermented, nonsalted, acidic and sun-dried leafy vegetable condiment staple to the Nepalese people [1, 2]. In India, gundruk is commonly prepared and consumed in Northeastern states, Darjeeling (West Bengal), Himachal Pradesh, Uttarakhand and other Nepali inhabited areas [3]. In Sikkim, the annual production of gundruk is about 3.2 kg/house with a consumption of 1.2 g daily per capita [4]. Generally, gundruk production technology is limited at household level for own consumption. Also, the verbal transmission of the producing technique is adopted from one era to the next [5, 6]. Thereby, the gundruk is locally sold in Sikkim and Darjeeling hills (West Bengal) of India and for this reason it is not commercialized on large scale [7].

Traditionally fermented vegetable products are indigenous and are very popular among the Northeastern states such as Mizoram, Manipur, Sikkim, Tripura, Meghalaya, Assam, Nagaland, and Arunachal Pradesh. The indigenous people of these states acquire deep knowledge about the environment and depend upon the plants, forests and their food products to meet their daily needs [8]. Traditionally, people have created and consumed fermented foods for a very long time. This is due to their strong roots in their tradition and culture. Such dependence exhibits the abundance of ethnic residents in terms of their abilities to prepare microbiological fermented goods in addition to drinks and food for a variety of occasions [9]. An investigation of traditional food allows greater and significant awareness into a dietary pattern. Also, one can know how they alter with time [10]. Vegetables of these states that have been fermented and partially fermented include gundruk, ganang tamdui, khalpi, anishi,

A. Barman (✉) · N. Yedi · N. M. Sangma
Department of Horticulture, North Eastern Hill University, Shillong 794002, Meghalaya, India
e-mail: arindamnehu@gmail.com

R. V. S. Uppaluri and L. Rangan (eds.), *Conservation of Biodiversity in the North Eastern States of India*, https://doi.org/10.1007/978-981-99-0945-2_8

narzi, hungrii, sinki, tsutuoice, inziangsang, bochu-mba, kosoi, ankamthu, simal tarul ko jaanr, anthur rep, galda gisi, goyang, and tam-um.

A gradual breakdown process called fermentation is conducted by microbes or enzymes through the production of organic acids and alcohols from readily accessible carbohydrates. This occurs through the anaerobic and partially anaerobic oxidation process. It is one of the ancient food processing methods that even exist till date and is regarded to be efficient with low energy requirements. Thereby, the process increases the shelf life and reduces the utility of other methods in the processing of foods [11]. Fermentation of fruits and vegetables are targeted so that the fermented products can be used during the time of scarcity, to impart desirable flavour, to improve the texture of the food, to reduce toxicity, to decrease the cooking time, to improve lifestyle through employment generation and well-being of minority and vulnerable group of people in the world [11, 12].

Vegetables and fruits can be targeted through the lactic acid fermentation for their better preservation. For this the targeted attributes include nutritional improvement and good sensory characteristics [13]. A large group of gram-positive do get included by lactic acid bacteria (LAB), acid-tolerant, anaerobic, lactic acid and non-sporulating producing bacteria. They are characterized on the basis of physiological, morphological and metabolic characters. However, based on the final product of fermentation, LAB can be categorized into two different types such as homofermentative LAB (it includes genera like *Pediococcus, Lactococcus* and *Streptococcus,* which after fermentation yield lactic acid as its primary constituent) and heterofermentative LAB (it includes genera like *Leuconostoc* and the Betabacteria a subgroup of the genus *Lactobacllius*, which produce ethyl alcohol, CO_2, acetic acid and lactic acid derivatives as fermentation products [14]. The Nepali women are major producers of gundruk [15].

The vast variation in the climatic and ecological condition of North-East India prompts the need to address the issue that how bio-diversity gets conserved in the lactic acid bacteria. Thereby, using relevant insights, the adaptability and threats for the microbes can be assessed and needful interventions can be sought for a healthy natural system. This is the central objective of this chapter.

8.2 Methods of Gundruk Preparation

Fermentation in container: A modern approach was followed in which the preparation involved the sequence of selection of fresh leaves, washing, wilting under the sun for 1–2 days, shredding with the knife, mildly crushing, then soaking in lukewarm water for some time, squeezing out the excess water and finally tightly packing the biomass into a plastic or glass container or an earthen pot to create an anaerobic condition. Then, after 7–10 days, fermentation occurs in a warm and dry place. At the time of fermentation, the room temperature was about 18 °C [16]. Finally, after another 3–4 days, fermented gundruk was sun-dried [15].

Fermentation in pit: A traditional method was followed. In this a pit of desired length, breadth and depth is dug and was dried by fire. Following this, paddy straw lining was allowed at the bottom and four corners of the pit were lined with a bamboo mat over the paddy straw layer. Then lining materials such as bamboo leaves, banana leaves, fern or onion leaves (*Galphylopteriopsis erubescens*), bhuletro leaves (*Dipterocarpus tuberculatus* Roxb.) and nevaro leaves (*Ficus palmata* forsk) were used to serve as a thick layer lining. Thereafter, *Brassica* leaves were wilted, shredded, and crushed and were placed over the lining. Further, the leaves were covered with a fine layer of lining materials and heavy materials such as stones or boulder. The stones were used on the top to compress it. In 15–22 days, in-situ-fermentation gets completed. Thereafter, sun-drying was allowed after another 2–4 days time period [2].

Fermentation in cow dung: The method is an indigenous method practiced by Bhutia people of Sikkim. The fresh leafy vegetables were crushed and placed inside polyethene bags. Thereafter, the bags were wrapped tightly to make them airtight. Eventually, the bags were kept in cow dung for about 20 days to complete their fermentation process. Subsequently, the bags were taken out and washed properly. Finally, the product was sun dried. Compared to other two methods, the fermentation rate is faster in this method [17]. Fully sun-dried gundruk with brownish to blackish colour has a storage life of about 2 years or more if it is stored at ambient room temperature [1, 2].

The conventional method of gundruk fermentation results in 90% loss of carotenoids during sun drying. This is a serious drawback of this method. However, the vitamin loss could be reduced by applying improved methods of gundruk drying [11]. A typical gundruk flavour and tangy taste is the determining factor for the evaluation of its quality [5]. Winter (December to February) is the main season for gundruk production. During this season, the availability and supply of perishable leafy vegetables for gundruk production is surplus. However, gundruk preparation during the rainy season is not encouraged due to rapid spoilage and undesirable flavour [18].

During gundruk fermentation, different lactic acid bacteria *(Lactobacillus)* species are engaged which include *L. casei, L. cellobiosus, Pediococcuspentosaceus, L. plantarum,* and *L. casei subsp. pseudoplantarum.* Both *L. cellobiosus* (a heterofermentative LAB) and *Pediococcuspentosaceus* (a homofermentative LAB) were responsible to initiate the gundruk fermentation process. Finally, *L. plantarum* (A homofermentative LAB) completes the fermentation [5]. Also, other LAB strains such as *L. brevis, L. paracasei, Leuconostoc fallax, Pediococcus acidilacticiand P. pentosaceus* have been reported in gundruk fermentation [1]. For gundruk production, lactic acid bacteria strains were used as a starter culture. These include *Lb. plantarum* GLn: R1 (MTCC 9483) and *P. pentosaceus* GLn:R2 (MTCC 9484) [3]. Various LAB strains obtained from gundruk have been characterized to possess high acidification capacity, probiotic attributes, antimicrobial properties and potentials to break down antinutritive elements [16].

Essential amino acids, minerals, and vitamin B are rich in gundruk [6, 19]. Therefore, the nutritive content of gundruk can be concluded as protein, carbohydrate, fat, ashes, moisture and food value and their corresponding values were 38.7%, 38.3%,

2.1%, 22.2%, 15% and 321.9 kcal/100 g DM (dry matter), respectively. The estimated chemical and mineral contents have been specified as pH, acidity, Ca, Na, and K and as 5.0%, 0.49%, 234.6, 142.2, and 677.6 mg/100 g, respectively [19, 20]. Gundruk is a good source of calcium, magnesium and iron. Hence, it acts as a source of essential minerals during off season to the consumers [21, 22]. Acetic and Lactic acid are the main acids present in gundruk [23]. The chief flavouring compound present in gundruk are esters, alcohols and phenylacetaldehyde such as Cyanate and isothiocyanate [5].

Gundruk is generally consumed and enjoyed as a soup (gundruk ko jhol), pickle, chutney, curry and dhedo [15, 18, 24]. Along with the main meal, it is generally served as a side dish. It is considered to have great appetizing attributes in a bland and starchy diet during off season. It also fulfils the daily mineral requirements due to its abundant mineral constitution [23]. Gundruk is known to possess immense health benefits. These include stimulation of bowl movement to check constipation, diarrhea, antimicrobial properties [17], aiding digestion, weight loss, memory booster, strong bone build up, reducing blood pressure and cholesterol level. Bhutia women uses gundruk soup to increase milking efficiency of the breast-feeding mothers. It is also consumed as a tonic by the elderly individuals [25].

Gundruk is known to be packed with essential vitamins, minerals and amino acids. For this reason, it can be used to overcome the malnutrition and overweight issues which are presently relevant to our country. Generally, the traditional production technology is very vast. However, it can be further improved by using essential microbes through biotechnological methods. Thereby, the characteristics of fermented foods can be enhanced. Fermented foods have a direct impact on human health. However, they are consumed directly on a daily basis in the Northeast region without the approval of FSSAI, or any other concerned authorities. The major problem for the marketing of relevant products is the lack of study, research and development. Thereby, systematic research on indigenously fermented food from a biological science perspective can assist in interesting discoveries that are beneficial for improved metabolism and comfort of humankind.

8.3 Materials and Methods

About five fermented sun-dried gundruk samples were collected randomly from Northeast India [three gundruk specimens were obtained from two different places of Arunachal Pradesh; one from Meghalaya; one from Sikkim]. For further investigation, Gundruk specimens were placed in sterilized clean bags and preserved at 4 °C. The study was accomplished randomly with four replicates.

8.3.1 Biochemical Analysis of Traditionally Fermented *Brassica* Leaf Product Gundruk

Anthrone method was used for the estimation of carbohydrates as described by Hedge and Hofreiter [26]. Lowry method was used for protein estimation [27]. Ascorbic acid was examined as described by Ranganna using titration method [24]. Folin-ciocalteau method was used for the estimation of total phenols [28, 29]. The estimation of total sugars was conducted with the Lane and Eyon method [30]. Determination of reducing sugar was achieved with Lane and Eyon method [30]. Crude fibre was estimated using AOAC [31] procedure. The technique used to calculate the total ash content was given by Ranganna [24].

8.3.2 Lactic Acid Bacteria (LAB) Isolation from Traditionally Fermented *Brassica* Leaf Product Gundruk

For the isolation of LAB, MRS (de Man, Rogosa, and Sharpe) agar plates were used [32]. MRS agar plates were supplemented with 1% $CaCO_3$ for distinguishing bacteria from another bacteria-producing acid. Gundruk (2 g) sample was weighed, crushed with 20 mL of 0.75% NaCl solution and was followed by volume makeup up to 40 mL with prepared NaCl solution. By adding 9 mL of autoclaved double distilled water sample (1 mL) extracts were diluted into 10^{-1} dilution. By taking 1 mL of well-vibrated suspension, stepwise dilution was done up to 10^{-4} dilution and then 9 mL of water was added into blank tubes. From the $10^{-1}$to 10^{-4} dilution, 100 µL of the prepared sample extract was taken and was pipette out satirically on the agar plates. Then, it was thoroughly spread with the help of sterile L-shaped spreader. For three to five days, the plates were kept incubated in an anaerobic environment at 30 °C. A clear zone around colony identified colonies of acid-producing bacteria, followed by random selection and transferring into sterile MRS broth medium.

8.3.3 Isolated Lactic Acid Bacteria for Molecular Identification

Extraction of DNA. 1 mL of 0.89% NaCl solution was transferred into the centrifuge tubes. Thereby, the bacterial smear from streaked plates were transferred into the centrifuge tubes. At 10,000 rpm, the tubes were centrifuged for half a minute. Bacterial DNA pellets were seperated after discarding the supernatant. 100 µL of prepared TE buffer solution was pipetted and transferred into the tubes containing bacterial DNA pellet. Following this, gentle tapping was allowed to dissolve bacterial DNA pellet in the TE buffer solution. Later, centrifuge tubes containing bacterial DNA pellet were stored at −20°C.

8.3.4 Sequence Search

The 16s rDNA genes were amplified using two bacterial-specific primers namely 16S F27, forward 5'...AGA GTT TGA 124 TC(AC) TGG CTC AG...3' and 16S R 1492, reverse 5'...TAC GG(CT) TAC CTT GTT 125 ACG ACT T...3'. Utilizing BLAST (Basic Local Alignment Search Tool) algorithm, the resulting 16S rRNA genomes were verified with the library of the National Center for Biotechnology Information (NCBI), which contains identified 16S rRNA genomes [33].

8.3.5 Construction of Phylogenetic Tree

By collecting the partial 16s rRNA entire genome for various species of *Lactobacillus* from NCBI database, phylogenetic relatedness of the isolated strain was investigated. The acquired dataset underwent iterative sequence alignment using the MAFFT algorithm version 7.427 by [34] to indicate the strains' evolutionary interrelationship using the unweighted pair group method with an arithmetic mean method. Using maximum likelihood method based on the Tamura-Nei model, the tree was constructed. MEGA7 [35] was utilised for this. For tree viewing and editing iTOL web interface [36] was used.

8.3.6 Probiotic Activity Analysis of Isolated Lactic Acid Bacteria

Acid tolerance test. From an agar plate, pure colonies of the culture were isolated and placed into a tube containing 5 mL MRS broth medium (de Man, Rogosa, and Sharpe). Further, for 24 h, the broth culture was grown at 37 ºC. The turbidity of inoculums was maintained with 0.5 McFarland standards (contained 10^7–10^8 CFU/mL). 10 mL MRS broth with pH adjusted to 2.0, 3.0, 4.0, 5.0, 6.0 of each isolate was transferred into and 8.0 respectively and after 24 h of incubation at 37 ºC, OD at 620 nm was measured [37].

Bile tolerance test. One day old pure bacterial colonies of the isolated culture were selected and 5 mL of MRS broth medium was transferred into test tubes and for 24 h it was incubated at 37ºC. Using McFarland standards (containing 10^7–10^8 CFU/mL), the turbidity of inoculums was maintained at 0.5. Each test tubes containing 10 mL of prepared MRS broth with different bile salt constitution (i.e., 0.05%, 0.10%, 0.15%, 0.30% and 0.50% w/v) was inoculated with 0.2 mL of each inoculum suspension. As a control sample, MRS broth without bile salts was prepared and incubated for 24 h at 37ºC. After incubation, readings were taken at 620 nm OD in a spectrophotometer [38].

Activity of Bile salt hydrolase (BSH): MRS agar plates supplemented with taurodeoxycholic acid sodium salt (0.5%) and calcium chloride (0.3 g/L) were prepared. 10 μL of overnight culture was spread plated on prepared media MRS agar dishes. Anaerobically, the incubation of the dishes was carried out at 30 °C. The existence of an opaque halo around colonies (occurred due to precipitation of bile acid) indicated deconjugation of bile salts and thereby confirmed them as positive for the bile salt hydrolase (BSH) activity [39].

Tolerance of isolate to stimulated gastric and intestinal juice: The potential lactic acid bacteria isolate was subjected to stimulated intestinal and gastric juice tolerance analysis. The cells of 10^8 CFU/mL were washed twice with saline water and subjected to artificial gastric juice with pH 3.0 (having 3 mg/mL pepsin) and for various time intervals (0, 90 and 180 min). The cells were then exposed to different intestinal juice with pH 8.0 (having 0.3% bile and pancreatin (1 mg/mL)) and were then incubated for 90 (270) and 180 (360) min at 37°C. The cells suspended in normal saline at pH 7.0 have been treated similarly and were used as the control case. The CFU was estimated by serially diluting the cells from each time period and plating them on MRS agar [40].

8.3.7 Statistical Assessment of the Data

Using SPSS statistical software Version 15.0 (SPSS Inc., Chicago, IL, USA) was carried out the analysis of variance (ANOVA) [41].

8.4 Results and Discussion

8.4.1 Biochemical Characterization of Gundruk (Fermented **Brassica** *Leaf Product)*

Among the gundruk treatments, the carbohydrate content altered from 14.15% to 28.77%. The treatment T-1 was found to have highest carbohydrate content (28.77%). On the other hand, the treatment T-5 was found to have the lowest carbohydrate content (14.15%). The grand mean observed for 5 treatments was 18.12%. Therefore, the significant mean differences between the gundruk treatments for carbohydrate can be observed by comparing the mean values of different treatments with the critical difference value (CD) at 5% from Table 8.1. In the past research, Tamang (2006) reported 38.3% of carbohydrate content in fermented sun-dried gundruk [42]. The findings of the study revealed that the significant variation existed among the different gundruk treatments. This could be due to the quality of raw materials used, alternate utility of brassica crops, age of the raw materials, harvesting period and time, environmental conditions, processing methods, drying methods, storage conditions and

Table 8.1 Data summary of the eight mean biochemical characters of five gundruk samples

Sample (treatment)	Carbohydrates (%)	Protein (%)	Ascorbic acid (mg/100 g)	Total phenols (mg/100 g)	Total sugar (%)	Reducing sugar (%)	Crude fibre (%)	Total ash content (%)
1	28.77	23.36	20.70	1.05	1.55	0.79	16.87	8.15
2	14.94	20.21	12.30	0.76	1.47	0.81	13.75	3.00
3	17.71	18.72	16.20	1.36	1.34	0.88	15.62	7.36
4	15.04	17.63	9.60	0.83	1.67	0.80	13.87	5.28
5	14.15	22.53	17.70	1.15	1.40	0.76	11.35	6.29
Grand mean	18.12	20.49	15.30	1.03	1.48	0.80	14.29	6.01
CV	2.72	1.45	8.19	9.29	3.77	4.25	5.85	5.62
Sem	0.23	0.14	0.62	0.04	0.02	0.03	0.41	0.16
CD (1%)	1.03	0.64	2.70	0.20	0.12	0.13	1.81	0.73
CD (5%)	0.73	0.45	1.93	0.14	0.08	0.07	1.29	0.52
Range (max)	28.77	23.36	20.70	1.36	1.67	0.88	16.87	8.15
Range (min)	14.15	17.63	9.60	0.76	1.34	0.76	11.35	3.00

fermenting containers. The lower carbohydrate content in the present investigation could be due to the age of the raw material used for gundruk preparation. Among the studied gundruk treatments, the protein content altered from 17.63% to 23.64%. The treatment T-1 was found to have the highest protein content (23.36%), whereas, the treatment T-4 was found to have the lowest protein content (17.63%). The grand mean observed for 5 treatments was 20.49%. Therefore, the significant mean differences between the gundruk treatment for protein can be observed by comparing the mean values of different treatments with the critical difference value (CD) at 5% from Table 8.1. Mishra and Kharel (2013) reported 29.76% of protein content in Sisnu gundruk [43]. The findings of the study revealed that significant variation existed among the different gundruk treatments. This could be due to the use of alternate brassica crops for gundruk preparation, environmental conditions, quality of raw materials, age of the raw materials, harvesting period and time, processing methods and drying methods. Among the studied gundruk treatments, the ascorbic acid content altered from 9.60 to 20.70 mg/100 g. The treatment T-1 was analyzed to contain the largest amount of ascorbic acid (20.70 mg/100 g), whereas, the treatment T-4 was analyzed to contain the lowest amount (9.60 mg/100 g). The grand mean observed for 5 treatments was 15.30 mg/100 g. Therefore, the significant mean differences between the gundruk treatment for ascorbic acid content can be observed by comparing the mean values of different treatments with the critical difference value (CD) at 5% from Table 8.1. The study revealed that in this findings, significant variation existed among the different gundruk treatments for ascorbic acid content. This might have occurred due to the genetic makeup of raw materials, drying method used, age of the raw materials and processing methods as well as environmental conditions. From the past study, Cheigh et al. (1994) reported 21 mg/100 g of ascorbic acid content in kimchi [44]. Among the studied gundruk treatments, the total phenol content varied as 0.76–1.36 mg/100 g. The treatment T-3 was found to have highest total phenol content (1.36 mg/100 g), whereas, the treatment T-2 was possessed lowest amount of total phenol content (0.76 mg/100 g). The grand mean observed for 5 treatments was 1.03. Therefore, the significant mean differences between the gundruk treatment for total phenol content can be observed by comparing the mean values of different treatments with the critical difference value (CD) at 5% from Table 8.1. Similarly, Deb and Jamir (2020) reported the highest total phenolic content in anishi with 1.44 mg/100 g [45]. The variation in total phenol content among gundruk treatments might have occurred due to the genetic makeup of raw materials and environmental conditions. Among the studied gundruk treatments, the total sugar content altered from 1.34–1.67% among the gundruk treatments studied. The treatment T-4 was found to have highest total sugar content (1.67%), whereas, the treatment T-3 was obtained with lowest total sugar content (1.34%). The examined grand mean for five treatments was 1.48. Therefore, the significant mean differences between the gundruk treatment for total sugar content can be observed by comparing the mean values of different treatments with the critical difference value (CD) at 5% from Table 8.1. Lairenlakpam (2004) reported that the 30 days old fermented gundruk possessed 2.2% of total sugar content and 40 days old fermented gundruk possessed 2.1% [2]. The results affirmed that the significant variation in the total sugars among different gundruk treatments could be occurred due to the use of different fermenting containers, genetic makeup of raw materials, age of raw materials, environmental conditions and microorganisms responsible for fermentation.

Among the studied gundruk treatments, the reducing sugar content ranged from 0.76–0.88%. The treatment T-3 was analyzed to contain highest reducing sugar content (0.88%), whereas, the treatment T-5 was inferred to contain lowest reducing sugar content (0.76%). The grand mean observed for five treatments was 0.80. Therefore, the significant mean differences between the gundruk treatment for reducing sugar can be observed by comparing the mean values of different treatments with the critical difference value (CD) at 5% from Table 8.1. In the previous study of Lairenlakpam (2014), 1.0% of reducing sugar content in 30 days old fermented sinki was reported [2]. On the other hand, in the present study the variation in reducing sugar content among the gundruk treatment might have occurred due to the usage of variant fermenting containers, genetic makeup of raw materials, age of raw materials, environmental conditions and microorganisms.

The crude fibre content altered from 11.35 to 16.87% among the studied gundruk treatments. The treatment T-1 was found to have highest crude fibre content (16.87%). On the other hand, the treatment T-5 was found to have the lowest crude fibre content (11.35%). The grand mean observed for 5 treatments was 14.29. Therefore, the significant mean differences between the gundruk treatment for crude fibre content can be observed by comparing the mean values of different treatments with the critical difference value (CD) at 5% from Table 8.1. The study's results showed that the significant variation existed among the different gundruk treatments. This might be due to the quality of raw materials used, different brassica crops used, age of the raw materials, harvesting period and time, environmental conditions, processing methods and drying methods. Ibrahim et al. (2011) revealed in locally fermented *Hibiscus sabdariffa* seeds that 17.00% of crude fibre content existed [46].

The total ash content among the gundruk treatments ranged from 3 to 8.15%. The treatment T-1 was found to have highest total ash content (8.15%), whereas, the treatment T-2 was found to have the lowest total ash content (3.00%). The grand mean observed for 5 treatments was 6.01. The mean difference between the gundruk treatment T-1 (8.15%) and T-3 (7.36%) was found 0.79, which was more than the CD at 5% (0.52) level of significance, indicating that the gundruk treatment T-1 and T-3 has not been par for the total ash content. Therefore, the significant mean differences between the gundruk treatment for total ash content can be observed by comparing the mean values of different treatments with the critical difference value (CD) at 5% from Table 8.1. The conclusions drawn from the investigation demonstrated that the significant variation existed among the different gundruk treatments. This could be due to the quality of raw materials used, different brassica crops used, age of the raw materials, harvesting period and time, environmental conditions, processing methods and drying methods. Ibrahim et al. (2011) revealed that in laboratory fermented *Hibiscus sabdariffa* seeds, total ash content was 9.65% [46].

8.4.2 Principal Component Analysis

In the present study, PCA was conducted based on biochemical observation of five gundruk treatments to determine variation in the different quantitative characters among and between the treatments. In the data set, a portion of genetic distance was observed. By only taking into account variables having an eigen value >1.0, the number of critical components in the dataset was determined. Thereby, the principal component with an eigen value less than one was considered to be non-significant [47].

A sum of three principal components PC1, PC2 and PC3 with eigen values 3.686%, 2.076% and 1.778% respectively was reported form principal component analysis (PCA) as shown in Fig. 8.1. This altogether accounted for the significant variation and affirms genetic variability among the gundruk treatments (Table 8.2). A greater positive coefficient was considered to be significant for each main component. Therefore, indicated by their respective loadings, the levels of relationship between every variable and component may be calculated and all variables included in the study and their degree of similarity to one another is given in Table 8.2.

The three Principal Components accounted for 94.255% of total variance with the first principal component (PC-1) contributing the highest (46.070%), followed by PC-2 contributing 25.954% and PC-3 contributing 22.231% respectively. The relative discriminating power of the principal axes with 3.686% for PC-1, 2.076% for PC-2 and as low as 1.778% for PC-3 as per the eigen value.

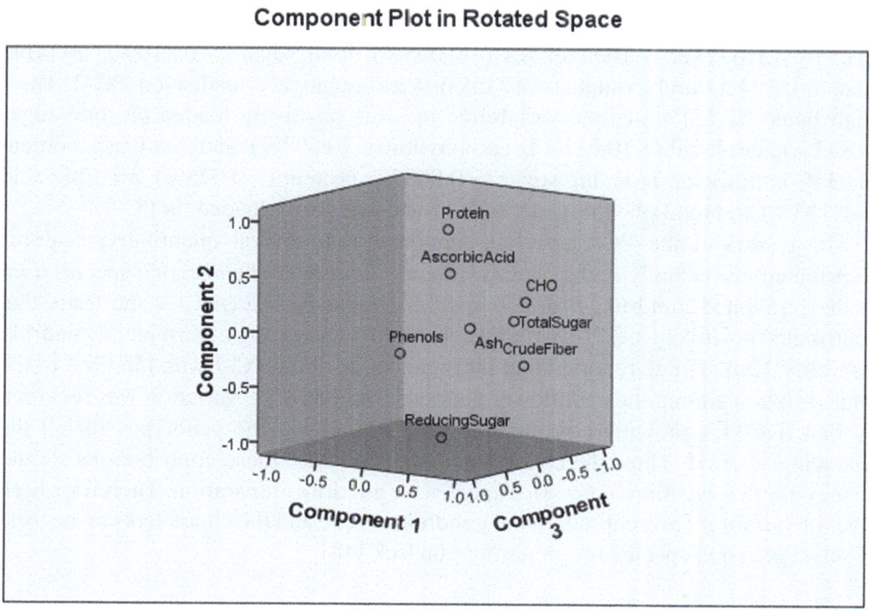

Fig. 8.1 Variables distribution of for PC1, PC2 and PC3 cases during PCA analysis

Table 8.2 Data summary of the principal components analysis of eight alternate biochemical characters in five gundruk samples

Characters	PC-1	PC-2	PC-3	Final communality
Carbohydrates	0.798	−0.008	0.593	0.989
Protein	0.689	−0.715	−0.025	0.986
Ascorbic acid	0.962	−0.233	−0.133	0.997
Total phenols	0.678	0.427	−0.569	0.965
Total sugar	−0.354	−0.150	0.843	0.859
Reducing sugar	−0.117	0.990	0.003	0.993
Crude fibre	0.534	0.512	0.611	0.920
Total ash content	0.874	0.255	0.034	0.830
Eigen value	3.686	2.076	1.778	
Variability (%)	46.070	25.954	22.231	
Cumulative (%)	46.070	72.024	94.255	

A total variability of 94.255% was recorded where PC-1 contributed 46.070%. The PC-1 was positively loaded with respect to ascorbic acid (0.962%), total ash content (0.874%), carbohydrates (0.798%), protein (0.689%), total phenols (0.678%), and crude fibre (0.534%) but was negatively loaded for reducing sugar (−0.117%). PC-2 contributed 25.954% of total variability and was positively loaded on reducing sugar (0.990%), crude fibre (0.512%), total phenols (0.427%) and total ash content (0.255%). However, carbohydrates (−0.008%), total sugar (−0.510%), ascorbic acid (−0.233%) and protein (−0.715%) were negatively loaded on PC-2. PC-3 contributed 22.231% of total variability and was positively loaded on total sugar (0.843%), crude fibre (0.611%), carbohydrates (0.593%) and total ash content (0.034%). However, reducing sugar (−0.003%), protein (−0.025%), ascorbic acid (−0.133%) and total phenols (−0.569%) were negatively loaded on PC-3.

The results of the PCA analysis supported that several quantitative features contributed individually to the overall variance along with the *brassica* species used in the fermented gundruk. *Brassica* species should be selected for the traits that contributed positively to 3 PCs without considering the quality of fermented gundruk. Similarly, Kumar et al. reported that the eigen value of the PC-1 was 45.696% of the total variation among the cauliflower genotypes and 81.72% variation was recorded by first five PCs showing considerable diversity among the genotypes for all the characters studied. Thus, the characters associated with these contributions should be used for the selection of *brassica* species for gundruk preparation. Therefore, high level of variability present within the gundruk sample and the characters can be used to select *brassica* species for preparing gundruk [48].

8.4.3 Isolation of Lactic Acid Bacteria (Lab) from Traditionally Fermented Brassica *Leaf Product Gundruk*

The G-7 isolate retained purple colour after gram staining and thereby confirmed it as a gram-positive bacterium. The G-7 isolate exhibited rod-shaped morphology under 10X, 40X and 100X magnification (Fig. 8.2). The colony shape was large in size, the elevation of colony was raised, was shiny creamish in colour, and cleared out zones with no colour pigment (Table 8.3). This type of colony is generally formed by the lactic acid bacteria. Angmo (2014) discovered that many microorganisms produce enzyme called catalase, that breaks down (H_2O_2) into water and release oxygen as gas bubbles. Thereby, the released oxygen affirms upon the presence of catalase enzyme in the microorganism being tested [49]. The G-7 isolate showed no gas bubbles after being subjected to catalase test. Thereby, it fulfilled the criteria of being catalase negative. Based on morphology and biochemical test, Karki et al. reported that 47 isolates were selected out of 464 strains isolated strains from 8 gundruk samples [5]. Similarly, Nguyen et al. (2013) identified 881 g positive, catalase negative lactic acid bacteria strains from 21 samples of dua muoi (fermented mustard and beet) and ca muoi (fermented eggplant) [50]. Later, G-7 isolate was subjected to alternate temperature conditions and to affirm upon their growth at those temperatures. The maximum growth was observed at 30 °C followed by moderate growth at 15 °C and 37 °C. However, no growth was observed at 4 °C (Table 8.4). Similarly, Goswami

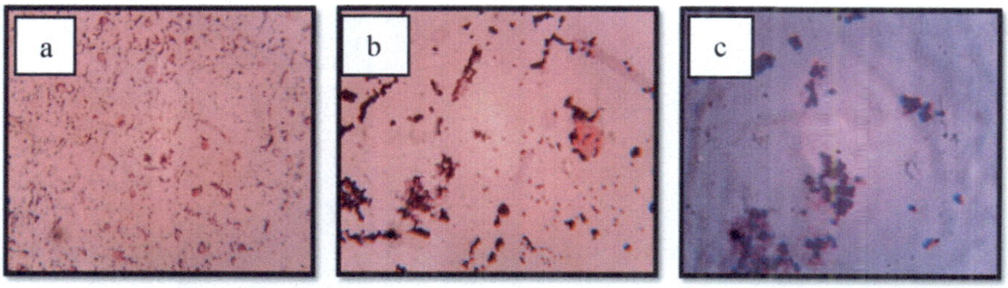

Fig. 8.2 Microscopic images of gram positive bacteria at various magnifications (A.10X, B. 40X, C. 100X)

Table 8.3 A list of ten potential lactic acid bacteria isolated from gundruk

Isolate	Colony
G-1	Small, off white colour, flat elevation, no colour pigment
G-2	Small, white colour, flat elevation, no colour pigment
G-3	Large, white shiny colour, raised elevation, no colour pigment
G-4	Medium, creamy white, convex, no colour pigment
G-5	Medium, off white colour, flat elevation, no colour pigment
G-6	Small, white, convex elevation, no colour pigment
G-7	Large, shiny creamish, raised elevation, no colour pigment
G-8	Large, white shiny, raised elevation, no colour pigment
G-9	Medium, off white, flat elevation, no colour pigment
G-10	Small, off white shiny, raised elevation, no colour pigment

Table 8.4 Growth data of G-7 isolate at various temperatures

Temperature	4 °C	15 °C	30 °C	37 °C
Growth	–	+	++	+

et al. (2017) observed that the growth of isolated lactic acid bacteria was maximum at 30 °C [51].

8.4.4 Molecular Identification of Isolated Lactic Acid Bacteria

The isolated LAB was identified by using 16 s rRNA gene sequencing. Using BDT v3.1 cycle sequencing kit on ABI 3730xl genetic analyzer, the forward and reverse RNA sequencing reaction of PCR amplicon was conducted with forward primer F27 and reverse primer R1492. The consensus sequence of 16S rRNA gene was generated

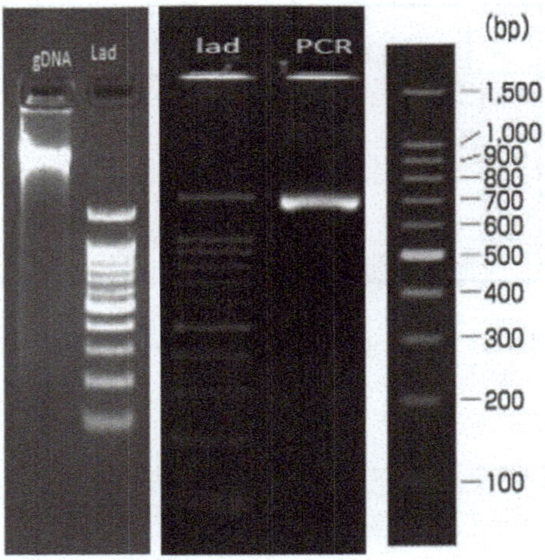

Fig. 8.3 Image depicting molecular identification of isolated lactic acid bacteria through the PCR analysis

from forward and reverse sequence data using aligner software. In the present study, a single band of high-molecular weight DNA and a single discrete PCR amplicon band of 1500 bp was observed upon resolution in 1.5% agarose gel (Fig. 8.3). The PCR amplification and 16Sr RNA gene sequencing of G-7 isolate showed 100% similarity with *Lactococcus lactis* subsp. *tructae* strain. Pheng et al. (2020) reported by the 16S rRNA gene sequence showed that the strain S-13T isolated from kimchi had the highest similarity to *Lactococcus taiwanensis* 0905C15T (97.9%), *Lactococcus lactis* subsp. *tructae* L105T (97.6%), *Lactococcus lactis* subsp. *cremoris* NCDO 607 T (97.5%), *Lactococcus lactis* subsp. *hordniae* NBRC 100931T (97.2%), and *Lactococcus lactis* subsp. *lactis* JCM 5805T (97.2%) [52]. Similarly, Chen et al. (2014) reported that *Lactococcus formosensis* was isolated from fermented broccoli stems [53]. A single band of high-molecular weight DNA was observed on 1.5% agarose gel. A single discrete PCR amplicon band of 1500 bp was observed when resolved on agarose gel. The isolated strain showed high similarity with *Lactococcus lactis subsp. tructae* strain based on nucleotide homology and phylogenetic analysis.

8.4.5 Sequences Producing Significant Alignments

The sequence homology was analyzed by using BLAST analysis to observe the similarities with closely related known sequences from NCBI database. The maximum score and total score ranged from 1007 to 1018. *Lactococcus lactis* subsp. *tructae* strain L105 *Lactococcus lactis* subsp. *hordniae* strain NCDO 2181, *Lactococcus lactis* strain NCDO 604 and *Lactococcus lactis* subsp. *cremoris* strain NCDO 607 recorded maximum score and total score (1018). On the other hand, *Lactobacillus nenjiangensis* strain 11,102 and *Lactobacillus equi* strain YIT 0455 showed least among maximum and total score (1007).

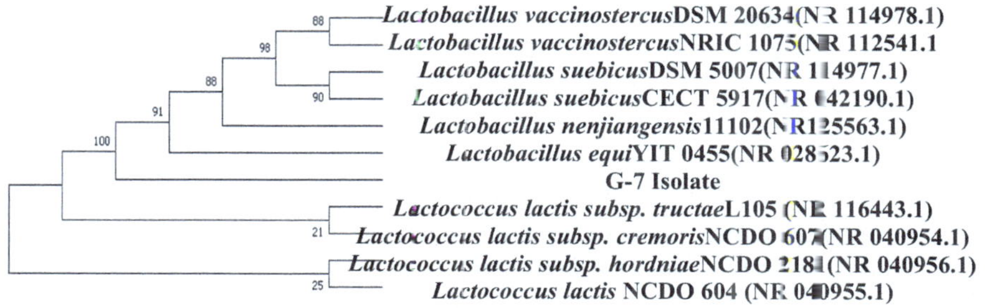

Fig. 8.4 Molecular phylogenetic analysis through the maximum likelihood method

The query cover ranged from 95 to 88%. *Lactobacillus suebicus* strain DSM 5007 *and Lactobacillus suebicus* strain CECT 5917 showed highest query cover (95%) and *Lactobacillus equi* strain YIT 0455 showed lowest (88%). The E. value was recorded '0' for all ten selected sequences. The percentage identification ranged from 81.32 to 82.15%. *Lactobacillus suebicus* strain DSM 5007 *and Lactobacillus suebicus* strain CECT 5917 showed lowest percentage identification of 81.32% but highest was displayed by *Lactobacillus equi* strain YIT 0455 (82.15%). The accession length ranged from 1413 to 1573. *Lactobacillus suebicus* strain CECT 5917 showed highest accession length (1573) and *Lactobacillus equi* strain YIT 0455 showed the lowest (1413). However, *Lactococcus lactis* subsp. *tructae* strain L105 with accession number NR_116443.1 recorded maximum score (1018), total score (1018), query cover (94%), E. value (0), percentage identification (81.39%) and accession length (1541%). Therefore, based on nucleotide homology, the G-7 isolate showed high similarity with *Lactococcus lactis* subsp. *tructae* strain. The phylogenetic tree was constructed with the sequences of representative bacteria using MEGA7. This revealed that the G-7 isolate was clustered with both Lactobacillus and Lactococcus group (Fig. 8.4) and multiple alignment software program Clustal W was used for selecting and aligning first ten sequences. This was based on the maximum identity score. According to Chen et al. (2014) the 16S rRNA gene sequencing results confirmed that the strain 516^{T} had 98.9% sequence similarity to that of the type strain *Lactococcus garvieae* NBRC 100934^{T} in fermented broccoli stem [54].

8.4.6 Probiotic Activity Analysis of Isolated Lactic Acid Bacteria

Probiotic is a fermented product or a preparation composed of living beneficial microbes or bacteria's in required numbers that adjusts the gut microflora of the host and provides health benefits to the host. Probiotic microorganisms are known to have beneficial effects on human host gut micro flora and thereby exhibits certain benefits

Table 8.5 Acid tolerance data of G-7 isolate at 620 nm (OD)

pH	2.0	3.0	4.0	5.0	6.0	8.0
G-7 isolate	0	0.148	0.557	0.235	0.162	0.354

such as antimicrobial activity, potential to reduce cholesterol level, diarrhea treatment, overcome lactose intolerance, antimutagenic activity and improves immune system functioning [14]. In the present study, the isolated G-7 has been subjected through four different probiotic tests and they have been discussed below.

Acid tolerance test. Acid tolerance is one of the most essential characteristics for selecting a probiotic bacterium with ability to survive various acidic levels in different regions of human gastro-intestinal tract in which the pH can be low as 1.5. The isolated G-7 isolate should be able to survive under these acidic conditions so that it can be used as a probiotic strain. In the present study, the isolated G-7 isolate was subjected to acid tolerance test in different pH mediums ranging from 2.0, 3.0, 4.0, 5.0, 6.0, and 8.0 represented in Table 8.5. It was found that the G-7 isolate showed highest growth and resistance at pH 4.0 with OD value (625 nm) of 0.557 followed by pH 8.0 (0.354), pH 5.0 (0.235), pH 6.0 (0.162), pH 3.0 (0.148). This was lowest at pH 2.0 (0). Patel and Parikh, (2016) reported that the SS-2 strain (*Lactobacilli*) isolated from local dairy products exhibited highest resistance and tolerance at pH 3.0 [55].

Bile tolerance test. Bile tolerance is a typical criterion for the selection of a probiotic bacterium. Human gastro-intestinal tract possess different bile salt concentrations in different regions of the tract. The human bile salt concentration ranges from 0.1 to 0.5% (Patel and Parikh, 2016). The ability of the isolate G-7 to survive or tolerate under different bile salt concentrations will determine its probiotic potential. In the present study, bile salt concentration ranged from 0.05%, 0.10%, 0.15%, 0.30% and 0.50% and the bile tolerance findings have been presented in Table 8.6. G-7 isolate showed highest growth and survivability based on OD (620 nm) at 0.05% bile salt concentration (0.336), then at 0.50% (0.228), 0.10% (0.214), 0.30% (0.175) and lowest at 0.15% (0.116) was observed at 620 nm. Similarly, Patel and Parikh (2016) observed that the highest growth and survival of SS-2 strain (*Lactobacilli*) isolated from local dairy products was at 0.05% bile salt concentration [55].

Bile salt hydrolase (BSH) activity. The isolate G-7 from gundruk treatment T-1 was placed on MRS agar plates supplemented with taurodeoxycholic acid sodium salt and calcium chloride. Thereby, their ability to hydrolyze the sodium salts of the taurodeoxycholic acid was targeted and assessed. The G-7 isolate displayed no opaque halo around colonies after incubation at 30°C. This conveyed the result to be negative to the bile salt hydrolase (BSH) activity even though it is a gram-positive

Table 8.6 Bile tolerance data of G-7 isolate at 620 nm (OD)

Bile salt concentration	0.05%	0.10%	0.15%	0.30%	0.50%	Control (without bile salt)
G-7 Isolate	0.336	0.214	0.116	0.175	0.228	1.024

Table 8.7 Tolerance data of G-7 isolate to stimulated gastric and intestinal juice

Isolate	Conditions	Cell survival [Log cfu/mL (%)]				
		Gastric juice			Intestinal juice	
		0 min	90 min	180 min	270 min	360 min
G-7	pH 3.0	7.6	7.4	7.5	–	–
	pH 7.0	7.7	7.9	7.7	7.1	7.1
	pH 8.0	–	–	–	6.4	6.3

Where, – = no growth

bacterium. Thus, the case is an exception among gram positive bacteria due to its negative result against the BSH activity [55].

Stimulated gastric and intestinal juice for tolerance of isolate. Improving gastric and intestinal juice for the tolerance of isolate is one of the important characteristics to select the probiotic strain. At different pH conditions ranging from pH 3.0, pH 7.0 and pH 8.0, G-7 isolate was subjected to different time periods (gastric juice) from 0 min, 90 min, 180 min, 270 min and 360 min (intestinal juice). The findings have been presented in Table 8.7. The G-7 isolate considerably survived the gastric juice transition in pH 7.0 from 0 to 90 min and up to 180 min. However, a marginal reduction in the survival of cell in intestinal juice transition was observed from 270 to 360 min. Therefore, the isolate was enabled to survive at gastro-intestinal nature and colonised efficiently in the gut. Similarly, Sadishkumar, (2017) reported that the isolates survived considerably from 0 to 90 min and then up to 180 min during the gastric transit. However, a marginal reduction in survival was observed during continuous exposure of cells in the intestinal juice at 270 min and 360 min [56]. Also, Savedboworn et al. (2014) reported that the strains KMUTNB 5–9, KMUTNB 5-36 and KMUTNB 6-21 representing *Pediococcus pentosaceus* (obtained from fermented vegetables) were found to be tolerant to stimulated gastric juice for 180 min at pH 2.0. However, the authors reported reduction in viable cells at pH 8.0 in intestinal juice [57].

8.5 Conclusions

Firstly, 8 biochemical characters namely carbohydrate, protein, ascorbic acid, total phenols, total sugar, reducing sugar, crude fibre and total ash content have been evaluated for 5 gundruk samples being collected from three North eastern states. The experiments were conducted in completely randomized design (CRD) and with 4 replications. Treatment T-3 recorded highest in total phenol content (1.36 mg/100 g); treatment T-4 recorded highest in total sugar content (1.67%); treatment T-3 recorded highest in reducing sugar content (0.88%). Based on the exhibition of optimal combinations of biochemical characters namely carbohydrates content (28.77%), protein (23.36%), ascorbic acid (20.70 mg/100 g), crude fibre (16.87%) and total ash content

(8.15%), the treatment T-1 was found best to be the best. Hence, further bacterial analysis was carried out with treatment T-1. TreatmentT-1 was prepared from radish, cabbage and broccoli leaves collected from Itanagar, Arunachal Pradesh.

PC-1 contributed 46.07% to the total variability of 94.25%. The biochemical characters which contributed positively to PC-1 were ascorbic acid (0.96), total ash content (0.87), carbohydrates (0.79), protein (0.68), total phenols (0.67) and crude fibre (0.53). PC-1 contributed maximum eigen value of 3.68 and accounted for 94.25% of the total variability. Incidentally, PC-2 contributed 25.95% to the total variability. The biochemical characters which contributed positively to PC-2 were reducing sugar (0.99) and crude fibre (0.51). PC-2 contributed eigen value of 2.07 and accounted for 94.25% of the total variance. PC-3 contributed 22.23% to the total variability. The biochemical characters which contributed positively to PC-3 were total sugar (0.84), crude fibre (0.61) and carbohydrates (0.59). PC-3 contributed an eigen value of 1.77.

Based on the visual appearance, colony morphology and growth on MRS agar plates supplemented with $CaCO_3$, the maximum bacterial growth was observed in 10^{-1} dilution factor followed by 10^{-2} and 10^{-3} but no growth was observed on 10^{-4} dilution factor after 3 to 5 days of incubation at 30 °C. A sum of ten potential lactic acid bacteria isolates i.e., G-1, G-2. G-3, G-4, G-5, G-6, G-7, G-8, G-9 and G-10 were obtained from gundruk treatment T-1. Later, upon being analyzed for biochemical tests such as gram staining, catalase test and growth at different temperatures, the isolate G-7 retained violet/purple colour after gram staining and thereby confirmed to be a gram-positive bacterium. The G-7 isolate exhibited rod shape structure upon observation under microscope at alternate magnifications namely 10X, 40X and 100X. There was no occurrence of bubbles during catalase test. Thereby, it was confirmed to be negative for catalase test. The maximum growth of G-7 isolate was observed at 30 °C, moderate at 15 °C and 37 °C, but no growth was at 4 °C.

An absorbance ratio of 1.75 was observed and DNA concentration of 0.81 μg/μL was recorded in UV spectrophotometer.

In the present study, G-7 isolate was put through molecular means of characterization (PCR) using 16s rRNA gene sequence analysis. A single band of high-molecular weight DNA was observed on 1.5% agarose gel. Fragment of 16S rRNA gene was amplified by 27F and 1492R primers. A single discrete PCR amplicon band of 1500 bp was observed upon resolution on Agarose gel. The G-7 isolated was compared with the already existing bacterial data recorded in NCBI. *Lactococcus lactis* subsp. *tructae* strain L105 with accession number NR_116443.1 recorded maximum score (1018), total score (1018), query cover (94%), E. value (0), percentage identification (81.39%) and accession length (1541%). It showed 100% similarities with the G-7 isolate based on nucleotide homology and phylogenetic analysis.

Selected isolate was tested for different probiotic attributes such as acid tolerance, bile tolerance, bile salt hydrolase (BSH) activity and tolerance to stimulated gastric and intestinal juices. During acid tolerance test, it was found that the G-7 isolate showed highest growth and resistance at pH 4.0 (OD of 0.557 at 620 nm) and lowest at pH 2.0 (0). The study on the resistance of G-7 isolate to bile salts at 0.05%

concentration recorded highest growth and survivability. However, during bile salt hydrolase (BSH) activity studies, it exhibited negative response as no opaque halo clear out zones were observed around the colonies. Again, when the G-7 isolate was put through stimulated gastric and intestinal juice tolerance test considerable survivability was observed in the gastric juice in pH 7.0 from 0 to 90 min and then up to 180 min. However, a marginal reduction in the survival of cell was observed in intestinal juice transition from 270 to 360 min. Therefore, the isolate was able to survive gastro-intestinal nature and colonised efficiently in the gut.

This study revealed that the gundruk treatmentT-1 prepared from leaves of different *brassica* species (radish, cabbage and broccoli) resulted in higher composition of certain biochemical characters like carbohydrates, protein, ascorbic acid, crude fibre and total ash content. The G-7 isolate isolated from T-1 was found to be gram positive and catalase negative, and had maximum growth at 30 °C. The BLAST with the database of NCBI genebank database was carried out using 16S rRNA gene sequence and the isolated strain showed high similarity with *Lactococcus lactis* subsp. *Tructae* strain on the basis of phylogenetic analysis and nucleotide homology. However, the presence of *Lactococcus lactis* subsp. *Tructae* has been reported for the first time in gundruk. To determine the probiotic attribute, the G-7 isolate was put through certain test like acid tolerance test, bile tolerance test, bile hydrolase (BSH) activity and tolerance to stimulated gastric and intestinal juice. Highest growth and resistance at pH 4.0 (0.557) was observed during acid test. At 0.05% bile salts concentration, highest growth and survivability was recorded. However, negative response exhibited to bile salt hydrolase (BSH) activity and the system survived both artificial gastric and intestinal juice environment.

By carrying out proper in vitro investigation related to the probiotic attributes of the isolated strain, it can be used to produce probiotic drinks and other products. Further studies can be carried out to infer upon the role of the isolated lactic acid bacteria for increasing the biochemical characters of the raw materials used for the preparation of gundruk or other fermented products. It can help in standardizing the traditional methods of fermented vegetable product preparation by reducing the time period. It can be further studied for bacteriocin production, which can be used as a biopreservative. It can be used as starter culture to produce different fermented vegetable products. Thus, utilization in green biorefinery, agriculture sector and entrepreneurial development can be targeted.

From bio-diversity conservation perspective, the regional alterations in phenotypic and genotypic characteristics can be targeted in all states of North-East India and by adopting traditional gundruk preparation methods. Thereby, possible scope for commonalities and alterations can be sought.

Acknowledgements Authors express their sincere thanks to Department of Horticulture, NEHU, Tura Campus, Meghalaya for the laboratory facilities.

References

1. Tamang, J.P., Tamang, B., Schillinger, U., Franz, C.M.A.P., Gores, M., Holzapfel, W.H.: Identification of predominant lactic acid bacteria isolated from traditionally fermented vegetable products of the Eastern Himalayas. Int. J. Food Microbiol. **105**(3), 347–356 (2005)
2. Lairenlakpam, S.: Microbial and Biochemical Assesment of Fermented Vegetable Food (Gundruk and Sinki) with Special Reference to Nepali Community of Manipur. Department of Life Sciences, Manipur University, Imphal, Ph.D.thesis (2004)
3. Tamang, J.P.: Himalayan Fermented Foods: Microbiology, Nutrition, and Ethnic Values, pp. 1–255. CRC Press/Taylor & Francis Group, New York (2010)
4. Tamang, B., Tamang, J.P.: Role of lactic acid bacteria and their functional properties in Goyang, a fermented leafy vegetable product of Sherpas. J. Hill Res. **20**(2), 53–61 (2007)
5. Karki, T., Okada, S., Baba, T., Itoh, H. and Kozaki, M.: Studies on microorganisms and their role in Gundruk fermentation. Studies on the microflora of Nepalese pickles Gundruk. Nippon Shokuhin Kogyo Gakkai Shi=J. Japan. Soc. Food Sci. Technol. **30**(6), 357–367 (1983)
6. Karki, T., Ojha, P., Panta, O.P.: Ethnic fermented foods of Nepal. In: Tamang, J.P (Ed.) Ethnic Fermented Foods and Alcoholic Beverages of Asia, pp. 91–117. Springer, New Delhi, India (2016)
7. Das, A.J., Deka, S.C.: Mini review on fermented foods and beverages of the Northeast India. Int. Food Res. J. **19**(2), 377–392 (2012)
8. Jaiswal, V.: Culture and ethnobotany of Jaintia tribal community of Meghalaya, Northeast India-A mini review. Indian J. Tradit. Knowl. **9**(1), 38–44 (2010)
9. Sekar, S., Mariappan, S.: Usage of traditional fermented products by Indian rural folks and IPR. Indian J. Tradit. Knowl. **6**(1), 111–120 (2007)
10. Trichopoulou, A., Stravroula, S., Vasilopoulou, E.: Traditional foods: A science and society prespective. Trends Food Sci. Technol. **18**(8), 420–427 (2007)
11. Battock, M., Azam-Ali, S.: Fermented fruits and vegetables: a global perspective, food and agriculture organisation of the United Nations, Rome, Italy. FAO Agric. Serv. Bull. **134** (1998)
12. Rolle, R., Satin, M.: Basic requirements for the transfer of fermentation technologiesto developing countries. Int. J. Food Microbiol. **75**(3), 181–187 (2002)
13. Karovicova, J., Kohajdova, Z.: Lactic acid fermented vegetable juices palatable and wholesome foods. Chem. Pap. **59**(2), 143–148 (2005)
14. Kumar, M.: Characterization of *lactobacilli* isolated from indigenous fermented food and human guts with special reference to their probiotic attributes. Ph.D. thesis, Department of biotechnology and Environmental sciences, Thapar University, Patiala, Punjab (India) (2012)
15. Tamang, J.P., Tamang, B.: Traditional knowledge of bio preservation of perishable vegetables and bamboo shoots in Northeast India as food resources. Indian J. Tradit. Knowl. **8**(1), 89–95 (2009)
16. Steinkraus, K.H.: Fermentations in world food processing. Compr. Rev. Food Sci. Food Saf. (1)23–32 (2002)
17. Khadka, B.D., Lama, P.J.: Traditional fermented food of Nepal and their nutritional and nuetraceutical potential, Nutritional and Health Aspects of Food in South Asian Countries, pp. 165–194. Academic Press/Elsevier (2020)
18. Tamang, J.P., Sarkar, P.K., Hesseltine, C.W.: Traditional fermented foods and beverages of Darjeeling and Sikkim—a review. J. Sci. Food Agric. **44**(4), 375–385 (1988)
19. Tamang, J.P., Tamang, N., Thapa, S., Dewan, S., Tamang, B., Yonzan, H., Rai, A.K., Chettri, R., Chakarabarty, J., Kharel, N.: Microorganisms and Nutritional value of Ethnic fermented foods and alcoholic beverages of North East India. Indian J. Tradit. Knowl. **11**(1), 7–25 (2012)
20. Tamang, B.: Role of lactic acid bacteria in fermentation and bio preservation of traditionalvegetable products. Ph.D. thesis, Food Microbiology Laboratory, Sikkim Government College, Gangtok (North Bengal University) (2006)
21. Moser, P.B., Reynolds, R.D., Acharya, S., Howard, M.P., Andon, M.B.: Calcium and magnesium dietary intakes and plasma and milk concentrations of Nepalese lactating women. Am. J. Clin. Nutr. **47**(4), 735–739 (1988)

22. Shrestha, R., Bhattarai, R.R. andKatawal, B.S.: Effect of different fermentation containers on the quality of gundruk. Int. J. Curr. Res. **4**(11), 315–320 (2012)
23. Karki, T., Itoh, H., Nikkuni, S.: Improvement of gundruk processing by selected lactic strains. Nippon Shokuhin Kogyo Gakkaishi **33**(10), 734–739 (1986)
24. Ranganna, S.: Hand Book of Analysis and Quality Control for Fruit and Vegetable Product, p. 2240. Tata McGraw-Hill Publishing Co. Ltd, New Delhi (1997)
25. Singh, A., Singh, R.K., Sureja, A.K.: Cultural significance and diversities of ethnic foods of Northeast India. Indian J. Tradit. Knowl. **6**(1), 79–94 (2007)
26. Hedge, J.E., Hofreiter, B.T.: Determination of reducing sugars and carbohydrates. In: Methods in Carbohydrate Chemistry, vol. 1, p. 380. Academic Press, New York (1962)
27. Lowry, O.H., Rosebrough, N.J., Randall, R.J.: Protein measurement with the Folin phenol reagent. J. Biol. Chem. **193**, 265–275 (1951)
28. Swain, T., Hillis, W.E.: The phenolic constituents of *Prunus domestica,* The quantitative analysis of phenolic constituents. J. Sci. Food Agric. **10**(1), 63–68 (1959)
29. Walter, W.M., Purcell, A.E.: Evaluation of several methods for analysis of sweet potato phenolics. J. Agric. Food Chem. **27**(5), 942–946 (1979)
30. Lane, J.H., Eynon, L.: Determination of reducing sugar by Fehling's solution with methylene blue as indicator. J. Soc. Chem. Ind. (42), 32–37 (1923)
31. AOAC: Official Methods of Analysis. Association of official analytical chemists, Inc.17th ed. Arlington, Virginia, USA (1990)
32. Abbas, M.M., Mahasneh, A.M.: Isolation of Lactobacillus strains with probiotic potential from camels milk. Afr. J. Microbiol. Res. **8**(15), 1645–1655 (2014)
33. Johnson, M., Zaretskaya, I., Raytselis, Y., Merezhuk, Y., McGinnis, S., Madden, T.L.: NCBI BLAST: a better web interface. Nucl. Acids Res. **36**(WebServer), W5–W9 (2008)
34. Katoh, K., Kuma, K., Toh, H., Miyata, T.: MAFFT version 5: improvement inaccuracy of multiple sequence alignment. Nucl. Acids Res. **33**(2), 511–518 (2005)
35. Kumar, S., Stecher, G., Tamura, K.: MEGA7.: molecular evolutionary genetics analysis version 7.0 for bigger datasets. Mol. Biol. Evol. **33**(7), 1870–1874 (2016)
36. Letunic, I., Bork, P.: Interactive Tree Of Life (iTOL) v4: recent updates and new developments. Nucl. Acids Res. **47**, W256–W259 (2019)
37. Sieladie, D.V., Zambou, N.F., Kaktcham, P.M., Cresci, A., Fonteh, F.: Probiotic properties of lactobacilli strains isolated from raw cow milk in the western highlands of Cameroon. Innov. Rom. Food Biotechnol. **9**, 12–28 (2011)
38. Vinderola, C.G., Reinheimer, J.A .: Lactic acid starter and probiotic bacteria:a comparative "in vitro" study of probiotic characteristics and biological barrier resistance. Food Res. Int. **36**(9–10), 895–904 (2003)
39. Dashkevicz, M.P., Feighner, S.D.: Development of a differential medium for bile salt hydrolase-active *Lactobacillus* spp. Appl. Environ. Microbiol. **55**, 11–16 (1989)
40. Ripamonti, B., Agazzi, A., Bersani, C., De Dea, P., Pecorini, C., Pirani, S., Rebucci, R., Savoini, G., Stella, S., Stenico, A., Tirloni, E., Domeneghini, C.: Screening of species-specific lactic acid bacteria for vealcalvesmulti strainprobiotic adjuncts. Anaerobe **17**, 97–105 (2011)
41. Thorat, P.P., Sadawarte, S.K., Sawate, A.R., Machewad, G.M.: Studies on effect of fermentation physiochemical properties of vegetables and preparation of sauce. Int. J. Curr. Microbiol. Sci. **6**(8), 3537–3545 (2017)
42. Tamang, B.: Role of lactic acid bacteria in fermentation and bio preservation of traditional vegetable products. Ph.D. thesis, Food MicrobiologyLaboratory, Sikkim Government College, Gangtok (North Bengal University) (2006)
43. Mishra, A., Kharel, G.P.: Preservation and Quality Evaluation of Sisnu (*Urtica plaviflora*) by making Gundruk like Fermented Product. J. Food Sci. Technol. Nepal **6**, 114–117 (2013)
44. Cheigh, H., Park, K., Lee, C.Y.: Biochemical, microbiological, and nutritional aspects of kimchi (Korean fermented vegetable products). Crit. Rev. Food Sci. Nutr. **34**(2)175–203 (1994)
45. Deb, C.R., Jamir, B.: Ethnic fermented food products of Nagaland, India. J. Food Chem. Nanotechnol. **6**(1), 18–27 (2020)

46. Ibrahim, A.D., Sani, A., Shinkafi, S.A.: Production, microbial and physio chemical evaluation of 'dawadawan botso' (a condiment) produced by the fermentation of *Hibiscus sabdariffa* seeds. Int. J. Biol. Chem. Sci. **5**, 2481–2490 (2011)
47. Nwangburuka, C.C., Kehinde, O.B., Ojo, D.K., Denton, O.A., Popoola, A.R.: Morphological classification of genetic diversity in cultivated okra *Abelmoschus esculentus* using principal component analysis (PCA) and single linkage clusteranalysis (SLCA). Afr. J. Biotech. **10**(54), 11651–21172 (2011)
48. Kumar, V., Singh, D.K., Panchbhaiya, A., Singh, N.: Genetic divergence studies in mid-season cauliflower (*Brassica oleracea* var. *botrytis* L.) through Principal Component Analysis (PCA) and D2 analysis. Int. J. Chem. Stud. **5**(4), 989–993 (2017)
49. Angmo, K.: Isolation and probiotic characterization of Lactic Acid Bacteria from traditional fermented foods and beverages of Ladakh. Ph.D. Thesis, Department of Biotechnology, Himachal Pradesh University Summer Hill, Shimla, India (2014)
50. Nguyen, D.T.L., Van Hoorde, K., Cnockaert, M., De Brandt, E., Aerts, M., Binh Thanh, L., Vandamme, P.: A description of the lactic acid bacteria microbiota associated with the production of traditional fermented vegetables in Vietnam. Int. J. Food Microbiol. **163**(1), 19–27 (2013)
51. Goswami, G., Bora, S.S., Parveen, A., Chandra Boro, R., Barooah, M.: Identification and functional properties of dominant lactic acid bacteria isolated from Kahudi, a traditional rapeseed fermented food product of Assam, India. J. Ethnic Foods **4**(3), 187–197 (2017)
52. Pheng, S., Han, H.L., Park, D.S., Chung, C.H., Kim, S.G. *Lactococcus kimchi* sp. nov., a new lactic acid bacterium isolated from kimchi. Int. J. Syst. Evol. Microbiol. **70**, 505–510 (2020)
53. Chen, Y.S., Otoguro, M., Lin, Y.H., Pan, S.F., Ji, S. H., Yu, C.R., Liou, M.S., Chang, Y.C., Wu, H.C., Yanagida, F.: *Lactococcus formosensis* sp. nov., a lactic acid bacterium isolated from yan-tsai-shin (fermented broccoli stems). Int. J. Syst. Evol. Microbiol. **64**, 146–151 (2014)
54. Patel, S.A., Parikh, S.C.: Isolation and screening of probiotic potential lactic acid bacteria from local dairy products. Int. J. Curr. Microbiol. Appl. Sci. **5**(11), 490–498 (2016)
55. Begley, M., Hill, C., Gahan, C.C.M.: Bile salt hydrolase activity in probiotics. Appl. Environ. Microbiol. (72), 1729–1738 (2006)
56. Sadishkumar, V.: Characterization of lactic acid bacteria isolated from idli batter fermented with piper betle leaves, their probiotic properties and antimicrobial peptides. Ph.D. thesis, Department of Biochemistry and Molecular Biology, School of Life Sciences, Pondicherry University, Puducherry (2017)
57. Savedboworn, W., Riansa-ngawong, W., Sinlapacharoen, W., Pajakang, S.,Patcharajarukit, B., Tipkanon, S.: Assessment of probiotic properties in lactic acid bacteria isolated from fermented vegetables. KMUTNB. Int. J. Appl. Sci. Technol. **7**(4), 53–65 (2014)

Chapter 9
An Online Survey to Gain Insights into Respondents' Awareness About Mithun, a Unique Biota of the NE Region of India and Its Husbandry Practices

Khriengunuo Mepfhuo and Mihir Sarkar

9.1 Introduction

The North-East region of India boasts of a plethora of exotic flora and fauna. These add to the biodiversity of the region and that of the country. Among the diverse species of animals that are found in the region, a bovine species namely the mithun (*Bos frontalis*) is unique to the NE states of Arunachal Pradesh, Nagaland, Mizoram, and Manipur. According to the report of 20th Livestock Census [1], the population of mithun is India is 3.90 lakh and 3.5 lakh, 23,123, 9059 and 3287 in Arunachal Pradesh, Nagaland, Manipur and Mizoram states of India respectively. Mithun are also found albeit in small numbers in Myamar, Bhutan, Bangladesh, and China [2]. This bovine species occupies a place of pride for the people of North-East especially the states of Arunachal Pradesh and Nagaland. In these states, it is revered for playing an important role from time immemorial in social, economical, and cultural aspects.

Mithun is traditionally reared in the free range system. The animals graze openly on their own in the deep forests as reported by Heli [3]. Kiba in the book, Pictorial documentation of Naga mithun stated that the mithun was domesticated more than 8000 years ago in the plains of the NE region. The author cites many interesting aspects of mithun rearing in the mentioned book where information conveys that the mithun rearers do spread across the villages of Nagaland. Kiba also relates to the rearing of the mithun to the conservation of the biodiversity of the region. Thereby, preserving its importance in the social, cultural, and economic tapestry of the communities can be achieved by those who are involved in rearing this animal and for many generations. It is also mentioned that mithuns belonging to different members of the community are also kept together and allowed to graze freely in a fenced area [4]. This system of rearing mithun is a unique husbandry practice associated with the animal. Dorji et al. in their review study highlighted that due to the

K. Mepfhuo (✉) · M. Sarkar
ICAR—National Research Centre on Yak, Dirang, West Kameng, Arunachal Pradesh, India
e-mail: k.mepfhuo@icar.gov.in

R. V. S. Uppaluri and L. Rangan (eds.), *Conservation of Biodiversity in the North Eastern States of India*, https://doi.org/10.1007/978-981-99-0945-2_9

importance of the animal among the ethnic communities, it lead to forest conservation [5]. However, in this system, only natural breeding is followed as reported by Pal and Rajkhowa [6]. This involved instances of mating with selected superior bulls for the purpose of obtaining progenies with desired traits as noted by Kiba [4]. Only salt is fed to the mithuns as supplementary feeding by the owner and its rearers to establish a bond whenever they visit the animal herd in the forests according to the findings of Mepfhuo [7]. Kiba [4] recorded that generally mithuns are not milked and the milk is meant only for consumption by the calves. Pal and Rajkhowa [6] also reported that there was a great demand for mithun meat as consumers felt the meat to be tender and of high quality in comparison to that of other species apart from pork. The hide is not processed and is consumed along with the meat. However, the horns are used as drinking cups and as a decorative item as per the finding of Kiba [4]. For the care and management of the mithun herd, it was found out by Mepfhuo [7] that practices like vaccination, deworming and approaching a veterinarian for treating sick animals are only followed partially and the traditional method of applying local herbs to wounds and applying salt to detach leech attached to the bodies of mithuns which is a major problem during the rainy season are still practiced. These findings were collected through the interaction with the mithun rearers or herders using an interview schedule and covered three districts of Nagaland. Nyodu [8] reported that mithun is reared for varied reasons and purposes. These include generation of economic benefit, financial security, for household meat requirements, to serve as a financial security, the bovine is seen as a source of animal power and is slaughtered during traditional ceremonies. While Kiba [4] revealed there was a strong social and religious significance in some communities, others opined that there was little or no traditional belief to their cultural ethos but simply reared as a means of livelihood and status symbol. Mepfhuo inferred that mithun in Nagaland were reared mainly for meat purpose and as a part of tradition and for economic purpose. However, among certain tribes including Nyshi and Apatani of Arunachal Pradesh, mithuns have been reported to be sacrificed for ritual purposes (Fig. 9.1) [9].

However, with the setting up of the ICAR-National Research Centre on Mithun at Nagaland in the year 1988 by the Government of India, many more potentials of mithun has been explored and revealed (Fig. 9.2).

Scientific studies conducted by scientists at ICAR-National Research Centre on Mithun located at Jharnapani, Dimapur, Nagaland has proved the superiority of different products of Mithun, namely, its meat, milk and hide. Mondal et al. [10] noted that mithun's milk has been more nutritious due to its higher content of fat, protein and solids not fat content (SNF). In the same study, mithun milk has been also found to have double the energy value in comparison to cow's milk. The fat and protein content in 100 g of mithun milk are 8–13% and 5–7%. Comparatively, in cow's milk it is 3.4% and 3.3% respectively. Dhali et al. [11] reported the conception of healthy mithun following artificial insemination of three mithuns using cryopreserved semen. Lalchamliani et al. [12] reported that the mithun meat was highly preferred and relished by the consumers due to scoring higher in the sensory attributes in a study that was conducted to evaluate the physico-chemical characteristics of mithun meat. Das et al. [13] reported that the leather quality of mithun was found to

Fig. 9.1 A mithun herd seen with its herder in its natural habitat

Fig. 9.2 Mithuns reared in the semi-intensive system (in sheds) at ICAR-National Research Centre on Mithun, Medziphema, Nagaland

be superior to that of the cattle. However, mithun was also found to be susceptible to foot and mouth disease (FMD) which could have a devastating impact on the animal as reported by Borah et al. [14]. Efforts made by ICAR-NRC on mithun has shown that mithun can be successfully reared under the semi-intensive system with scientific inputs (Fig. 9.3). This can prove to be a viable solution to the constraint in rearing mithuns in the traditional system of free ranging due to the dwindling forest area. However, the challenge is the level of awareness about the potentials of mithun and its products, the scientific practices that can sustain this livestock rearing in the long run amongst the people. For the second case especially young people are to be educated in the states where this majestic animal is found and is known to play a significant role-socially, economically, and culturally. Due to the strides that has been made in the best practices in mithun husbandry, there is no lack in the available information. However, there is paucity in studies assessing this knowledge level of the rearers and owners. Therefore, keeping these views in mind, this present study was visualised and undertaken. The works that have been extensively quoted in this study are those related to the mithun rearing practices. Further, the socio-cultural and socio-economic relevance to its rearers and the communities among whom it is reared have been gathered through direct observation and interaction with the mithun farmers themselves. The scientific husbandry practices followed are based on the scientific works that have been conducted by the researchers in the laboratories and farms.

Fig. 9.3 Mithun bulls reared in the semi-intensive system (in sheds) at ICAR-National Research Centre on Mithun, Medziphema, Nagaland

9.2 Materials and Methods

The present study was conducted among respondents spread across the four mithun rearing states of North-East India—Arunachal Pradesh, Nagaland, Manipur, and Mizoram. For the purpose of the study, a questionnaire containing questions related to the study was prepared for the purpose of data collection. This was done in consultation with available literature and the scientists of ICAR-NRC on Mithun (Fig. 9.4). After finalising the interview schedule, it was decided that due to the ongoing COVID 19 pandemic, the data would be collected using the Google forms. The respondents for this study was randomly selected and the questionnaire in the form of a Google form link was randomly send to the respondents through WhatsApp along with a cover letter requesting to fill the questionnaire on a voluntary basis while assuring anonymity for the individual respondent's personal details and responses.

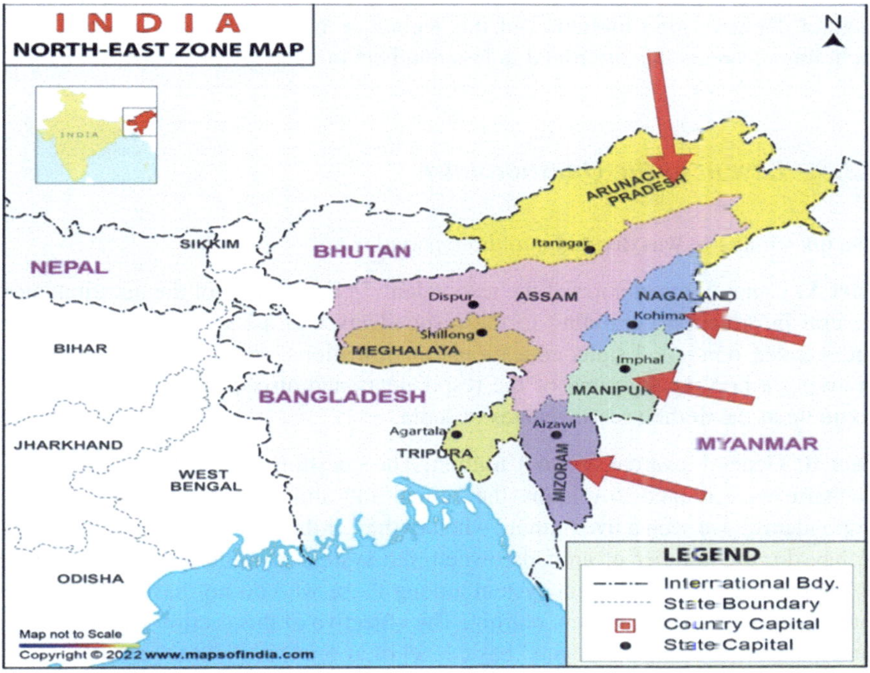

Fig. 9.4 Arrows marks depicting of North-East India in which mithuns are found. *Source* www.mapsofindia.com, 2022 [15]

9.2.1 Study Locale

The prepared questionnaire in the Google form was sent out randomly to the respondents in the four mithun rearing states of North-East India—Arunachal Pradesh, Nagaland, Manipur, and Mizoram.

9.2.2 Sampling Technique

For the purpose of the study, the Google form link was randomly sent to the respondents of the mithun rearing states. The selection of the states was deliberate with the condition been that it should be one where mithuns are found. The respondents' selection followed the simple random sampling. This method was applied so that the selection will be bias free which is one of the apparent advantages of this sampling method. However, the limitation of this technique is the decreased probability of including elements that are found in less numbers in the population [16].

9.2.3 Layout of the Questionnaire

The questionnaire was divided into three parts:

Part A: General information of the respondent: In this section of the questionnaire, general information pertaining to the respondents was asked which includes the age, gender, domiciled state, educational qualifications, and occupation. It helped in giving a brief background of the respondents and also ensured that they were domicile of the mithun rearing states of India.

Part B: General awareness about mithun: The questions in this part of the questionnaire were related to mithun, the animal and aimed at knowing whether the respondents have seen a live mithun, whether they or their families own mithun, and if they do, the number of animals owned, the system of mithun rearing followed or knowledge of the prevalent system among those who do not have mithuns, the purposes or reasons for mithun rearing. The objective of this section was to gain an understanding of the respondents' basic level of awareness about the animal.

Part C: Awareness related to mithun husbandry: For this section, on the basis of available literature and experts advices, statements related and relevant to the established mithun husbandry were prepared and respondents were asked to cite if they were aware of the practices. The objective was to garner an understanding of the respondents' awareness with respect to mithun husbandry.

For the purpose of this study, the definition of awareness as a derivative of the word aware, 'as having knowledge of a situation or fact' was considered.

The online link of the questionnaire was activated on the 25th of July 2020. Responses from the respondents were recorded from 25th July 2020 to 6th August 2020.

The data collected were quantified, compiled, and properly tabulated and were analysed by suitable statistical methods.

The following statistical were applied for analysis of data:

- Descriptive statistics: i.e., frequencies, percentage, means, standard deviation, and range.
- The study was purely descriptive in its scope and extent.

9.3 Results and Discussion

9.3.1 General Information About the Respondents

A total of 138 respondents responded to the online survey. Among these, 53% were male and 47.10% were female (Fig. 9.5).

The average age of the respondents was found to be 28.76 years and the age varied from 18 to 56 years. Out of the 138 respondents, 58.69%, 17.39%, 15.94%, and 7.97% were from Nagaland, Arunachal Pradesh, Manipur, and Mizoram respectively (Table 9.1).

As can be noted from Table 9.2, 0.74% of the respondents studied up to primary class, 17.39% possessed high school degree, while 51.45% were graduates, and 30.43% possessed post graduate and above degrees.

It can be noted from Table 9.3 that 42.75% of the respondents were students, 27.54% were in Govt. Service, 4.35% in private service, and 3.62% engaged in business with regard to the occupation of the respondents.

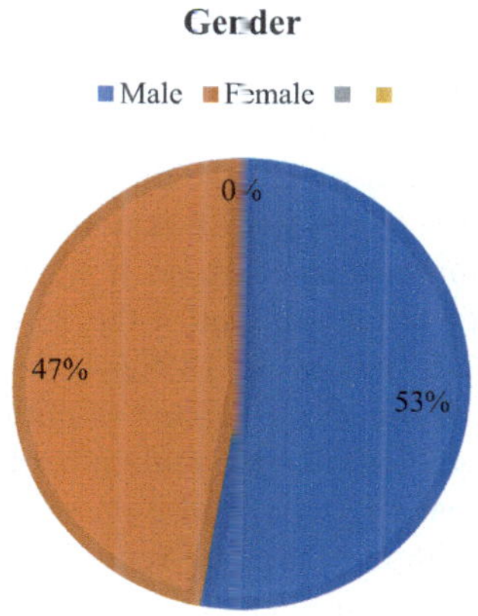

Fig. 9.5 Gender distribution of the respondents

Table 9.1 Data summary of the distribution of the respondents, and as per their domiciled state

State	Number of respondents
Nagaland	81 (58.69%)
Arunachal	24 (17.39%)
Manipur	22 (15.94%)
Mizoram	11 (7.97%)
Total	138

Table 9.2 Data summary of respondents' qualifications

Educational qualifications	Respondents
Primary school	1 (0.74%)
High school	24 (17.39%)
Graduate	71 (51.45%)
Post graduate and above	42 (30.43%)

Table 9.3 Data summary of respondents' occupation

Occupation	Respondents
Agri and allied	8 (5.80%)
Govt. service	38 (27.54%)
Private service	6 (4.35%)
Business	5 (3.62%)
Student	59 (42.75%)
Unemployed	15 (10.87%)
Self-employed	7 (5.07)

9.3.2 General Awareness of Respondents Related to Mithun

It can be seen from Table 9.4 that 84.05% of the respondents have seen a live mithun and 15.94% expressed in the non-affirmative classification. From the same table, it can also be noted that out of the 138 respondents, 23.91% were found to own mithun which also implies that the family of the respondent owns mithun. However, it is to be noted that none of these respondents were involved in looking after the mithuns as a herder. It was also recorded that 76.09% of the respondents did not own mithun. Out of the 33 respondents (23.91%) who own mithun, 39.39% own 1–3 mithun, 18.18% own 4–6 mithun, 15.15% own 7–9 mithun, and 27.27% own 10 and above number of mithun. This finding is similar to that of Mepfhuo [7] who reported that the mithun herd size range from 1 to 12. With respect to the system of rearing among those possessing mithun, it was observed that 81.81% were rearing mithun in the free-range system, 3.03% and 15.15% were rearing in the intensive and semi-intensive system respectively. This finding is in resonance to those of Heli [3], Kiba [4], and Mepfhuo [7] that mithun are reared in the free range system.

Table 9.4 Data summary of distribution of respondents' responses with respect to general awareness related to mithun

General awareness of respondents	Responses of respondents
Seen a live mithun	
(a) Yes	116 (84.05%)
(b) No	22 (15.94%)
Self or family own mithun	
(a) Yes	33 (23.91%)
(b) No	105 (76.09%)
Number of mithun own by self/family (response from those owing mithun)	
(a) 1–3	13 (39.39%)
(b) 4–6	6 (18.18%)
(c) 7–9	5 (15.15%)
(d) 10 and above	9 (27.27%)
System of rearing followed by those who or whose family own mithun	
(a) Free range	27 (81.31%)
(b) Intensive	1 (3.03%)
(c) Semi-intensive	5 (15.15%)
Response of respondents to the states where mithun is found in India	
(a) Correct answer	83 (60.14%)
(b) Incorrect answer	55 (39.85%)

Amongst the respondents, 60.14% in the study gave correct response to the question on the states in which mithun are found (Table 9.4).

A close appraisal of Table 9.5 revealed that for the respondents who or whose family own mithun, 28% of the responses were for the aforementioned five reasons or purposes which include income/livelihood, meat (nutrition source), symbol of status and prosperity, for slaughter during festivals and religious ceremonies and, to serve as a bride price. These results are in resonance with the ones reported by Nyodu [8] that stated the reasons for rearing mithun. These include economic venture, meat purpose, and slaughter during ceremonies. For those not owning mithun, 29.10% chose this response as their perceived idea on why mithun were reared.

9.3.3 Awareness Related to Mithun Husbandry

An examination of Table 9.3 gave an overview of the respondents' awareness about mithun husbandry. It can observed from Table 9.3 that 48.48% of the respondents who own mithun were aware that mithun can be reared in sheds, 65.71% of the respondents who do not own mithun were aware of this practice. This fact corresponds

Table 9.5 Data summary of responses of the respondents for reasons and purposes of mithun rearing

Reasons/purposes for rearing mithun	No. of responses of those who don't own mithun	No. of responses of those who/whose family owns mithun
Income/livelihood	28 (20.90%)	10 (20%)
Meat (additional nutrition source)	17 (12.69%)	6 (12%)
Symbol of status and prosperity	19 (14.18%)	6 (12%)
Slaughter during festivals or religious ceremonies	7 (5.22%)	9 (18%)
To serve as bride price	8 (5.07%)	4 (8%)
All the above	39 (29.10%)	14 (28%)
None of the above	4 (2.99%)	1 (2%)
Don't know	12 (8.96%)	0 (0%)

to the rearing practice followed at ICAR-National Research Centre on mithun where the animals are reared in sheds as per Annual Report (2019) [17] of the institute.

It can be seen that 42.42% of the respondents who own mithun and 66.67% of the respondents who don't own mithun were aware that mithun could be milked. Mithun milk is suitable for human consumption was known to 66.67% of the respondents who own mithun and 77.14% of the respondents who do not own mithun.

From Table 9.6, it can be noted that 78.79% of the respondents who own mithun and 63.80% of those who do not own mithun were aware that mithun's meat is tender and tastes better in comparison with other bovine species. These findings are similar to the ones by Pal and Rajkhowa [6] who cited that there was a great demand in mithun meat due to its high tenderness. Lalchamliani et al. [12] also reported that consumers relished mithun meat.

It can be concluded from the same table that 69.70% of the respondents who own mithun and 81.90% of the respondents who don't own mithun have awareness about the possibility of processing and developing mithun's hide into good quality leather items. Though it was reported by Kiba [4] that mithun hide is consumed along with the meat. A high number of respondents can be seen to be aware of the potential of the processing of mithun hide to quality leather. This superiority of mithun hide was cited by Das et al. [13].

A scrutiny of the table also revealed that 27.27% of the respondents who own mithun and 54.29% of the respondents who do not own mithun were privy to the potential of mithun for draught purpose. The use of mithun as a potential draught animal was reported by ICAR-NRC on Mithun in its Annual report (2019) [17].

It can be seen that 69.69% of the respondents who own mithun and 72.38% of the respondents who do not own mithun had knowledge that apart from salt, mithun can

Table 9.6 Data summary of frequency distribution of responses of respondents for the awareness on mithun husbandry

Practices	Who own mithun		Who don't own mithun		Pooled response	
	Yes	No	Yes	No	Yes	No
(1) Mithun can be reared in sheds	16 (48.48%)	17 (51.51%)	69 (65.71%)	36 (34.29%)	85 (61.59%)	53 (38.41%)
(2) Can be milked	14 (42.42%)	19 (57.58%)	70 (66.67%)	35 (33.33%)	84 (60.87%)	54 (39.13%)
(3) Milk suitable for human consumption	22 (66.67%)	11 (33.33%)	81 (77.14%)	24 (22.86%)	103 (74.64%)	35 (25.36%)
(4) Meat is tender and taste better	26 (78.79%)	7 (21.21%)	67 (63.80%)	38 (36.19%)	93 (67.39%)	45 (32.61%)
(5) Hide can be processed into leather goods	23 (69.70%)	10 (30.30%)	86 (81.90%)	19 (18.10%)	109 (78.99%)	29 (21.01%)
(6) Mithun can be used for draft purpose	9 (27.27%)	24 (72.73%)	57 (54.29%)	48 (45.71%)	66 (47.83%)	72 (52.17%)
(7) Apart from salt, mithun need to be fed mineral mixture	23 (69.69%)	10 (30.30%)	76 (72.38%)	29 (27.62%)	99 (71.74%)	39 (28.26%)

(continued)

Table 9.6 (continued)

Practices	Who own mithun		Who don't own mithun		Pooled response	
	Yes	No	Yes	No	Yes	No
(8) AI can be done for breeding purpose	25 (75.76)	8 (24.24%)	79 (75.24%)	26 (24.76%)	104 (75.36%)	34 (24.63%)
(9) Identification can be done with ear tags and microchips	27 (81.82%)	6 (18.18%)	93 (88.57%)	12 (11.43%)	120 (86.96%)	18 (13.04%)
(10) Age can be detected by dentition pattern	25 (75.76%)	8 (24.24%)	85 (80.95%)	20 (19.05%)	110 (79.71%)	28 (20.29%)
(11) Mithun suffers from infections and parasitic infestation	29 (87.88%)	4 (12.12%)	97 (92.38%)	8 (7.62%)	126 (91.30%)	12 (8.70%)
(12) Mithun can needs to vaccinated (against FMD, HS etc)	28 (84.85%)	5 (15.15%)	96 (91.43%)	9 (8.57%)	124 (89.86%)	14 (10.14%)
(13) Mithun can needs to be dewormed (against parasitic infestation)	26 (78.79%)	7 (21.21%)	87 (82.86%)	18 17. 14%)	113 (81.88%)	25 (18.11%)
(14) Visited NRCM	8 (24.24%)	25 (75.76%)	30 (28.57%)	75 (71.43%)	38 (27.54%)	100 (72.46%)
(15) Interested in rearing mithun in the future	29 (87.88%)	4 (12.12%)	53 (50.48%)	52 (49.52%)	82 (59.42%)	56 (40.58%)

also be fed with mineral mixture. Mepfhuo reported that the salt feeding is practiced by the mithun rearers.

From the table it can be also observed that 75.76% of the respondents who own mithun and 75.24% of the respondents who do not own mithun know that artificial insemination could be done in mithun. Dhali et al. [11] reported the success of Artificial Insemination in mithuns.

It may also be noted that 81.82% of the respondents who own mithun and 88.57% of the respondents who do not own mithun were aware that animal identification methods like ear tags and micro-chips could be employed in mithuns. The findings of Heli [3] revealed that the mithun owners identified their mithun based on colour, body markings and ear notching (marked during calfhood) and also by the structure of the horns. Also Kiba [4] reported that in some mithuns, owners recognized the mithun through its voice.

In the same table, it can observed that 75.76% of those who owned mithun and 80.95% of those who don't know mithun were aware that age of mithun could be determined by noting the dentition pattern.

From the table it can be observed that 87.88% of the respondents who own mithun and 92.38% of the respondents who do not own mithun were aware that the Mithun suffer from infections like foot and mouth disease (FMD), haemorrhagic septicemia (HS) and parasitic infections. The susceptibility of mithun to FMD was reported by Borah et al. [14].

It can also be observed that 84.85% of the respondents who own mithun and 91.43% of the respondents who do not own mithun know that Mithun can be vaccinated to prevent FMD and HS. From Table 9.6 it can be seen that 78.79% of the respondents who own mithun and 82.86% of the respondents who do not own mithun expressed their awareness in the affirmative for the regular deworming of mithun to prevent parasitic infections.

Table 9.6 also revealed that 87.88% of the respondents who own mithun expressed their interest to continue rearing mithun in the future. While 50.48% of the respondents who do not own mithun were seen to be interested in rearing mithun in the future. This finding is similar to that of Mepfhuo [7]. The authors revealed that all interviewed respondents expressed their willingness to continue the rearing of mithun.

It can be seen from Table 9.7 that 23.5% of the responses cited internet as source of information on mithun and mithun husbandry. Further, 12.5% of the responses were for print media, 35% for family and friends, 3.5% for trainings and seminars, 3.5% for radio and television and 8% from a visit to ICAR-NRCM. Six percent of the recorded responses were for all mentioned sources and 8% responses mentioned that their sources of the information were not cited in the questionnaire. Among the respondents, as being conveyed in Table 9.4, 23.90% of the respondents owned mithun in their families (also as seen from Tables 9.3 and 9.2). Further, 42.75% and 51.45% respondents in the survey were students with graduate degrees respectively. This could explain family and friends, and the internet ranked very high as the sources

Table 9.7 Data summary of frequency distribution of responses on the sources of information

Sources of information related to mithun and mithun husbandry	Pooled (including mithun owners and non mithun owners) responses of respondents
Internet	47 (23.5%)
Print media	25 (12.5%)
Family and friends	70 (35%)
Training, seminars etc	7 (3.5%)
Radio and television	7 (3.5%)
Visit to ICAR-NRCM	16 (8%)
All of the above	12 (6%)
None of the above	16 (8%)

of information for the topic of the study. The qualification and the occupation of the respondents could further explain the high level of awareness related to the scientific mithun husbandry practice among the respondents. These can further be validated by the findings of Biam et al. that family and friends, television and internet are the main sources of information for the mithun rearers and owners [18].

9.4 Conclusions

Based on the findings of the present study, it was noticed that many aspects of scientific mithun husbandry are yet to reach the doorsteps of the mithun owners. This is crucial so that mithun apart from maintaining its status as an animal of pride can be successfully transitioned into a remunerative source of income for its rearers.

One hundred thirty eight (138) people responded to this survey and among this 23.91% belonged to families who own mithuns. A good percentage of respondents were from Nagaland (58.69%) followed by Arunachal Pradesh (17.39%). It was analyzed that 51.45% of the respondents were graduates. The present study indicated that mithun is being reared to act as a source of income or livelihood and meat, status of symbol and prosperity confirmed by (28% of the respondents who have mithun and 29.10% of those who don't own mithuns). The study also indicates that the respondents do have a good degree of awareness related to mithun husbandry. Also, the internet and print media sources ranked high in the source of information for the respondents apart from friends and family. While this study was mainly restricted to those with access to the internet and knowledge. As it was conducted through an online survey.

The method employed in this study remains distinctive in the way that no such surveys employing online mode has been reported on the aspects that being covered in the study. The study was primarily aimed at knowing the awareness level of the respondents in the states where mithuns are found irrespective of whether they own mithun or not. For this the conception was that higher awareness about this animal is indicative of its reverence in the communities attributable to the place of importance it occupies in the biodiversity of the places in which it is found. However, limitations include the reach of the survey which may have restricted the participation of only those with access to smart phone and internet, and a certain degree of literacy to partake in the survey and also the bias of the respondents who may have received the questionnaire but chose not to participate in the survey due to no direct interaction with the respondents. This study which used online means to gather information had the advantage of collecting information in the shortest possible time with less effort in comparison to the previously conducted works that involves travelling to the different places and having discussion with groups of farmers. All these require time, effort and financial aid. The responses were solely on the discretion of the respondents. There is also a possibility that the respondents while replying to the survey may had accessed the internet to get answers to certain questions. However, such surveys can be taken to trigger the curiosity of the respondents and prompt a person to seek answer to the question being asked. This is a major take-away from this study. While this study was mainly restricted to those with access to the internet and knowledge, it may be mention that such surveys amongst the youth may also catalyze them to check out information to respond to the surveys and may act to fulfill their curiosity. At such institutes, researchers or other stakeholders involved in mithun husbandry could be a source of tested and proven scientific information and can play crucial roles to provide better access to such knowledge to the public.

The future of the majestic mithun remains in the hands of the younger generation and the current stakeholders. Therefore, all efforts should be made by the government and other concerned parties to make mithun rearing as lucrative and sustainable to retain the interest and attention of its owners and rearers who have expressed their willingness to continue rearing this unique bovine species despite the challenges. However, efforts as revealed from this study refer to the definite need to create more awareness about the relevant husbandry practices that can be successfully followed by the rearers to yield more profit. Due to the small population of the mithun, maximizing conservation endeavours will be a key to secure the place of the animal in the biodiversity of the North-East region. The methodology applied in this study which entailed the use of the online means to assess awareness level can be further deployed among different groups to generate interest within the community members as well as those outside to learn more about this rare species which has its home in the North-East states of India.

The conducted research involving the driving of awareness of mithun through the online survey is useful towards bio-diversity conservation emphasis in North-East India. Further public forums, books, journals, and conferences that advocate for such noble cause can enhance human engagement in scientific studies that are involved in creating awareness about the bio-diversity of North East India.

Acknowledgements The help extended by the personnel of ICAR-National Research Centre on Mithun, Jharnapani, Nagaland during the course of this study is highly solicited and appreciated.

References

1. Anonymous: Basic Animal Husbandry Statistics, p. 85. Published by the Ministry of Fisheries, Animal husbandry & Dairying, Department of Animal Husbandry & Dairying, Government of India, Krishi Bhawan, New Delhi (2019)
2. Bhattacharyya, H.K., Goswami, B.K., Biswas, R.K., Deka, B.C.: First record case of seminal vesiculitis in Mithun (*Bos frontalis*): research and reviews. Zool. Sci. **2016**(4), 38–39 (2016)
3. Heli, T.: Studies of some physical characteristics and meat production in mithun under the prevailing managemental practices in Arunachal Pradesh. M.V.Sc Thesis, Assam Agricultural University, Guwahati-22 (1993)
4. Kiba, H.Z.: Pictorial documentation of Naga Mithun. Issued in commemoration to the 1st "Mithun Festival" 12th to 14th Jan 2012. Organised by the Department of Veterinary and Animal Husbandry Nagaland, NRC on mithun (ICAR), NLDB and NABARD (2012)
5. Dorji, T., Wangdi, J., Shaoliang, Y., Chettri, N., Wangchuk, K.: Mithun (Bos frontalis): the neglected cattle species and their significance to ethnic communities in the Eastern Himalaya—a review. Anim. Biosci. **34**(11), 1727–1738 (2021)
6. Pal, D.T., Rajkhowa, C.: Evaluation and improving locally available feed resources and developing feeding system for improving livestock production. Annual Report 2001–2002, National Research Centre on mithun, Jharnapani, Nagaland, India, pp. 19–22 (2001)
7. Mepfhuo, K.: Traditional and scientific rearing of Mithun (Bos frontalis) in Nagaland. M.V.Sc Thesis, Assam Agricultural University, Guwahati-22 (2012)
8. Nyodu, M.: Role of animal husbandry in tribal communities of Pasighat developmental block of Arunachal Pradesh. M.V.Sc Thesis, Assam Agricultural University, Guwahati-22 (2002)
9. Tayo, T., Heli, T., Bengia, A., Gama, N.: Mithun production and management system in Papum Pare District of Arunachal Pradesh. J. Krishi Vigyan **2**(1), 36–40 (2013)
10. Mondal, M., Singh, R.K., Baruah, K.K., Kumar, A., Singh, V.: Mithun: a promising milch animal of the northeastern hill region of India. Res. Rev.: J. Dairy Sci. Technol. **3**(3), 17–21 (2015)
11. Dhali, A., Karunakaran, M., Mech, A., Nath, N., Prakash, B., Rajkhowa, C., Mishra, D.P.: Birth of the first mithun (Bos frontalis) calf through artificial insemination. Anim.: Int. J. Anim. Biosci. **2**, 879–882 (2008)
12. Lalchamliani, Chauhan, G., Mitra, A., Hanah, S.S., Chamuah, J.K.: Quality evaluation of meat from adult male mithun (Bos frontalis). Int. J. Curr. Microbiol. Appl. Sci. **8**(5), 137–149 (2019)
13. Das, K.C., Mukherjee, G., Baruah, K.K., Khate, K., Rajkhowa, C.: Study of growth performance and leather quality in mithun (Bos frontalis) and its comparison with local cattle (Bos indicus) fed on tree leaves based ration. Livest. Res. Rural. Dev. **23**(5) (2014)
14. Borah, P., Deka, B., Sharma, K., Baro, S., Hazarika, A.K., Das, C., Garam, G.B., Boro, P., Ltu, K.: Isolation, identification and retrospective study of foot-and-mouth disease virus from affected mithun (Bos frontalis) in North-Eastern India. Transbound. Emerg. Dis. **65**(1), e63–e69 (2018)
15. Maps of India Homepage. https://www.mapsofindia.com/maps/northeast/sevensisters.htm. Last accessed 9 July 2022
16. Ray, G.L., Mondal, S.: Research Methods in Social Sciences and Extension Education, Third Revised and Enlarged Edition. Kalyani Publishers, New Delhi (2011)
17. Annual Report: ICAR-National Research Centre on Mithun (NRCM) Medziphema, Dimapur, Nagaland, India (2016)
18. Biam, K.P., Devi, L.S., Khate, K., Singh, N.U., Paul, P., Gowda, C.: Information sources, their utilization pattern vis-à-vis mithun (Bos frontalis) husbandry in Arunachal Pradesh. Indian J. Ext. Educ. **58**(3), 14–17 (2022)

Part III
Value-Added Product Development and Benefit Sharing in Bio-Diversity for the Bioprospecting Perspective

Chapter 10
A Study on Various Factors Affecting the Pigmented Rice Value Chain of North East India with a Focus on Black Rice

Rubeka Idrishi, Siddhartha Singha, and Latha Rangan

10.1 Introduction

The NER of India is in the Indo-Burmese biodiversity hotspot which constitutes around 45% of the total biodiversity of India. For a long time, Indian ecology is known for its biodiversity. This diversity in the biological systems differs greatly from state to state. This region of India comprises of the states namely Arunachal Pradesh, Assam, Meghalaya, Manipur, Tripura, Mizoram, Nagaland and Sikkim which can be physio-graphically categorized into the Eastern Himalayas, the Northeast hills, the Brahmaputra river region and the Barak Valley plains [1]. The biodiversity of flora fauna in this region is highly impacted by the Indo-Malaysian, Indo-Chinese and other Indo-biogeographical realms influence. This region is in itself is quite unique for providing a mix of different habitats with high endemism and diversity in the prevalent biota. It is also home to approximately 135 tribal communities of India which is out of 450 in total. The exchange in their customs, traditions and culture also have an important role in the diversity of the black rice found in this region [2].

Among various bio diversities in the north east, the black pigmented rice is counted to be also one of the most important crops [3]. A rough estimate had been made in a relevant prior art [4] about the cultivars which is grown in the north eastern and eastern part of India. Approximately 30,000 varieties are cultivated which also

R. Idrishi · S. Singha
School of Agro and Rural Technology, Indian Institute of Technology Guwahati, Guwahati 781039, Assam, India
e-mail: rubeka1995@iitg.ac.in

S. Singha
e-mail: siddharthafp@iitg.ac.in

L. Rangan (✉)
Department of Biosciences and Bioengineering, Indian Institute of Technology Guwahati, Guwahati 781039, Assam, India
e-mail: lrangan@iitg.ac.in

R. V. S. Uppaluri and L. Rangan (eds.), *Conservation of Biodiversity in the North Eastern States of India*, https://doi.org/10.1007/978-981-99-0945-2_10

includes many strains of black rice, with many being aromatic/scented [4]. Thus, this whole region is practically a treasure reserve for rice germplasms with wide genetic variability. Unlike other rice varieties which contains white pericarp, black rice contains a dark brown to black pericarp due to high anthocyanin content [5]. It is highly nutritious as it contains high level of iron, fiber, zinc, vitamins and other micro nutrients such as antioxidants and magnesium which helps in preventing and curing many types of diseases. It is also good for diabetic patients as the low glycemic content of red rice helps in regulating the insulin level [6]. In spite of being highly nutritious and organic in nature, this rice was not cultivated commercially due to low yield. However, with the increase in demand for the organic products, the demand for this rice too has increased and some farmers have started cultivating this pigmented rice on a commercial basis. In India, black rice is mostly grown in the north eastern and southern states.

Indian Council of agricultural Research (ICAR) identified NER as one of the hubs for rare rice germplasms where the National Bureau of Plant Genetic Resources (NBPGR), India enlisted the emphasized regions in their reports. In these reports it was highlighted that the region hosts many wild varieties of black rice and other varieties from NER of India. Notable among these are the tallest rice in Nagaland, pigmented rice like Chak-hao of Manipur. However, a proper repository is still needed to enlist all kinds of pigmented rice of North-East India. Tripura being one of the regions which has pretty high agro-biodiversity with 47 indigenous types of rice [2], does not yet report on the many black rice varieties. The genetic diversity of black rice of few regions of north India was studied by Vanlalsanga and colleagues [7]. The authors highlighted that the cultivars found here having considerable diverse genes of indigenous origin might be used as good sources for various crop improvement programs and thereby cultivate genetically sound and improved varieties. Similar findings were highlighted in a study done on rice varieties of the Eastern Himalayan region of Northeast India [8]. In a report by UNDP and EXIM [9], pigmented rice specifically red and black were highlighted as top exportable products amongst 7 others which are eri silk, jute, ginger, jackfruit, pine apple, lac and mushroom.

10.2 Methodology

In the chapter, the undertaken review-based study focused primarily on various external factors that affects the pigmented rice value chain. To do so, a systematic literature review has been followed to review articles from AGRIS, CAB abstracts, PubMed, Scopus and Web of science. Various reports of government organizations working in North East India specifically has also been fetched and reviewed for the raw data and the outcome of the projects.

10.3 Factors Affecting the Diversity of Black Rice

Here, various external factors have been discussed which affects the properties of pigmented rice specifically black rice this is due to the pertinent differences in cultivation, genetic and physiochemical properties. These factors contribute to the biodiversity and land specific varieties of black rice of NER of India.

10.3.1 Landcover of North East India

The diverse landcover of north eastern India is one of the major factors for variation of traits in crops such as pigmented rice. Cultivation of black rice is currently dominated by the inorganic high yielding non-pigmented rice varieties which has led to the reduction of the yield and genetic diversity of black rice significantly. The land specificity of black rice has been increased with the reduction in yield. Some important gene pools of pigmented rice have been found from the landcover diversity. The agroclimatic diversity leads to the diversity of the landcover which differs from tropical to alpine, thereby making it habitable for wide variety of pigmented rice [10]. The surface water also contributes along the landcover. This part of India has largest forest covers. Area viz after Madhya Pradesh and upto Arunachal Pradesh contributes to the second largest forest covers in India followed up by Chhattisgarh, Odisha and Maharashtra [11]. From forest cover perspective, as the percentage of total geographical area in the north eastern regions of India, Mizoram leads the list (85.41%) followed up by Arunachal Pradesh (79.63%), Meghalaya (76.33%), Manipur (75.46%), and Nagaland (75.31%) of Fig. 10.1. Few pigmented rice cultivars were studied by [12] who highlighted that these rice were found to more tolerant in the diverse land covers such as mountain lands, plains, deep water lands and unfertile soil covered lands.

Rice cultivation in North East India experiences diverse agroclimatic and topographic conditions. Thereby, the cultivation which is usually categorized on the basis of the season of crop, grain type and quality and habitat of the location.

There are three types of cultivation patterns found in north East India for pigmented rice. The *Sali, Boro* and *Jhum* type. The *Sali* type is the one which constitutes majority of the black rice cultivation. This is generally followed in the lowland lying flood plains, the areas which have these kinds of patterns are the Brahmaputra river valley and Barak valley regions. The other type is the *Boro* cultivation which is the winter cultivation of rice done from November to May. This is also done in lowland areas of sufficient water availability during the colder and dry months of the year. Therefore, the varieties which is grown under this cultivation patterns are generally cold shock proof, genetically stronger and suitable for cold climate adaptation. The third type of cultivation followed in North East India is the *Jhum* type. This is also called shifting or slash and burn cultivation. This is done on dryland and hilly areas for cultivation. This particular cultivation can adapt to a wide range of ecological conditions and climate which ranges from dry and low moisture level

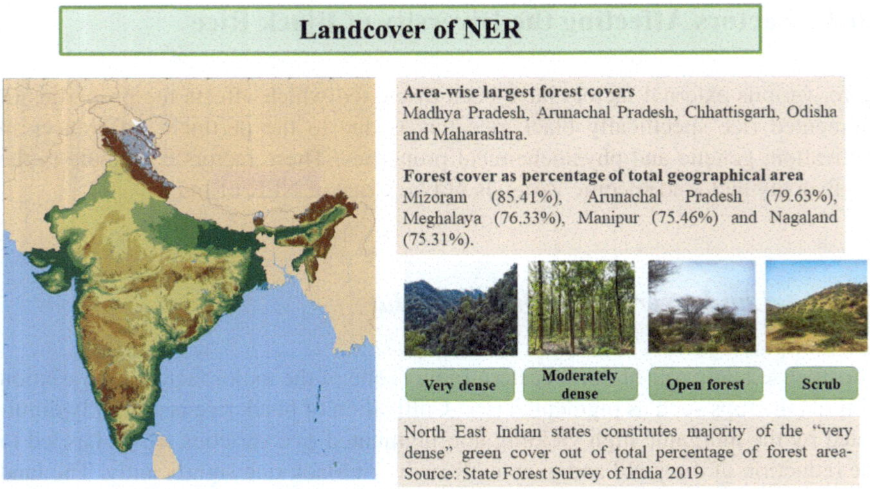

Fig. 10.1 The landcover of North East India

areas to high altitudes even inclining more above 3000 m above the level of sea [8]. This type of cultivation generally grows the glutinous kind of black rice and other rice varieties such as sticky and Joha which are commonly found in NER of India. This type of rice is a source of grain for the preparation of sweet dish such as Pitha, desserts such as pudding/kheer and breakfast options. In various tribal and ethnic groups of the region, it is consumed in the form of traditional recipes as desserts, flakes, cakes, bread and special traditional preparations. In Manipur a recipe named Utong Chak is prepared traditionally with black rice and in bamboo after smoking or simmering effects. It also finds various medicinal use where traditional medical practitioners of Manipur use black rice for treating various ailments such as dysentery, muscular sprain, dog bite, reduced eyesight, mouth ulcer, hypersensitivity of teeth, pre- and post-partum feeding and dandruff. In Meghalaya, it is used to treat skin allergy and tonsils [4]. Adding to these types of rice, various other naturally wild species of rice strains are also grown in this part of India which includes *O. rufipogon* (said to be wild ancestor of black rice [13]. *O. granulata, O. officinalis, O. nivara, O. meyeriana, Hygrorhiza aristata, Leersia hexandra and Zizenia latifolia* [14].

10.3.2 Water Resources

The water resources of the north east India are quite diverse due to its perennial rivers and its tributaries. The NE India is divided into four distinct units on a physiographic basis. These are (i) the Himalayan region at the East, (ii) the mountains at the East, (iii) the Meghalaya- Mikir table land and (iv) the Brahmaputra valley. The rainfall

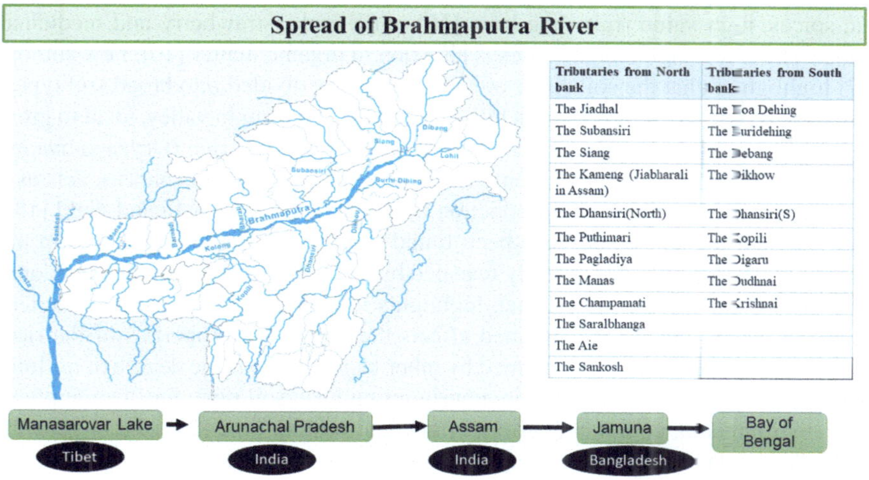

Tributaries from North bank	Tributaries from South bank
The Jiadhal	The Noa Dehing
The Subansiri	The Burideling
The Siang	The Debang
The Kameng (Jiabharali in Assam)	The Dikhow
The Dhansiri(North)	The Dhansiri(S)
The Puthimari	The Kopili
The Pagladiya	The Digaru
The Manas	The Dudhnai
The Champamati	The Krishnai
The Saralbhanga	
The Aie	
The Sankosh	

Fig. 10.2 The dispersion of Brahmaputra alluvial plains with its tributaries based on the review of the authors

and climatic zones of these areas also affects the black rice cultivation of these regions. The regions which receive highest rainfall amongst all is the Himalayan zone which is covered with thick and dense vegetation areas and is suitable for black rice cultivation as well. The Terai region of Himalaya houses many varieties of black rice. The mountain region experiences relatively lesser rainfall than that of Himalayan region which ranges from 1000 m towards the Brahmaputra valley to 5000 m towards China. The valley regions of NE India vary with altitude where the physiographic divisions are divided into Brahmaputra valley, Barak valley and Terai region of North Bengal. These regions of NE India are further divided into three parts which are (i) The Himalayan mountain, (ii) Subdued peninsular block and (iii) Great Brahmaputra alluvial plains (Fig. 10.2). These regions constitute various physiographic sections. Majority of these are based on the water resources they have such as mountains, flood plains, piedmont slopes, flood plains, longitudinal valleys, river terrains and terraces, rivers and their tributaries and slip off slopes [15]. Moving from the Tibet region around the origin of Himalayas to the lower plain lands of Brahmaputra basin, a huge variation exists in the cultivation pattern, types of pigmented rice, anthocyanin content of black rice and the yield of the same.

10.3.3 Soil Fertility

The type of soil fertility found in North East India is also one of the major reasons of the crop being diverse from other parts of the country. The loamy and red soil of north east is found to be rich in minerals and suitable for cultivating tuber crops [9], citrus, tea, spices, high value fruits like kiwi [16], pine apple, strawberry and medicinal plants. The soil found in eight states is also rich in organic matter [10]. Few authors [17] highlighted that the soil cover of Manipur can be divided into broad soil types. These are (i) red ferruginous soil in hill area and (ii) alluvium in valley, ideal to grow the black rice (Oryza sativa cv. Chakhao Poireiton) and red rice (Oryza sativa cv. Chakhao Amubi). Low mineral content in soil and especially phosphorus, acts as a major constraint for black rice production by affecting plant growth and yield [18]. The seed and the soil quality has been found to have a positive correlation in the yield of black rice [19]. This clearly revealed by a detailed analysis of the nutritional content of the pigmented rice which highlights the fact that the nutritional content of soil in which the rice is cultivated affects the nutritional properties of the rice. These factors are further accompanied by other factors such as the degree of milling, method of preparation of rice before consumption [20], soil type, fertilizer application, environmental condition, variety of rice and milling time [21, 22]. In general, the black rice landraces possess higher stress tolerance towards relatively stressful (poor soil nutrient and water) conditions in comparison to hybrid and other traditional rice varieties. Chakhao poreiton [23] which is the mostly preferred black rice of Manipur is found to be the most stress and drought tolerant among the cultivated landraces of Manipur [24].

10.3.4 Type of Cultivation

During monsoon season in some areas of North East India, the black rice is cultivated in Jhum cultivation system, also well known as the shifting cultivation. The cultivars of rice grown in this kind of farming have demonstrated higher resistance to long season of rainless weather and droughts. The dark color of Chak-hao of Manipur is also said to be one of the major outcomes of this cultivation along with others such as disease resistance especially blast and gall midge. Few Assam varieties like Baon show deep water tolerance and the Mizoram Hmawrhang variety shows good drought resistance [14, 25, 26].

The cultivation of black rice along with Jhum cultivation (seen in slash or burn fields) can also be seen in other farming methods such as wet rice cultivation (known as Pani Kheti and dry rainfed terrace cultivation). The pani kheti method is generally followed on plain/flat lands whereas on the hills, tilling method is done by hands. On hills, it is almost entirely done by hands due to lack of mechanical tools specialized for hill farming of rice. In Arunachal Pradesh, few traditional ploughing tools named as Khampti and Sherdukpen are used. The introduction of shallow tube wells or canals help to reduce the dependency on rain. This has been recently introduced to some areas of Assam plains. For ploughing and tilling, tractors and power tillers are presently used. Since past few years in some parts of this region where Jhum cultivation is constantly discouraged due to inputs from other parts of India. However, in hills, Jhum cultivation is a way of livelihood for rural people [27].

Over 200 land owners in the Goalpara area of Assam, India, under the guidance of Upendra Rabha (a farmer, known for introducing black rice in Assam from one single paddy plant) have chosen to grow black rice rather than other various kinds of rice. Black rice may be grown on a variety or combination of diverse soil types, such as sedimentary, topsoil, and rock. This variety is equally resistant to acidic conditions in the soluble soils. Black rice harvests are said to grow best on loamy or clayey soils that easily pool into mud and form cracks when dried. This kind of farming may be applied in a variety of areas with a high yield, including stream valleys, floodplains, deltas, and shoreline fields. Fast-growing rice varieties may be grown on lighter soils with significant levels of topsoil. Rice may grow well in dark magma soil as well. Due of the diversity of different climates and terrain, black rice growing in India is referred to be "cultivator culture" and is not a good candidate for mechanization. Setting up the seed bed, transmitting seeds, transplanting plants from nurseries to the fields, gathering, and winnowing are the main tasks that are completed by hand. Therefore, there is a scope to make these steps easy by some tool, equipment or machinery interventions as it needs a considerable insight to effectively address development through a moderate effort-based approach. However, in Punjab and Haryana, the rice production is mostly dependent on temporary laborers from eastern U.P. and Bihar. A relevant research group [28] studied the black rice cultivation with respect to farming practices and success stories of black rice farmers of India and highlighted various cultivation practices opted by the farmers from Telangana, Assam and Manipur in India.

10.3.5 Agroclimatic Zones

The agroclimatic zone varies significantly in different areas of North East India. The ecosystems differ from being higher in altitudes in Sikkim and Arunachal Pradesh to flood prone areas of Barak valley and verges of Brahmaputra river. The Assam is majorly upland and long terrains of greens provide an ideal agroclimatic zone for black rice farming. In other states, the steep terraces, rainfed areas, irrigated lands and deep-waterbodies encourage the Jhum and tilla land ecologies that support the production of Black rice. However, they have but has not been practiced to a larger extent. In NER of India, the rice self-sufficiency is around 80%. In states such as Manipur, Assam and Tripura, surplus rice cultivation takes place. However, there is a need for others to improve their production potential and productivity with suitable interventions [4]. The altitude, temperature and the relative humidity are one of the major factors that contribute to these varying agroclimatic zones.

Altitude: Meghalaya is the third major contributor of black rice cultivation [29]. The state attempted to cultivate black rice under mid hill altitude. Of Meghalaya, in few experimental studies, it was highlighted that the integrated nitrogen management of organic and inorganic sources is beneficial for black rice cultivation in Meghalaya. This can be addressed and studied for other states as well. Due to the fact that

Table 10.1 Average temperature, relative humidity (RH), and rainfall of eight states of North East India

S.no.	State	Temperature range (°C)	Average RH (%)	Average rainfall (mm)	Reference
1	Arunachal Pradesh	20–20	79–80	750–6250	[32]
2	Assam	10–31.7	57–83	65.9–3000	[4]
3	Manipur	5–27	36–85	2375	[17]
4	Meghalaya	3–27	NA	NA	[29]
5	Mizoram	12–30	NA	1660	[32]
6	Nagaland	1–13	NA	NA	[4]
7	Sikkim	9–22	75–92	NA	[32]
8	Tripura	10–34	81.5	2700	[33]

the anthocyanin content of black rice is the most prized component amongst all, a research group [30] studied black rice plantations on high altitudes. Thereby the authors inferred that the black rice anthocyanin content alters with altitude and higher altitude favor better anthocyanin content.

Temperature, relative humidity and rainfall: The altitudinal differences should be coupled with varied physiography of NER of India in order to understand the factors that contribute to great climatic variations of the region [4]. Table 10.1 enlists the average temperature, relative humidity (RH) and the average rainfall of all the states of North East India. Among all of these eight states, black rice is majorly grown primarily in Assam and Manipur and in few regions in Mizoram and Meghalaya [31]. Furthermore, research is needed in studying the relation between these three parameters with respect to black rice cultivation.

The suggested climate for black rice cultivation was suggested by a research group [34] to sow it in the month of June–July as it prefers a warm and longer growing conditions of 3–6 months. The temperature for germination was suggested to be around 21 °C with a low temperature maintenance during maturity for better quality and yield of the final crop. The conditions of the seeds should be healthy and to remove the inferior ones, 1% brine solution can be used. Around 3–5 kg seed is enough for 1 acre of the land, with two seedlings per hill and wider spacing. These enhance the longevity of the plant. Soaking the seeds of black rice before and after pre-germination is often carried out. Removal of weeds, usage of green manures like *Sesbania* spp and plants like Anabena and Azolla are used to enhance the nutrient needs of the crop. Though in North East India black rice is usually produced by organic farming, fertilization with other methods is however being practiced in few parts of India to enhance the yield.

10.4 Possible Strategies to Enhance the Production of Black Rice in India

There is a huge potential to commercialize the black rice of North Eastern states of India. This is due to only few other states of India that produce the black rice such as Himachal Pradesh, Himalayan Terai region, Odisha, Tamil Nadu, Kerala and Karnataka. The advantage of the NER black rice is that majorly it is organically produced. However, there are few gaps that needs to be addressed to synchronize the disrupted value chain of black rice. These have been enlisted as follows:

1. Development of state specific variety packages based on the variations in agro-climatic conditions and other factors.
2. The infrastructural facilities such as irrigation, seed production inclusion of all types of cropping pattern such as Jhum cultivation can be developed and promoted with the participation of government, private and non-governmental sectors.
3. Research studies are needed to enlist all black rice of India, protection of existing germplasms, their characterization, value addition and their variety traits studies in the context of their agriculture practices.
4. The policies centered to the promotion of pigmented rice of north east India can serve as a boosting factor for the marketing of black rice.
5. Once there is enough production of black rice to fulfill the demands on Indian market, further efforts can be made to target international markets.
6. Adequate capacity building and training for farmers is needed, to develop skilled manpower for the farming of black rice in other regions as well.
7. The Amguripara, Goalpara farmer Upendra Rabha in Assam can be a good example and template for other districts of Assam where approximately 12 tonnes of black rice were produced in 13.2 ha. He was also applauded by the government [34] for maintaining and hybridization efforts of black rice in Assam.
8. The realization of economic importance of the black rice can generate many employment opportunities for people of the major growing regions of India which are the North East followed by South India.
9. To ensure high rice grain quality, Kumar and group [35] suggested three types of interventions: Environmental, Genetic/molecular/Biochemical and Policy interventions.
10. Strengthening of Institutional framework that devotes towards the research on the North Eastern Region (NER) biodiversity of the black rice cultivated in North east India. Thereby, facilitate a synchronized black rice value chain in the region. Figure 10.3 is an institutional framework (conceptualized by authors) of NER which can be linked in the similar manner.

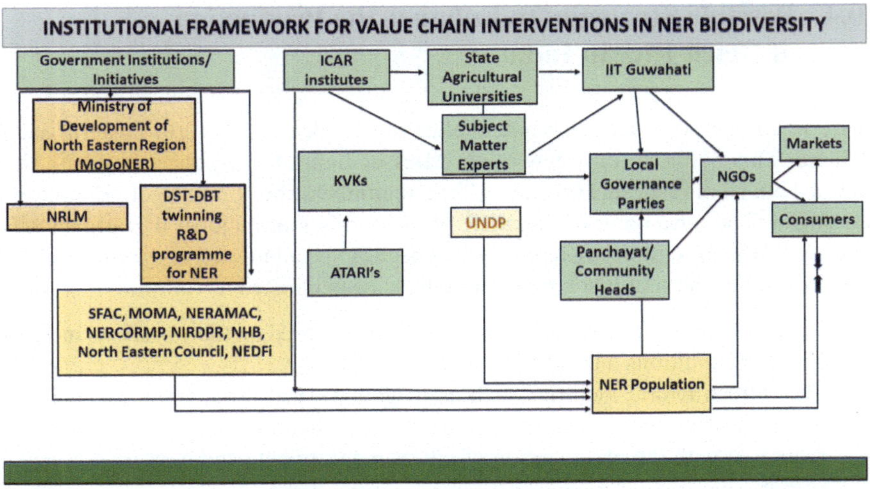

Fig. 10.3 Institutional framework for the posing of key intervention areas to ensure increased black rice production in North East India (SFAC: Small farmers agribusiness consortium; MOMA: Manipur organic mission agency; NERAMAC: North Eastern Regional Agricultural Marketing Corporation Ltd; NERCORMP: North Eastern Region Community Resource Management Project; NIRDPR: National Institute of Rural Development an Panchayati Raj; NHB: National Horticultural Board; NEDFi: North Eastern Development Finance Corporation Ltd.)

10.5 Conclusions

Factors which affect the cultivation, production and quality of black rice can be intrinsic and extrinsic. The extrinsic factors or the external ones are the major contributors of the biodiversity of black rice. In this chapter, five external factors have been discussed with respect to the agro-biodiversity of pigmented rice with a focus on black rice. The landcover of the north east India is full of greens and dense forest covers which when accompanied with the other factors such as the soil type, rainfall, temperature, humidity, altitude, water resources. All these together form an ideal combination to cultivate pigmented rice. The black rice found in these agroclimatic regions have been found to be more stress tolerant and disease resistant than that of black rice produced from other states. Despite of the fact that there is a difference in cultivation pattern and season of sowing and harvesting the black rice in each state, these regions have been able to grow a significant amount of black rice. This production scenario needs to be scaled up without the exclusion of complexities in their cultivation systems. The organic production of the pigmented rice of these regions is what that makes them different from other hybrid and usual rice varieties. Hence, the undertaken review study highlights upon the importance of inclusivity of the factors which are responsible for the biodiversity of black rice. These findings can be used to explore further research that targets the inclusion of black rice that strengthens the black rice value chain of north east India. In a nutshell, biodiversity conservation of north east India is most important to achieve sustained growth in black rice cultivation in this region and its associated economic growth.

References

1. Umakanth, B., Vishalakshi, B., Kumar, P.S., Devi, S.J.S.R.: Diverse rice landraces of North-East India enables the identification of novel genetic resources for Magnaporthe resistance. **8**(August), 1–13 (2017). https://doi.org/10.3389/fpls.2017.01500
2. Chatterjee, S., Saikia, A., Dutta, P., Ghosh, D., Pangging, G., Goswami, A.K.: Background Paper on Biodiversity Significance of North East India, pp. 1–71. World Wide Fund for Nature-India (2006)
3. Ito, V.C., Schnitzler, E., Demiate, I.M., Eusébio, M.E.S., Lacerda, L.G., Castro, R.A.E.: Physicochemical, thermal, crystallographic, and morphological properties of biodynamic black rice starch, and of residual fractions from aqueous extraction. Starch/Staerke **70**(11–12) (2018). https://doi.org/10.1002/star.201700348
4. Mohanty, S.V.N.A.K., Pattanayak, A.: Rice in North East India Status Paper on Rice in North East India, pp. 1–82 (2019)
5. Hu, C., Zawistowski, J., Ling, W., Kitts, D.D.: Black rice (Oryza sativa L. indica) pigmented fraction suppresses both reactive oxygen species and nitric oxide in chemical and biological model systems. J. Agric. Food Chem. 5271–5277 (2003). https://doi.org/10.1021/jf034466n
6. Kushwaha, U.K.S.: Book review: black rice: research, history, and development. Adv. Plants Agric. Res. **5**(1), 15406 (2016b). https://doi.org/10.15406/apar.2016.05.00165
7. Vanlalsanga, Singh, S.P., Singh, Y.T.: Rice of Northeast India harbor rich genetic diversity as measured by SSR markers and Zn/Fe content. BMC Genet. **20**(1), 1–13 (2019). https://doi.org/10.1186/s12863-019-0780-6
8. Choudhury, B., Khan, M.L., Dayanandan, S.: Genetic structure and diversity of indigenous rice (Oryza sativa) varieties in the Eastern Himalayan region of Northeast India. **2**(1), 1 (2013). https://doi.org/10.1186/2193-1801-2-228
9. UNDP and Export-Import Bank of India: Value chain studies (2010)
10. SFAC: Value Chain Analysis of Select Crops in North Eastern States (2012)
11. Sudhakar Reddy, C., Jha, C.S., Manaswini, G., Alekhya, V.V.L.P., Vazeed Pasha, S., Satish, K.V., Dadhwal, V.K., et al.: Nationwide assessment of forest burnt area in India using Resourcesat-2 AWiFS data. Curr. Sci. **112**(7), 1521–1532 (2017). https://doi.org/10.18520/cs/v112/i07/1521-1532
12. Ahuja, U., Ahuja, S.C., Chaudhary, N., Thakrar, R.: Red rices—past, present, and future. Asian Agri-Hist. **11**(4), 291–304 (2007)
13. Roy, S.C., Shil, P.: Black rice developed through interspecific hybridization (O. sativa x O. rufipogon): origin of black rice gene from Indian wild rice. BioRxiv 1–35 (2020). https://doi.org/10.1101/2020.12.25.423663
14. Hore, D.K.: Rice diversity collection, conservation and management in northeastern India. Genet. Resour. Crop Evol. **52**, 1129–1140 (2005)
15. Sarkar, A.N.: Resource potential and biodiversity of north eastern region. Indian J. Hill Farming **7**, 11–21 (1994)
16. Yadav, H., Shalendra, A.: Kiwi value chain analysis and market assessment for lower subansiri district Arunachal Pradesh. Report by CCS National Institute of Agricultural Marketing (A Govt of India organization under Ministry of Agriculture and Farmers welfare) (2017)
17. Asem, I.D., Imotomba, R., Mazumder, P.B.: Black scented rice (Chakhao) of Manipur. In: Advances in International Rice Research, pp. 125–136 (2017). https://doi.org/10.5772/67193
18. Roy, S., Verma, B.C., Banerjee, A., Kumar, J., Ray, U.S.: Genetic diversity for drought and low-phosphorus tolerance in rice (Oryza sativa L.) varieties and donors adapted to rainfed

drought-prone ecologies. Sci. Rep. **11**, 1–10 (2021). https://doi.org/10.1038/s41598-021-93325-2
19. Rajendran, M., Chandran, K.R.: Grain dimension, nutrition and nutraceutical properties of black and red varieties of rice in India. Curr. Res. Nutr. Food Sci. **8**(3), 903–923 (2020). https://doi.org/10.12944/CRNFSJ.8.3.20
20. Rathna Priya, T., Raeboline, A., Eliazer, L., Ravichandran, K., Antony, U.: Nutritional and functional properties of coloured rice varieties of South India: a review. **3**, 1–11 (2019)
21. Verma, D.K., Srivastav, P.P.: Bioactive compounds of rice (Oryza sativa L.): review on paradigm and its potential benefit in human health. Trends Food Sci. Technol. **97**(November 2019), 355–365 (2020). https://doi.org/10.1016/j.tifs.2020.01.007
22. Yu, S., Nehus, Z.T., Badger, T.M., Fang, N.: Quantification of vitamin E and gamma-oryzanol components in rice germ and bran. J. Agric. Food Chem. **55**, 7308–7313 (2007)
23. Saikia, S., Dutta, H., Saikia, D., Mahanta, C.L.: Quality characterization and estimation of phytochemicals content and antioxidant capacity of aromatic pigmented and non-pigmented rice varieties. Food Res. Int. **46**(1), 334–340 (2012). https://doi.org/10.1016/j.foodres.2011.12.021
24. Borah, N., Athokpam, F.D., Semwal, R.L., Garkoti, S.C.: Chakhao (black rice; Oryza sativa L.): a culturally important and stress tolerant traditional rice variety of Manipur. Indian J. Tradit. Knowl. **17**(4), 789–794 (2018)
25. Singh, K.I., Singh, M.P., Sharma, P.R., Singh, T.R.: Screening of certain promising local rice varieties for resistance to the Manipur gall midge (Orseoliaoryzae wood-Mason) biotype. Indian J. Entomol. **68**(3), 265–268 (2006)
26. Mahender, A., Swain, D.M., Gitishree, D., Subudhi, H.N., Rao, G.J.N.: Molecular analysis of native Manipur rice accessions for resistance against blast. Afr. J. Biotechnol. **11**, 1321–1329 (2012)
27. Anon: Package of Practices for Rabi Crops. Department of Agriculture, Govt. of Assam, Guwahati (1997)
28. Santhosh, K.: Black rice cultivation and forming practices: success story of Indian farmers. J. Rice Res. **9**(2) (2021). ISSN 2375-4338
29. Menaka, P., Lala, S.: Indian journal of hill farming performance of black rice with different nitrogen management practice under mid hill of Meghalaya. **31**(1), 115–119 (2018)
30. Kushwaha, U.K.S.: Black rice anthocyanin content increases with increase in altitude of its plantation. Adv. Plants Agric. Res. **5**(1), 469–472 (2016a). https://doi.org/10.15406/apar.2016.05.00170
31. Handique, A.K., Handique, G.K.: Nutritional parameters of unique black rice in Assam, India, and its prospect of boosting rural economy. Curr. Sci. **119**(10), 1613–1615 (2020)
32. Kumar, S., Devi, E.L., Sharma, S.K., Ansari, M.A., Phurailatpam, S., Ng, T.C., Kumar, A., et al.: Review article rice breeding strategies of North Eastern India for resilience to biotic and abiotic stresses: a review. **54**(1) (2017)
33. Anupam, A., Imam, J., Quatadah, S.M., Siddaiah, A., Prasad Das, S., Variar, M., Prasad Mandal, N.: Genetic diversity analysis of rice germplasm in Tripura state of Northeast India using drought and blast linked markers. Rice Sci. **24**(1), 10–20 (2017). https://doi.org/10.1016/j.rsci.2016.07.003
34. Saha, S., Vijayakumar, S., Saha, S., Mahapatra, A., Kumar, M., Sundaram, R.M.: Black rice cultivation in India—prospects and opportunities **6**(2), 044–048 (2022)
35. Kumar, N., Chhokar, R.S., Meena, R.P., Kharub, A.S., Gill, S.C., Tripathi, S.C., Gupta, O.P., Mangrauthia, S.K., Sundaram, R.M., Sawant, C.P., Gupta, A., Naorem, A., Kumar, M., Singh, G.P.: Challenges and opportunities in productivity and sustainability of rice cultivation system: a critical review in Indian perspective. Cereal Res. Commun. 1–29 (2021). https://doi.org/10.1007/s42976-021-00214-5

Chapter 11
Plant Bio-diversity Conservation in North-East India Through the Development of Mixed Non-leafy Vegetable Soups

Imdadul Hoque Mondal, Latha Rangan, and Ramagopal V. S. Uppaluri

11.1 Introduction

Horticultural produces constitute an important sector of the diverse plant systems. Their effective exploration for consumer products can enhance and better human thinking towards ecological sustainability and effective conservation of plant biodiversity. While such efforts may not be relevant for plant systems advocating non-consumer relevance, they are highly effective towards the stabilization of human thought process based on the effective utility of atleast few plant systems for human nutrition and metabolism. For the non-consumer characteristics of diverse plant systems, other approaches are inevitable such as their role in fostering plant and ecological sustainability and thereby promote conservation and human sustenance based plant-biodiversity. This is due to the obvious hypothesis that natural systems are

I. H. Mondal
Department of Food Technology and Nutrition, Lovely Professional University, Phagwara 144001, India
e-mail: imdadul.27343@lpu.co.in

I. H. Mondal · L. Rangan · R. V. S. Uppaluri (✉)
School of Agro and Rural Technology, Indian Institute of Technology Guwahati, Guwahati 781039, India
e-mail: ramgopalu@iitg.ac.in

L. Rangan
e-mail: lrangan@iitg.ac.in

L. Rangan
Department of Biosciences and Bioengineering, Indian Institute of Technology Guwahati, Guwahati 781039, India

R. V. S. Uppaluri
Department of Chemical Engineering, Indian Institute of Technology Guwahati, Guwahati 781039, India

R. V. S. Uppaluri and L. Rangan (eds.), *Conservation of Biodiversity in the North Eastern States of India*, https://doi.org/10.1007/978-981-99-0945-2_11

highly complex and hence the interaction between plant, animal and ecosystems is as well complex. Thus, this article aims to foster greater and renewed interest in the plant bio-diversity in the north-eastern region of India through the effective exploration of abundant vegetable produces that have a critical say on human nutrition. Thereby, grouping or integration of such produces into food consumer products such as soups, cookies, chips etc., can foster the philosophical inclination of the modern humankind towards valuing ecology and nature. Thereby, emphasis on bio-diversity conservation can be renewed in the contemporary and contextual drift in human centric science and technology endeavors. With such philosophical endeavor, the article next delineates on the science and technology of low cost mixed non-leafy vegetable based soup formulation based on the abundant vegetable produces of North-east India.

Complex demographic factors such as enhanced income, greater health awareness and extremely busy lifestyle prompt towards substantial variation in the human consumption habit. Rapid advances in science and technology further augment towards the mentioned factors and indicate an enhancing demand to develop and customize ready to cook/eat foods. These products have been developed from the perspective of nutrition, functionality and cost efficacy. Among various ready to consume food products, vegetable mix based value added soup products are often targeted due to their ease of preparation, longer storage and shelf life characteristics, higher nutritional characteristics etc. [1–3]. Thereby, they prompt renewed human thoughts and engagement for plant bio-diversity conservation.

Among abundant horticultural produces, non-leafy vegetables constitute a significant proportion. Only a limited set of vegetables such as tomato, carrot etc., have been considered till date for the preparation of a non-leafy vegetable based soup formulation. However, it is a fact that the non-leafy vegetables constitute a wide range of vegetables that are often underutilized. These include abundantly produced but underutilized vegetables such as Kaskal (*Musa splendida*), Komora (*Benesica hispida*), Kolphul (*Musa balbisiana* Colla blossom) and Posola (*Musa balbisiana* Colla pseudostem). Such vegetables are well known endemic produces of the NE region of India, an acclaimed bio-diversity hot spot region. The needful engagement of active research in such a domain of research that targets the customized soup product development using underutilized non-leafy vegetables is beneficial from consumption and economic competitiveness perspective.

A non-leafy vegetable mix based soup formulation is conventionally developed using appropriate constitutions of dried vegetable mix, spices and thickening agents. Drying inevitably influences the nutritional characteristics of the vegetable mix. Henceforth, the final vegetable soup mix product formulation must be addressed from the retention perspective of its nutritional characteristics. Simultaneously, the sensory characteristics need to be achieved as they also dictate the product acceptability of the consumer.

Till date, several optimum vegetable soup mix formulations have been reported. But, mostly, trial and error methodology based approaches have been followed to attain such formulations using few pre-defined formulations [4, 5]. The authors characterized the soup formulations based on nutritional and sensory parameters. On the other hand, very few research groups targeted optimization of non-leafy vegetable based soup mixes and with the response surface methodology (RSM). Through such approach, optimal soup premix with tomato was targeted based on varied tomato powder, whey powder and salt constitutions [6]. Thereby, based on the sensory characteristics, optimal soup mix formulation was identified by the authors. On the other hand, few other literature are also available for leafy vegetable soup formulations [3, 7]. The authors adopted trial and error and mathematical optimization techniques respectively to formulate leafy vegetable based soup products. The literature thus provide limited scope for bio-diversity conservation through the development of mixed vegetable soup formulations.

A detailed analysis of the available prior art indicates the non-consideration of a rational approach in the direction of non-leafy vegetable based soup mix development and characterization. Also, optimality of nutritional attributes related to dried non-leafy vegetables prior of the sensory and rheological evaluations has not been addressed. The associated rationale towards the pre-defined formulations and RSM based ingredient optimality is not very clear from a rational perspective. Also, among the reported vegetable soup formulations, very few devote towards the non-leafy mixed vegetable soup formulation. Hence, a non-linear programming (NLP) based matured mathematical optimization of nutritional components of the dried vegetables is very much needed to ensure novel paradigms and approaches towards the development of optimal vegetable soup mixes with nutritionally rich and desired sensory characteristics. Finally, the existing literature does not delineate upon any studies that targeted the chosen non-leafy (Kolphul, Posola, Kaskal and Komora) vegetables and thereby achieve optimal mixed vegetable soups.

Overcoming the above mentioned lacunae, the chapter targets an optimization (mathematical) approach being resolved through the solver of Microsoft (MS) Excel using RSM driven response model equations to develop a non-leafy vegetable mix formulation from widely available underutilized vegetables in the North-east India such as Komora, Kaskal, Posola and Kolphul. Thereby, nutritional parameters were optimized through the identification of the optimal non-leafy vegetable constitution followed with sensory and rheological evaluations.

11.2 Materials and Methods

11.2.1 Raw Materials and Sample Preparation

The non-leafy produces namely Posola, Komora, Kaskal and Kolphul were purchased from Shingimari area of Kamrup (Assam, India, with a latitude and longitude coordinate of 26.22° N and 91.62° E respectively). Thereby, the fresh vegetables obtained in polythene enclosed pouches were sorted and each vegetable was thoroughly washed to remove contaminants and was effectively drained. Other than Kolphul, all other vegetables were shaped and sliced to 1 mm thick samples and possessed an average diameter of 3.5 cm (Kaskal), 5 cm (Komora) and 8 cm (Posola). These sliced samples were placed as a single layer in a tray dryer and were subjected to heat based drying. The Kolphul samples for drying were prepared by first removing 2–3 outer layered bracts to pluck out the blossoms. Thereafter, the blossoms were cut as two halves (8.6 cm height, 0.1 cm length and 0.5 cm width) on a vertical basis and were subjected to single layer tray drying.

11.2.2 Drying Method and Selection of Model

A laboratory tray drying experimental set up supplied by International Commercial Traders, Kolkata, India was used to dry the samples. The system was customized with periodic lapse in the airflow. All samples were dried at an air rate of 4.5 m/s and equal time frame for stagnation and flow assurance (for each 20 s). The prior art provided detailed insights into the obtained experimental data and response characteristics (moisture, antioxidant activity and vitamin C characteristics) and also summarized upon the best fit quadratic model expressions to signify response variables in terms of drying temperature and time [8–10]. These model expressions have been used in the modelling effort reported in this work for the non-linear programming model being developed through the reported model expressions in the mentioned prior art.

11.2.3 Determination of Optimal Mix of Non-leafy Vegetables

The quadratic response variable model expressions have been included in the non-linear programming (NLP) model for the chosen models. Thereby, the probable constitution of the non-leafy vegetable soup mix formulations has been attempted. Since one can either dry vegetables on their own and later mix the dried vegetables or mix them prior to the drying and dry them together, both approaches have been followed. In this work, the former is termed as individual or stand alone drying and the later is termed as the co-drying. Thus, for both cases, two separate NLP optimization models have been formulated. These include an objective function and a set of constraints specified as equality and inequality constraints. The MS-Excel based solver was used to solve the NLP model and with generalized reduced gradient approach (GRG). Thereby, the optimal vegetable only constitution has been achieved for both cases. The inequality constraints for the NLP model have been based on the upper and lower constraints that were set as the maximum and minimum values of temperature and time. These were set based on both real world scenario and experimental research based inputs. A brief account of the mathematical model is as follows:

(a) Vegetable constitution as degrees of freedom: The target was to optimize the constitution of Komora (C_A), Kaskal (C_B), Kophul (C_C) and Posola (C_D). The following constraints are applicable for the constitution expressed in terms of percentage ratio of the mixed vegetables only recipe and on a dry basis:

$$10 \le C_{A-D} \le 80$$
$$C_A + C_B + C_C + C_D = 100$$

In the above, the first expression, refers to the minimum and maximum percentage ratio of each vegetable being set as 10 and 80 respectively. The second expression refers to the total constitution representing 100% of the percentage ratio.

(b) Drying time and temperature as degrees of freedom: Drying time (DTI, min) and drying temperature (DTE, °C) as key degrees of freedom have been specified for vegetables A–D and as DTI_{A-D} and DTE_{A-D}. Thereby, the following constraints are applicable:

$$50 \le DTE_{A-D} \le 70$$
$$390 \le DTI_A \le 720$$
$$240 \le DTI_B \le 450$$
$$240 \le DTI_C \le 420$$
$$360 \le DTI_D \le 630$$

In the above expressions, the first refers to specification of the maximum and minimum temperatures for drying based on experimental research conducted till date. The next four above expressions refer to minimum and maximum drying times set based on the mentioned prior art based experimental investigations.

(c) Objective function: The objective function (OBJ) involved a linear function based weighted representation of desired maximum (antioxidant activity and vitamin C) and minimum (moisture content) responses. It is conveniently expressed as

$$OBJ = WE_1CR_1 + WE_2CR_2 - WE_3CR_3$$

where CR_1, CR_2 and CR_3 refer to the weighted averages of dry vegetable mix system's antioxidant activity, vitamin C content and moisture content respectively. Further, the weights WE_{1-3} are positive weights that were used to manipulate the objective function. The negation of moisture content term in the objective function is due to the detrimental role of moisture content in affirming the shelf life of the dried vegetable mix system. Thereby, applying the theoretical ideal mixing rule [7], the CR_{1-3} can be evaluated using the expressions:

$$CR_1 = \frac{C_A AA_A + C_B AA_B + C_C AA_C + C_D AA_D}{C_A + C_B + C_C + C_D}$$

$$CR_2 = \frac{C_A VITC_A + C_B VITC_B + C_C VITC_C + C_D VITC_D}{C_A + C_B + C_C + C_D}$$

$$CR_3 = \frac{C_A MC_A + C_B MC_B + C_C MC_C + C_D MC_D}{C_A + C_B + C_C + C_D}$$

where AA_{A-D}, $VITC_{A-D}$, and MC_{A-D} refer to antioxidant activity, vitamin C and moisture content of vegetables A–D respectively. These can be evaluated using relevant literature reported quadratic models of drying temperature and time for the Kolphul, Kaskal, Komora and Posola [8–10].

(d) Supplementary constraints: The supplementary constraints for the NLP model have been set as follows:

$$0 \leq MC_{A-D} \leq 6$$
$$50 \leq AA_{A-D} \leq 90$$
$$100 \leq VITC_{A-D} \leq 300$$

In the above expressions, the first refers to maximum moisture content being set as 6 to ensure good shelf life of the dried vegetable mix. Similarly, the second expression has been defined to achieve a tighter representation of the antioxidant activity for each vegetable in the mathematical formulation and hence the dried vegetable mix. Similarly, the third expression accounts to the tighter specification of vitamin C content of each dried vegetable and henceforth for the entire mix.

The above set of expressions in sub-sections (a–d) of this section can only be applied for the individual or standalone case. For the case of co-drying, the NLP model was optimized by setting additional constraints and as follows:

$$50 \leq DTE_A = DTE_B = DTE_C = DTE_D \leq 70$$
$$240 \leq DTI_A = DTI_B = DTI_C = DTI_D \leq 720$$

In the above two expressions, the first refers to equal temperature specification for all individual vegetables due to the co-drying scenario. A similar case applies for the drying time expression presented as the second among the above two expressions.

Thus, expressions in sub-sections (a–d) of this section represent the NLP model for the individual drying case. Along with these expressions, the above mentioned two expressions account for the NLP model for the co-drying case.

11.2.4 Response Analysis

For the individual and optimized vegetable mix systems, the moisture content, vitamin C content and antioxidant activity, respectively, were determined using the AOAC 2010 method [11, 12], 2, 6-dichlorophenol indophenol (DCPIP) titrimetric method [8, 13] and 2, 2-Diphenyl-1-Picrylhydrazyl (DPPH) method [8, 14].

11.2.5 Sensory Analysis

The alternate vegetable mix based soup formulations have been prepared as follows. The optimized soup mix formulation indicated by the NLP based mathematical optimization (MS-Excel) has been realized as the F_{M1} formulation. With 5 g (100% weight in the vegetable only mix constitution) of each of the vegetables (Kolphul, A; Kaskal, B; Komora, C; Posola, D), $F_{A–D}$ formulations were achieved. Thereby, the sensory analyses based data of these formulations were used as an additional control set to infer upon the enhancement or decline of each of these vegetables after sensory optimization of F_{M1} formulation. Thereby, further refinement in F_{M1} was targeted. Thereafter, mixed dried vegetable to corn powder ratio was altered to achieve additional formulations based on F_{M1} recipe and these respectively refer to 20, 30, 40, and 50% of the mixed veg constitution in the veg and corn powder mix basis system. For reconstitution purpose, 1:20 of final soup mix to water was used. The soup was prepared through 10 min boiling for all mentioned formulations. Thus, individual formulations ($F_{A–D}$), mixed formulations (F_{M1}) and altered veg to corn powder formulations ($F_{5–8}$ based on F_{M1} formulation) were targeted for sensory studies.

Hedonic 9 point sensory scale system was used to evaluate the sensory characteristics in terms of appearance, texture, consistency, flavour, after-feel, mouth-feel, taste and overall acceptability. The scale system and methodology was as per the literature reported research methodology [1, 3, 15]. To conduct the sensory analysis, 5 male and 5 female experienced panelists in the age group of 25–50 years were approached. Further, these panelists were first provided few trails so that they can ensure confidence levels with respect to the reported sensory parameters due to adequate prior training in judging and providing sensory scores. With the likeliness bias, hedonic scale involving scores as 9 (like extremely), 8 (like very much), 7 (like moderately), 6 (like slightly), 5 (like nor dislike), 4 (dislike slightly), 3 (dislike moderately), 2 (dislike very much) and 1 (dislike extremely) have been used. Thereby, the qualitative representation of sensory scores has been mapped with the quantitative scores.

11.2.6 Rheological Behaviour of Optimal Soup Formulation

The best reconstituted soup identified from sensory evaluation was evaluated for rheological characteristics. To do so, a cone and plate geometry equipped rheometer (Physica MCR 301 supplied by Anton Paar) was used. The experimental investigations involved shear rate alterations upto 100 s^{-1} and for temperature upto 50 °C to experimentally measure steady state rheological parameters. Thereafter, the measured data of altered shear rate with shear stress were attempted for their fitness with Power and Herschel–Bulkley models to quantify the pertinent flow behaviour [7, 16]. These models have been presented as:

$$\tau = \tau_0 + K(\gamma)^n$$
$$\tau = K(\gamma)^n$$

where, τ denotes shear stress (Pa), K denotes flow consistency index (Pa·s^n), γ refers to shear rate (per second), τ_0 refers to yield stress (Pa) and n denotes flow behaviour index. The dynamic flow phenomena were evaluated for an upto value of 300 s duration and for 10 per second shear rate and at room temperature for both the desired responses (viscosity and shear stress).

Upto a maximum frequency of 10 Hz, the system's dynamic flow behaviour was assessed for 0.05% constant strain value and at room temperature. These studies involved the measurement of key parameters namely loss tangent (G′/G″), loss modulus (G″) and storage modulus (G′).

11.3 Results and Discussion

11.3.1 Response Variable Model Expressions

To represent the influence of process variables (time and temperature of tray drying) on the evaluated responses (vitamin C, antioxidant activity and moisture content), the literature reported best fit quadratic models have been referred. In other words, these models have been used in the modelling framework involving optimization studies in this work [8–10].

11.3.2 Optimization of Near Optimal Vegetable Soup Mix

Standalone drying: Both predicted and experimental data for various cases of standalone drying of the chosen vegetables were obtained (Table 11.1). Thereby, for Komora, Kaskal, Kolphul and Posola, the best proportions have been obtained as 45.91, 15, 29.09 and 10, respectively (1949.79 objective function value). For this system, the optimized responses have been 50%, 5.29% and 187.09 antioxidant activity, moisture content and vitamin C content, respectively. With a standard deviation value lower than 2.47, these are in corroboration with those obtained experimentally. Due to higher vitamin C constitution (187.09 mg/100 g), Komora has been in a higher quantity in the soup mix. Also, Posola with its poor nutritional properties has been conveyed to be the lowest in the soup mix by the mathematical tool. The literature reported soup mix also inferred in a similar way. In this literature, the authors conveyed that the higher antioxidant activity possessing vegetable has been in greater quantity in the optimized horticultural produce based mixed recipe [12].

Also, the Kolphul dried sample referred to the optimized values of 60.5 °C, 610.02 min, 5.08%, 89.41%, 87.97 mg/100 g for drying temperature, drying time,

Table 11.1 Predicted and experimental process parameter summary for mixed non-leafy vegetable soup recipes and for stand alone drying approach

		Drying technique	Optimal temp (°C)	Optimal time (Min)	Moisture content (%)	Vitamin C (mg/100 g)	Antioxidant activity (%)	Reference
Kolphul	Predicted	Tray drying	60.50	610.02	5.08	87.97	89.41	This work
	Actual	Do	60.50	610.02	6.13 ± 0.47	86.16 ± 2.04	88.29 ± 1.72	Do
	Prior art	Do	60	600	7.34	84.51	86.38	[12]
Kaskal	Predicted	Tray drying	57.90	386.32	6	91.1	72.81	This work
	Actual	Do	57.90	386.32	5.35 ± 0.39	88.76 ± 2.42	72.16 ± 0.97	Do
	Prior art	Do	60	600	7.34	84.51	86.38	[12]
Posola	Predicted	Tray drying	56.99	378.79	3.71	53.55	25.93	This work
	Actual	Do	56.99	378.79	4.55 ± 0.4	55.17 ± 1.84	23.38 ± 1.34	Do
	Prior art	Do	60	600	7.34	84.51	86.38	[12]
Komora	Predicted	Tray drying	62.07	532.35	5.53	310.34	22.82	This work
	Actual	Do	62.07	532.35	6.31 ± 0.38	313.07 ± 2.64	21.30 ± 0.63	Do
	Prior art	Do	60	600	7.34	84.51	86.38	[12]
M1 (29.09:15:45.91:10)	Actual	Do	62.07	532.35	6.31 ± 0.38	313.07 ± 2.64	21.30 ± 0.63	Do
	Actual	–	–	–	4.58 ± 0.36	190.52 ± 3.08	49.14 ± 1.74	Do
	Prior art	–	–	–	–	–	–	–
M2 (25.33:20:44.67:10)	Predicted	Tray drying	–	–	5.33	184.48	50	This work
	Actual		–	–	5.63 ± 0.52	183.25 ± 2.33	51.26 ± 0.86	Do
	Prior art			–	–	–	–	–
M3 (21.58:25:43.42:10)	Predicted	Tray drying	–	–	5.37	181.86	50	This work
	Actual	–	–	–	4.72 ± 0.32	184.38 ± 2.94	47.83 ± 2.14	Do
	Prior art	–	–	–	–	–	–	–

(continued)

Table 11.1 (continued)

		Drying technique	Optimal temp (°C)	Optimal time (Min)	Moisture content (%)	Vitamin C (mg/100 g)	Antioxidant activity (%)	Reference
M4 (17.83:30:42.17:10)	Predicted	Tray drying	–	–	5.41	179.24	50	This work
	Actual	–	–	–	5.08 ± 0.66	178.71 ± 3.22	48.26 ± 1.28	Do
	Prior art	–	–	–	–	–	–	–

moisture content, antioxidant activity and vitamin C respectively For Kaskal, Komora and Posola, the optimized values have been summarized similarly in the literature. With a standard deviation lower than 1.93, the measured and predicted values have been in corroboration with one another.

Co-drying: Compared to the standalone system, the co-drying performed poor and this is evident from the reported data in Table 11.2. The best set corresponds to 6.44%, 50% and 188.41 mg/100 g, 1749.32 for moisture content, antioxidant activity, vitamin C and objective function respectively. These findings have been in corroboration with the experimental data and the same has been assured with a standard deviation value lower than 2.02. On a 100% weight basis, the optimal quantities of Posola, Kaskal, Kolphul and Komora have been achieved as 10, 15.4, 34.88 and 40.12%, respectively. Thus, the optimal constitution had higher quantities of both Kolphul and Komora due to their very good combinations of vitamin C content and antioxidant activity. The achieved findings have been in corroboration with the literature reported findings for the vegetable mix recipe formulation [7, 17]. In addition, for all considered horticultural produces, 504.98 min and 61.41 °C have been the best temperature and time, respectively. Corresponding values for the optimal responses have been presented in the table. The experimental and predicted data for all cases confirmed a lower standard deviation of 2.78 and henceforth a good corroboration of the model values with those determined from experimental studies.

Further, it can be analysed that due to the drying time being not enough, the tray dried Komora and Kolphul samples possessed a little higher content of the moisture. Corresponding reduction in the dry matter translated into a comparably lower combination of the other two response variables. On the contrary, the long-term drying for Kaskal and Posola had a detrimental effect in terms of the significantly reduced values of the nutritional response variables. This is due to the negative role of prolonged heat on the vitamin C and antioxidant activity properties of the dried sample. However, the moisture content for both standalone and together drying cases has been similar for Kaskal and Posola cases. The equilibrium drying being achieved through the longer drying duration can be the most probable reason for such trend in these vegetables.

Near optimum dried vegetable mix: Based on standalone drying approach, the optimal dried vegetable constitution has been determined to be out of 100% as 15.45, 29.09, 10 and 45.91% for Kaskal, Kolphul, Posola and Komora, respectively. For this system, 187.09 mg/100 g, 50% and 5.29% have been the optimized values of responses namely vitamin C content, antioxidant activity and moisture content respectively. With this optimal content, the formulation F_{M1} was achieved. Also, F_{A-D} as control cases were also considered and subjected to sensory analyses. For all cases, other ingredients such as species and oil have been fixed as 3 g, 2 g, 0.5 g, 0.5 g, 0.5 g, 1 g, 0.5 g and 2 mL for salt, carrot flakes, onion powder, garlic powder, ginger powder, milk powder, black pepper powder and vegetable oil, respectively. For all considered formulations, the respective choices of other varied ingredients have been presented in Table 11.3. The fixed choice of spices and oil have been based on few trails and prior art data [2, 7, 18–20]. The literature reported findings confirmed

Table 11.2 Predicted and experimental process parameter summary for mixed non-leafy vegetable soup recipes and for co-drying approach

		Drying technique	Optimized temp (°C)	Optimized time (Min)	Moisture (%)	Vit C (mg/100 g)	Antioxidant activity (%)	Reference
Kolphul	Predicted	Tray drying with intermittent airflow	61.41	504.98	7.31	84.69	86.76	This work
	Actual	Do	61.41	504.98	8.24 ± 0.40	84.19 ± 2.13	89.03 ± 1.86	Do
	Prior art	Do	60	600	7.34	84.51	86.38	[12]
Kaskal	Predicted	Tray drying with intermittent airflow	61.41	504.98	4.09	75.50	60.08	This work
	Actual	Do	61.41	504.98	4.75 ± 0.52	77.82 ± 1.66	63.06 ± 1.38	Do
	Prior art	Do	60	600	7.34	84.51	86.38	[12]
Komora	Predicted	Tray drying with intermittent airflow	61.41	504.98	7.31	307.15	23.08	This work
	Actual	Do	61.41	504.98	6.74 ± 0.38	310.57 ± 2.70	21.56 ± 1.28	Do
	Prior art	Do	60	600	7.34	84.51	86.38	[12]
Posola	Predicted	Tray drying with intermittent airflow	61.41	504.98	3.77	36.11	14.65	This work
	Actual	Do	61.41	504.98	4.16 ± 0.29	39.18 ± 1.24	17.06 ± 0.83	Do
	Prior art	Do	60	600	7.34	84.51	86.38	[12]

(continued)

Table 11.2 (continued)

		Drying technique	Optimized temp (°C)	Optimized time (Min)	Moisture (%)	Vit C (mg/100 g)	Antioxidant activity (%)	Reference
Mix (37.28:10:47.72:5)	Predicted	–	–	–	6.44	188.41	50	This work
	Actual	–	–	–	7.82 ± 1.15	191.27 ± 2.36	51.48 ± 0.93	Do
	Prior art	–	–	–	–	–	–	–

Table 11.3 Composition of alternate mixed vegetable soups with non-leafy vegetables

Ingredients/formulation	Kolphul powder (g)	Kaskal powder (g)	Komora powder (g)	Posola powder (g)	Corn flour (g)
F_A	5	0	0	0	5
F_B	0	5	0	0	5
F_C	0	0	5	0	5
F_D	0	0	0	50	5
F_{M1}	1.45	0.75	2.30	0.50	5
F_{M2}	1.27	1.00	2.23	0.50	5
F_{M3}	1.08	1.25	2.17	0.50	5
F_{M4}	0.89	1.50	2.11	0.50	5
F_5	0.43	0.50	0.87	0.20	8
F_6	0.65	0.75	1.30	0.30	7
F_7	0.86	1.00	1.74	0.40	6
F_8	1.08	1.25	2.17	0.50	5

an equal ratio of corn flour and dried vegetable mix and was chosen accordingly in the recipes F_{A-D}, and F_{M1} [2, 7, 18–20]. Table 11.3 also mentioned additional formulations F_{M2}, F_{M3} and F_{M4} along with the above detailed formulations. Also, few cases of altered corn flour to dried mentioned horticultural produce mix ratio have been considered and referred to 20%, 30%, 40% and 50% choice of the corn flour in the respective formulations of F_5, F_6, F_7 and F_8. The linkage between additional formulations F_{M2-M4} and F_{5-8} will be further delineated in the following sub-sections of the article.

11.3.3 Organoleptic Properties of Various Soup Formulations

Optimality based on mix and control soup formulations: For the mentioned alternate formulations and control cases, Fig. 11.1 depicts a graphical representation of the obtained sensory scores. For the F_{M1}, the sensory scores of 7.8, 7.7, 7.6, 7.8, 7.6, 7.6, 7.8 and 7.9 have been obtained for texture, taste, flavor, appearance, mouth-feel, consistency, after-feel and overall acceptability respectively. Among control cases, F_B indicated the highest scores of 8.3, 7.9, 7.9, 8.2, 8, 8.3, 8.4 and 8.1 for texture, appearance, flavor, taste, mouth-feel, consistency, overall acceptability and after-feel respectively. The F_C affirmed marginally lower scores for the Komora based soup recipe. For F_B, higher scores could be due to the optimal constitution and their interaction and vice-versa for F_D (Posala based recipe).

These findings provide useful insights into further optimization of the soup mix formulations from the perspective of the sensory scores. The higher content of Kaskal

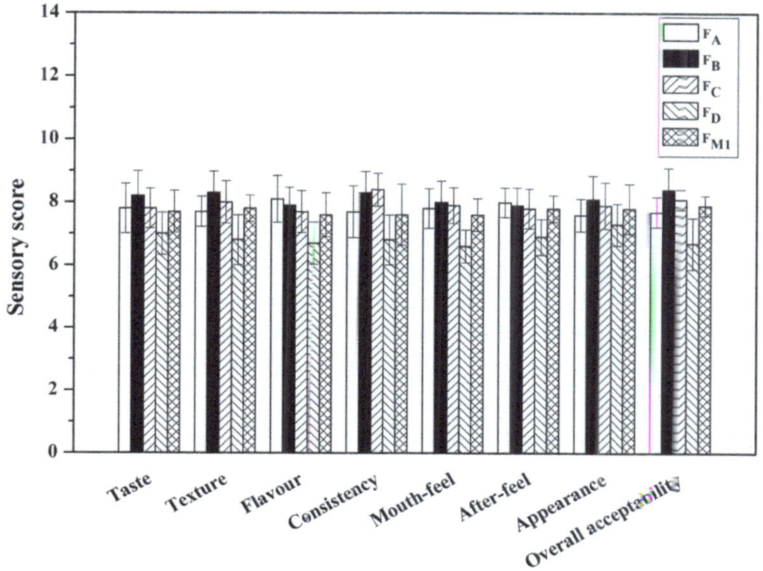

Fig. 11.1 Bar chart depicting the sensory properties of control and non-leafy vegetable based mixed soup recipes

in the F_{M1} is confirmed upon its higher overall acceptability. It can be further optimized through a simultaneous enhancement of Kaskal for a proportional reduction in Posala that achieved lowest sensory scores. However, since Posola constituted a very low content of 10% in the overall dry soup mix only recipe, its further reduction was not addressed. Also, the formulation F_A (Kolphul) had a comparatively lower sensory score. Therefore, Kaskal can be further explored in the best recipe mix that has been found so far based on the adopted research methodology.

Considering these very useful insights, it has been envisaged that Kaskal can be further enhanced and Kolphul can be further reduced in the best performing F_{M1} soup recipe. Accordingly, Kaskal constitution has been increased from the lower constitution of 15% in F_{M1} to 20% in F_{M2}, 25% in F_{M3}, 30% in F_{M4} in the dry soup mix formulation. Thereby, other dried vegetables constitution in the vegetable only mix recipe has been subjected to optimization using the non-linear programming approach (outlined in the materials and methods section) and with MS-Excel solver. In other words, for formulations $F_{M2\text{–}M4}$, the respective Kaskal constitution has been fixed and constitution of other vegetables (Komora and Kolphul) excluding Posola (10% fixed choice due to its lower content) has been targeted. Thereby, the optimal constitution of the F_{M2} has been achieved as 25.33, 20, 44.67 and 10% for Kolphul, Kaskal, Komora and Posola respectively. Similarly, the corresponding optimized formulation for the F_{M3} has been 21.58, 25, 42 and 10% and for the F_{M4}, the formulation has been 17.83, 30, 42.17 and 10%.

For these better optimized mixed vegetable recipes F_{M1-M4}, the sensory scores have been determined and depicted in Fig. 11.2. Among all cases, F_{M3} formulation achieved the best sensory score and indicated an optimal choice of 25% Kaskal in the mix. The non-linear programming based output confirmed 1897.14 for the objective function value and also affirmed 5.37%, 181.86 mg/100 g and 50% for the moisture content, vitamin C content and antioxidant activity, respectively. The corresponding sensory scores have been 8.3, 7.9, 8.2, 7.9, 8.1, 8.2, 8.3, 8.4 for texture, mouth-feel, taste, flavor, after-feel, appearance, consistency and overall acceptability, respectively. All these have been comparatively higher than the previously obtained sensory scores for the F_{M1} formulation. Thus, F_{M3} provided even better integration of constituent ingredients and their compatibility and affirmed optimal constitution of Kaskal and other ingredients in the vegetable only mix as well as the optimality of all other ingredients.

Optimality for varied mixed non-leafy vegetable to corn powder ratio: Based on the best achieved soup mix formulation F_{M3}, further alterations were targeted to evaluate the organoleptic characteristics of the mixed veg mix to corn flour ratio in the formulations. Accordingly, in the optimized soup mix formulation F_{M3}, dry veg mix to corn ratio was altered to achieve F_5 (20:80 veg mix to corn flour ratio), F_6 (30:70 veg mix to corn flour ratio), F_7 (40:60 veg mix to corn flour ratio) and F_8 (50:50 veg mix to corn flour ratio). Thereby, sensory analyses were conducted for the mentioned

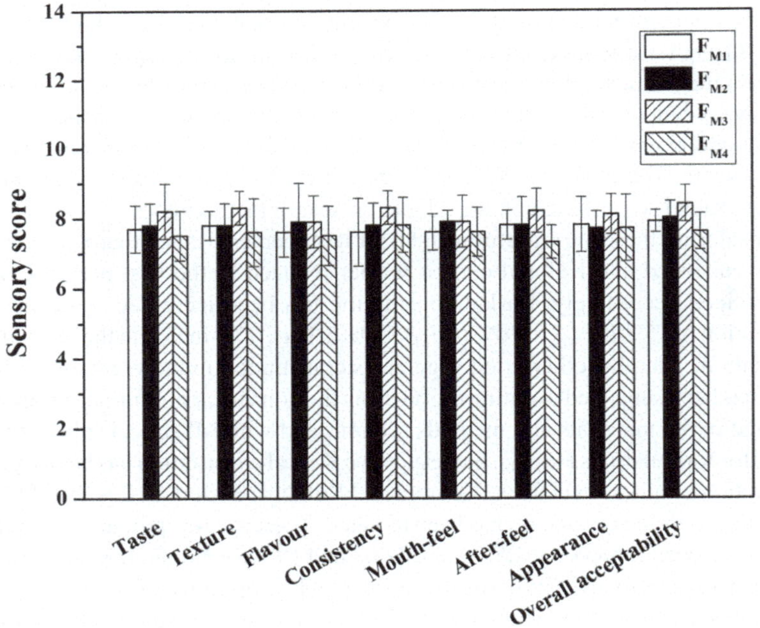

Fig. 11.2 Bar chart depicting the sensory properties of non-leafy vegetable based mixed soups for varied Kaskal constitution

soup mix recipes and have been depicted in Fig. 11.3. The figure confirmed that the corn flour content in the soup mix of the recipes highly influenced and altered the sensory characteristics. The best among all cases referred to the F_7 formulation (60:40 corn flour to dried mix veg ratio). For the case, the optimized sensory values referred to 8.3, 8.3, 8.4, 8.2, 8, 8.4, 8.3 and 8.6 scores out of 9 for texture, mouth-feel, taste, flavour, appearance, consistency, after-feel and overall acceptability respectively. Thereby, improved and better compatibility of the ingredients has been best achieved for the best recipe namely F_{M3}. Thus, further optimized compatibility of the chosen ingredients exists in the evaluated proportions for the achievement of the lowest sensory scores. Since only corn flour content has been altered, it is apparent that compared to the F_{M3} formulation, a higher corn flour content enabled the non-desired scores of the sensory parameters. Thus, beyond 60% of the corn flour in the veg to corn ratio percentage proved to be detrimental and the optimal choice for the dried veg mix to corn powder ratio is 40:60 (F_7 formulation). This could be due to the thicker constitution of the soup with increased ratio of corn powder to the veg mix. Also, such higher constitution of the corn powder in the soup mix is detrimental from the perspective of the nutritional properties of the developed formulations with the fact that higher corn powder content translates into lower dried veg mix content and hence forth reduced nutritional parameters such as antioxidant activity and vitamin C content.

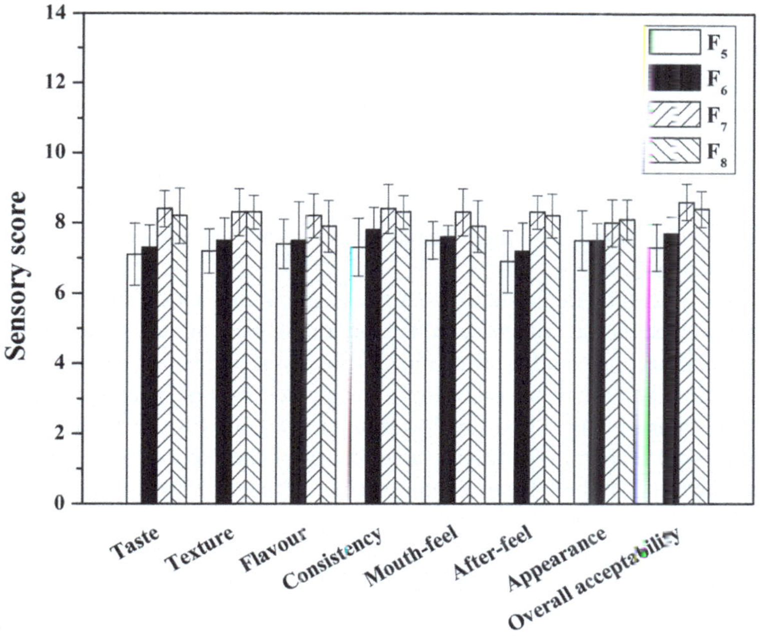

Fig. 11.3 Bar chart depicting the sensory properties of mixed non-leafy vegetable soup recipes with varied corn to dry mixed soup ratio

11.3.4 Flow Behaviour of Optimal Soup Sample

Steady state flow behaviour: The flow behavioural trends of the F_7 soup sample have been targeted in terms of the steady state flow phenomena. Thereby, shear stress alteration with shear rate and viscosity alteration with shear rate have been measured and have been respectively depicted in Fig. 11.4a, b. The shear stress figure corresponds to two distinct phase trends in which the first existed upto a shear rate of 10.8 s^{-1} and with shear stress rapid enhancement as 6.88–14.1, 4.97–11.5, 1.64–6.1 Pa at the lower temperature of 30 °C, moderate temperature of 40 °C and higher temperature of 50 °C, respectively. Thereafter, upto 100 s^{-1}, the corresponding shear stress altered gradually and respectively as 14.1–18.2, 11.5–14.7 and 6.1–9.48 Pa. The contrary viscosity alterations affirmed exponentially declining trends for the entire range of shear rate alteration and respectively as 74.5–0.18, 53.8–0.15 and 17.8–0.095 Pa·s. Relevant hypothesis for the phase wise alteration in the shear stress is as follows. The first phase of the shear stress for its characteristic rapid enhancement is very likely due to the sudden displacing of the large number of particles that constitute the aqueous soup mix system. Thus, there has been a significant reduction in the gradient of the velocities between any two adjoining layers of the fluid that prompted an abrupt reduction in the viscosity. The later effect of non-profound alterations in the shear rate dependent shear stress has been due to the stabilized and equilibrated displacing of the particles that constitute the aqueous soup system. Similar trend has also been apparent for the viscosity of the system. Thus, during mature application of the shear rate, it is apparent that the shear rate did not customize upon the restoration of the particles to their original state in the fluid. Further, beyond a particular context, the restored particles in their partial state could not be as well displaced. Also, the placement of the shear stress and viscosity trends with temperature accounts for the larger influence of the temperature on these variables. Thereby, weaker drag forces between adjoining layers of the fluid have been reasoned to foster the larger displacement of the particles that constitute the aqueous soup system.

The experimentally determined shear stress alterations with respect to shear rate have been studied for their fitness with standard rheological models such as Herschel–Bulkley and Power law models. Both models possess an index of the flow behavior to be lower than 1 and are hence relevant for the studied system. Thereby, fitness plots have been obtained (Fig. 11.5a for Power law model and Fig. 11.5b for Herschel–Bulkley model). Table 11.4 presents a summary of the model parameters of both models. The trends in both figures and associated fitness parameters do indicate the definite behavior in terms of the shear thinning flow of the soup mix sample after reconstitution. Similar inferences have been corroborated for reported soup products in the prior art [4, 5, 16, 21]. The model parameters reduced from 9.55 to 2.89 for Hershel-Bulkley model and from 9.66 to 3.02 for the Power model and with increasing temperature. These values are in corroboration with those reported for mango jam [22], tomato soup [22] and vegetable soups [7, 16].

Fig. 11.4 Variation of **a** shear stress and **b** viscosity with shear rate for the optimal mixed non-leafy vegetable soup recipe

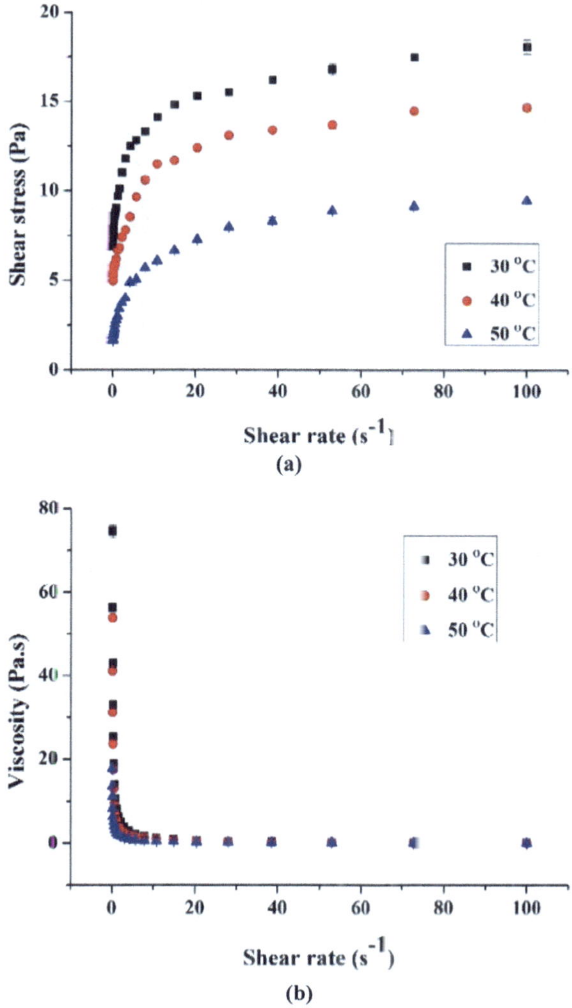

Time dependent rheological behaviour: For a constant shear rate, the shear stress and viscosity of the F_7 aqueous soup system has been studied and has been depicted in Fig. 11.6. The figure clearly conveys that apart from marginal alterations, both variables did not involve much variation and hence fairly constant trends (12.4–15.1 Pa for the shear stress and 1.25–1.50 Pa·s for the viscosity and upto a time of time 300 s). Thus, these time dependent evaluations confirmed a highly marginal role of time in influencing the mentioned responses. Thus, for the applied constant shear rate, the fluid reached back to its state prior to the experiment for the entire range of the time duration. Also, partial restoring affirms that for all time data sets, the restoration has not been possible and hence pertinent deviations from the exact

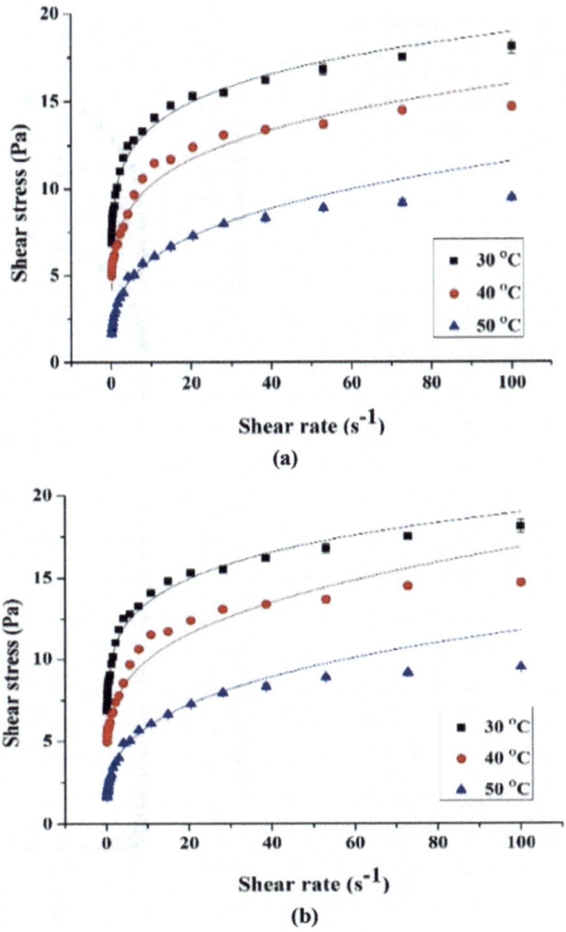

Fig. 11.5 Alternate rheological model fitness plots predicting shear stress values for the optimal mixed non-leafy vegetable soup recipe. **a** Power law **b** Herschel–Bulkley model

Table 11.4 A summary of rheological model fitness data for the optimal mixed non-leafy vegetable soup recipe

Rheological model	Model parameters	50 °C	40 °C	30 °C
Power model	n	0.29148	0.19389	0.14598
	K	3.0239	6.54921	9.6626
	R^2	0.99273	0.98306	0.99432
Herschel–Bulkley model	τ_o	0.12466	3.14283	0.1079
	n	0.30177	0.30373	0.1475
	K	2.89149	3.38448	9.55094
	R^2	0.99247	0.98946	0.99403

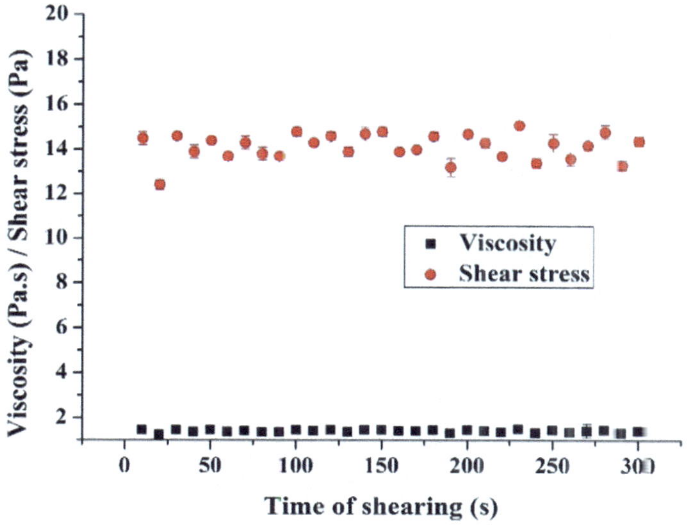

Fig. 11.6 Variation of viscosity and shear stress with shearing time for the optimal mixed non-leafy vegetable soup recipe

constant trend for both cases. These marginal fluctuations in both shear stress and viscosity with time are in corroboration with that reported by Chavan for tomato soup [16] and vegetable soups [7].

Dynamic flow behaviour: The alterations in three important variables from flow dynamics perspective (loss tangent expressed as (G'/G''), loss modulus expressed as G'' and storage modulus expressed as G') with frequency were considered. Figure 11.7 depicts the trends. The figure clearly illustrates that the loss moduli and storage moduli increased slowly from 6.86 to 2.17×10^2 and 5.9 to 13.4×10^2 respectively for the entire range of the studied frequency (0.01 to 10 Hz). Thus, the pertinent trends affirmed dominant storage moduli but not the loss moduli and do confirm upon the elastic nature of the formulated soup product. Thereby, it is confirmed that the system reached back to its state prior to the dynamic flow study after the application of each external force stroke. It is further hypothesized that the ingredients of the soup system fostered complex interactions and thereby assured upon a weak gel type structural formation that critically induced the observed elastic behaviour. Further, for a similar alteration of frequency, the loss tangent reduced from a value of 8.60 to a marginally reduced value of 6.14. These evaluated dynamic behavioural phenomena of the best soup formulation upon reconstitution have been in corroboration with the best findings reported for pollock [23], leafy vegetable [7] and tomato [16] soup systems.

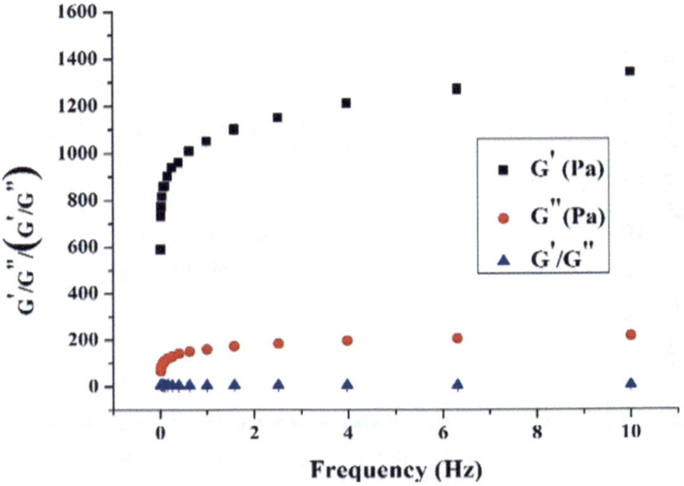

Fig. 11.7 Effect of frequency on loss factor, storage and loss modulus for the optimal mixed non-leafy vegetable soup recipe

11.4 Conclusions

The article provided many important conclusions in the specific research theme of non-leafy vegetable based soup mix products from North-east India's horticultural produces. These include the outcomes of the steps or hierarchy of nutritional optimization based on mathematical programming assessment followed by insights into further optimization of the soup recipe and corn to veg mix ratio optimization and rheological analysis of the best recipe. Thereby, detailed procedures summarized in the article provide ample scope for product based strategies for bio-conservation of plant species based on nutritional needs. This article also delineated on the general research pedagogy of mixed vegetable based soup development.

The listed conclusions of the article are as follows. The first is the successful integration of numerical optimization to sensory approaches that are often followed in the non-leafy vegetable based mixed soup formulations. In this, it was analysed that the separate drying has been better than the drying of mixed vegetables. For the separate drying case, best responses of 50%, 181.86 mg/100 g and 5.37% for antioxidant activity, vitamin C content and moisture content respectively have been obtained. The optimal vegetable mix ratio has been identified as 10:43.42:25:21.58 for Posola, Komora, Kaskal and Kolphul, respectively. Accordingly, the best soup mix recipe was identified as F_7 which obtained the highest overall acceptability score of 8.6 for the reconstitution of 20 g dry soup mix with water of 400 mL. Thirdly, this soup upon rheological analysis confirmed shear thinning and pseudoplastic behaviour. In summary, these findings strengthen the possibility for greater utility of underutilized non-leafy produces of North-east India in soup type commercial product development. Doing so, biodiversity can be conserved for wild and endemic species through human palatable applications. All these translate into newer hopes and horizons for the booming food processing industry in the North-eastern states of India. Such insights will be useful to serve as generic guidelines for the plant bio-diversity conservation based on the principles of value added consumer product development.

Acknowledgements Imdadul Hoque Mondal is thankful to School of Agro and Rural Technology, Indian Institute of Technology Guwahati for all the necessary infrastructural support. Further, he also sincerely thanks MHRD, Govt of India for the fellowship.

References

1. Anita, S., Sameshwari, T., Kalpana, B.: Formulation and evaluation of instant soup powder using millets. Int. J. Curr. Res. **8**, 35668–35673 (2016)
2. Chaudhari, S., Solanke, N.: Formulation and sensory quality of legume based soup powder. Life Sci. Leafl. **73**, 19–27 (2016)
3. Kokani, R.C., Mokashi, P.S., Shelar, Y.P.: Studies on development and standardization of moringa leaves instant soup mix powder incorporated with garden cress seeds. Int. J. Res. Rev. **6**(10), 242–246 (2019)
4. Abdel-Haleem, A.M., Omran, A.A.: Preparation of dried vegetarian soup supplemented with some legumes. Food Nutr. Sci. **5**(22), 2274 (2014)
5. Jayasinghe, P., Pahalawattaarachchi, V., Ranaweera, K.: Formulation of nutritionally superior and low cost seaweed based soup mix powder. J. Food Process. Technol. **7**(4) (2016)
6. Chavan, R., Kumar, A., Bhatt, S.: Tomato soup premix base: development and optimisation using response surface methodology. Ge-Int. J. Eng. Res. **3**(1), 35–48 (2015)
7. Mondal, I.H., Rangan, L., Uppaluri, R.V.: A robust and novel methodology for the optimal targeting of leafy vegetable mix soup formulations. LWT-Food Sci. Technol. **134**, 110152 (2020)
8. Mondal, I.H., Rangan, L., Uppaluri, R.V.: Parametric optimality of tray dried *Musa balbisiana* Colla blossom. J. Food Sci. Technol. **57**, 4599–4612 (2020)
9. Mondal, I.H., Rangan, L., Uppaluri, R.V.: Tray drying characteristics of Musa splendida and Musa balbisiana Colla psuedo-stem. J. Therm. Anal. Calorim. 1–14 (2022)
10. Mondal, I.H., Rangan, L., Uppaluri, R.V.: Process-product characteristics of tray-dried *Benincasa hispida*. J. Food Process. Preserv. **44**(9), e14697 (2020)
11. AOAC: Officials Methods of Analysis, 17th edn. Association of Official Analytical Chemists, Washington, D. C., U.S.A. (2010)
12. Mondal, I.H., Rangan, L., Uppaluri, R.V.: Effect of oven and intermittent airflow assisted tray drying methods on nutritional parameters of few leafy and non-leafy vegetables of North-East India. Heliyon **5**(11), e02934 (2019)
13. Anjali, K.M., Singh, N., Pal, K.: Effect of sulphur dioxide on plant biochemicals. Int. J. Pharma Prof.'s Res. **3**(2), 538–544 (2012)
14. Barimah, J., Yanney, P., Laryea, D., Quarcoo, C.: Effect of drying methods on phytochemicals, antioxidant activity and total phenolic content of dandelion leaves. Am. J. Food Nutr. **5**(4), 136–141 (2017)
15. Singh, J., Sindhu, S.C., Kumari, V.: Development and evaluation of Shiitake mushroom (*Lentinus edodus*) instant soup mixes. Int. J. Curr. Microbiol. Appl. Sci. **6**(5), 1232–1238 (2017)
16. Chavan, R., Kumar, A., Basu, S., Nema, P.K., Nalawade, T.: Whey based tomato soup powder: rheological and color properties. Int. J. Agric. Res. **5**(4), 301–313 (2015)

17. Larrosa, A.P., Cadaval, T.R., Jr., Pinto, L.A.: Influence of drying methods on the characteristics of a vegetable paste formulated by linear programming maximizing antioxidant activity. LWT-Food Sci. Technol. **60**(1), 178–185 (2015)
18. Dhiman, A.K., Vidiya, N., Surekha, A., Preethi, R.: Studies on development and storage stability of dehydrated pumpkin based instant soup mix. J. Appl. Nat. Sci. **9**(3), 1815–1820 (2017)
19. Srivastava, A., Attri, B., Verma, S.: Development and evaluation of instant soup premix using oyster mushroom powder. Mushroom Res. **28** (2019)
20. Sugumar, J., Guha, P.: Study on the formulation and optimization of functional soup mix of *Solanum nigrum* leaves. Int. J. Gastron. Food Sci. 100208 (2020)
21. Yilmaz, M.T., Sert, D., Demir, M.K.: Rheological properties of tarhana soup enriched with whey concentrate as a function of concentration and temperature. J. Texture Stud. **41**(6), 863–879 (2010)
22. Basu, S., Shivhare, U., Singh, T.: Effect of substitution of stevioside and sucralose on rheological, spectral, color and microstructural characteristics of mango jam. J. Food Eng. **114**(4), 465–476 (2013)
23. Kim, S.-G., Yoo, W., Yoo, B.: Effect of thickener type on the rheological properties of hot thickened soups suitable for elderly people with swallowing difficulty. Prev. Nutr. Food Sci. **19**(4), 358 (2014)

Chapter 12
Phytocompounds Analysis and Antimicrobial Potential of *Callistemon Viminalis* Essential Oil from North-East India

Riya Bhattacharya, Debajyoti Bose, Surjendu Maity, S. R. V. Siva Prasanna, and Srinu Nagireddi

12.1 Introduction

Recent advances in science have sparked renewed interest in the therapeutic potential of plants due to their low toxicity, pharmacological activity, and commercial feasibility [1]. In this kind of research, the positive effects of phytochemicals taken from plants on human health have been the primary focus. Plant-based additives might come in the form of individual chemicals, chemotypes, or essential oils [11]. More and more research on the benefits of using bioactive components in the food sector, either alone or in combination with other compounds, has been conducted in recent years [12]. Medicinal extracts and essential oils added directly to foods have been shown to have antioxidant or antimicrobial action [27]. Essential oils are a type of aromatic oily liquids derived from plants. They are sometimes referred to as volatile oils. Steam or hydro-distillation, a procedure that was pioneered in the middle ages by the Arabs, is frequently used to obtain essential oils that are sold on the market. The density of essential oils is frequently lower than that of water, and they are characterized by being liquid, volatile, lipid-soluble, rarely coloured, and dissolved in organic solvents [31]. Essential oils are incredibly complex natural mixtures that can contain somewhere between 20 and 60 various components, depending on the

R. Bhattacharya · D. Bose (✉)
School of Biotechnology, Faculty of Applied Sciences & Biotechnology, Shoolini University of Biotechnology & Management Sciences, Solan, Himachal Pradesh, India
e-mail: debajyoti1024@gmail.com

S. Maity
Terasaki Institute for Biomedical Innovation, Los Angeles, CA, USA

S. R. V. S. Prasanna
Department of Mechanical Engineering, Chandigarh University, Punjab, India

S. Nagireddi
Department of Chemical Engineering, Chandigarh University, Punjab, India

R. V. S. Uppaluri and L. Rangan (eds.), *Conservation of Biodiversity in the North Eastern States of India*, https://doi.org/10.1007/978-981-99-0945-2_12

oil [38]. Even though there are typically only two or three basic components present in essential oils, these tend to have significantly higher percentages (between 20 and 70%) than the other constituents. Essential oils get their name from the fact that they are extracted from plants. Due to the fact that essential oils are chemically formed from their oxygenated derivatives and terpenes, also known as terpenoids, which are phenolic compounds and aliphatic and aromatic acid esters, the amounts of their various components can vary greatly depending on the part of the plant and the species [39]. Hydrophobicity is an important attribute of essential oils and the constituents that make up essential oils because it enables these substances to break apart with the lipids that are present in the membranes of cells and mitochondria, which in turn makes these membranes more permeable by disrupting the structures of the cells [42].

Plant secondary metabolism played a crucial role in the evolution process that allowed plant species to successfully colonize terrestrial habitats. Furthermore, secondary metabolites provide an ecological link between plants and their environment. Pigments and aromatic compounds provide reproductive organs and fruit and vegetables their colour and aroma, which in turn attracts pollinators and encourages seed distribution by animals. Phytoalexins are broad-spectrum antibacterial metabolites, but volatile molecules can also deter phytophagous species like viruses and phytoplasma vectors [7]. Essential oils have a valuable role in nature by protecting plants from herbivores by lowering their appetite for plants with antibacterial, antifungal, insecticidal and antiviral activities. Essential oils have been deemed harmless by the Department of Health and Human Services Public Health Services, and some of them even have components that can be utilized as antibacterial additions. Various studies have reported that essential oils (EOs) are effective against viruses and food pollutants, which suggests that EOs could have uses in the food business. It has been observed that essential oils have potent antimicrobial, antioxidant, antiviral, antifungal, antiparasitic, and insecticidal properties [30]. Thus, essential oils can be an effective method to fight microbes.

Microorganisms, such as bacteria, viruses, fungi, and parasites, develop antimicrobial resistance (AMR) when they evolve to survive and even thrive in the presence of drugs that were once effective against them [15]. Each year, at least 23,000 individuals in the United States die from antibiotic-resistant diseases, and the CDC estimates that over two million people fall ill with such infections [34]. Also, antibiotic resistance will have a disproportionately negative effect on those who already suffer from chronic diseases like diabetes, asthma, and rheumatoid arthritis. Important classes of medicine, such as carbapenems and polymyxins, are not always easily accessible in developing countries, are expensive and have a wide variety of side effects; yet they are necessary because the efficiency of antibiotics will decrease due to persistent trends of AMR. Long-term efforts to combat infectious diseases including tuberculosis, HIV/AIDS, and malaria have been undermined by the rise of resistant strains of these microorganisms. Increases in drug-resistant HIV cases have been observed worldwide, although this trend is most pronounced in Sub-Saharan Africa, where 60% of HIV patients have become immune to HIV treatment [13]. A higher rate of death from HIV is expected among individuals who have developed resistance to HIV medications. Essential oils are gaining a lot of attention these days because they contain unique phytocompounds that can act as new drugs or help improve the potency of existing antibiotics.

The genus Callistemon also referred to as bottlebrush is a member of the Myrtaceae family and includes around 30 species. They are aromatic woody plants or shrubs (about 0.5–7 m in height) that were once widespread in the tropical region, particularly tropical Asia, South America, and Australia, but are currently found throughout the world [28]. *Callistemon* species feature slender, beautiful leaves and white, matte texture bark. *Callistemon* species are utilized for many other things such as essential oil production, forestry, degraded-land rehabilitation, and ornamental gardens [3]. Due to their attractive flowers, *Callistemon* species are cultivated in gardens, roadside areas and as ornamental plants in the North-Eastern states of India. In China, *Callistemon* species, particularly *C. viminalis* are used to cure hemorrhoids with traditional Chinese medicine supplements [12]. Widely recognized as weeping bottlebrush, *C. viminalis* is a short tree or shrub with distended leaves, but some types seem to be more pendulous than others. It attains a size of approximately 4 m in its natural environment, but in horticulture, especially in temperate zones, it is significantly smaller [38]. Furthermore, *C. viminalis* essential oils have been shown to exhibit antibacterial [27], antifungal [35], anti-quorum sensing [40] and anthelmintic properties in the limited research that has been conducted on them [2].

This chapter describes the extraction of *C. viminalis* essential oil (CVEO) by the Clevenger apparatus. Phytochemical analysis was performed using GC-MS. The GC-MS report of EO showed the presence of Eucalyptol, 2-Phenyl-4-Methylthiazolidine, and Amrinone. Compounds of this nature contribute to its therapeutic properties in terms of antimicrobial activity. This is investigated as the antibacterial and antifungal activity of CVEO by means of agar well diffusion and minimum inhibitory concentration (MIC). The current chapter claims for the first time the chemical components that contribute to antibacterial and antifungal activities of *C. viminalis* EO exist in the species found in the Gangtok district of Sikkim which is one of the North-Eastern states of India. A comprehensive understanding of the various biological and non-biological elements that influence the rate and degree to which antibiotic resistance develops is required for any method that can be considered rationally applicable to the problem of antibiotic resistance. Developing such methods through such analysis indicates a possible frontier for future research. This new generation of phytopharmaceuticals can provide useful insight into the design of new pharmacological regimes for the treatment of antibiotic resistance.

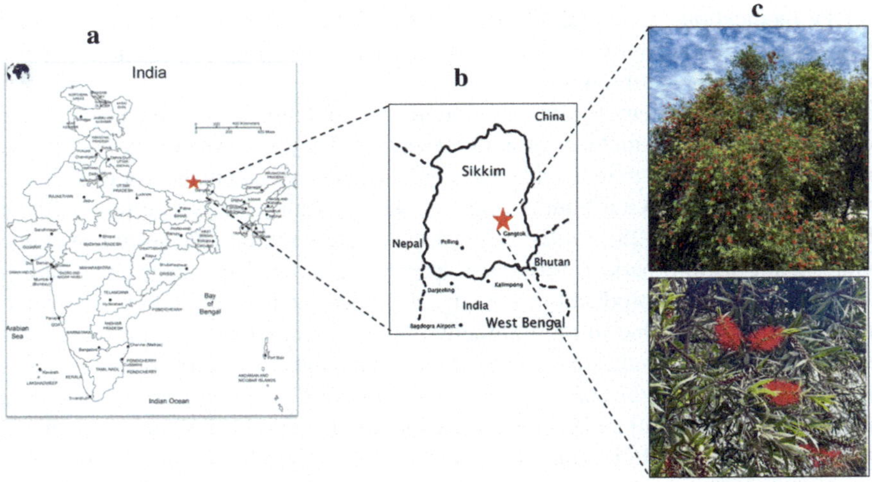

Fig. 12.1 **a** Map of India depicting Sikkim state with a star. **b** Map of Sikkim depicting the Gangtok district with a star. **c** *Callistemon viminalis* plant source being used to collect the leaves

12.2 Materials and Methods

12.2.1 Collection of Plant Material and Identification

Leaves were collected from the plant *C. viminalis* and the location for collecting the fresh leaves was Gangtok (27.3314°N, 88.6138°E), one of the districts of Sikkim, India is shown in Fig. 12.1. The place is located in an elevation of 5,600 feet (1,700 m) above sea level and is ideal for the growth of such plants. Leaves were collected in the month of February. The name of the plant has been confirmed from the online database of The Plant List. The Plant List aims to compile all of the information currently available about the world's plant species, namely those belonging to the vascular plant kingdom. This list is the product of a joint effort made by the Royal Botanic Gardens in Kew, United Kingdom, and the Missouri Botanical Garden in St. Louis, Missouri. This was accomplished by integrating multiple checklist data sets that were kept by each of these institutions as well as additional contributors [19].

12.2.2 Extraction of C. Viminalis Essential Oil

The Clevenger apparatus (JSGW, JSGW/476) was used to extract the essential oil (EO) from the *C. viminalis* leaves. In order to extract the oil, the Clevenger Apparatus goes through the stages of boiling, condensing, and decantation during the distillation process. The extraction of essential oils using this process is also the recognized standard method for quality control purposes. When this approach is taken, there is a smaller amount of waste in terms of the organic material from which the oil is extracted [43]. The fresh and young leaves of the plant were washed to eliminate the contaminants on their surfaces. With the help of a paper towel, the excess moisture was absorbed. In a round bottom flask (JSGW, JSGW/1196/6), 200 g of fresh leaves were added with distilled water (1000 ml) and then heated at 40 °C in the Clevenger apparatus. Essential oils (oils lighter than water) were extracted for 3–5 h before being collected and separated from the hydrosol with the help of anhydrous sodium sulphate (HiMedia, GRM419). The EO was put into storage at a temperature of 4 °C and kept in the dark.

Essential oil Yield percentage is calculated from Eq. 12.1 as given below.

$$\text{Yield}\ (\%) = \frac{\text{Volume of Essential oil (ml)}}{\text{Fresh weight of leaves (g)}} \times 100 \qquad (12.1)$$

This yield is critical in terms of commercializing such products as the market for essential oils has been continuously growing, and the number of people who are looking for these oils has been higher. After the distillation process has been finished, the essential oils of the plant are extracted from the water or steam that was used during the distillation process. This results in the production of a genuine essential oil fraction, which can be employed for a range of purposes, including aromatherapy, synergistic combination with standard drugs and the general improvement of health [23].

12.2.3 Characterization of Phytocompounds of C. Viminalis Essential Oil by GC-MS

The GC-MS analysis provides an understanding of the composition of the essential oil, which in turn enables comprehension of important components and implications associated with the usage of the oil [21]. This comprises the substance's purity, quality, and safety as well as its potential applications in clinical environments. To establish the authenticity and integrity of essential oil, a GC-MS analysis is necessary. In this case, the sample was prepared by dissolving pure essential oil in an n-hexane inappropriate amount (1% v/v). GC-MS (Thermo Fisher Scientific, USA) was evaluated at the analytical instrumentation facility, Shoolini University. The initial temperature for GC-MS was set at 80 °C for 1 min and then ramped at a rate of 10 °C/min to 240 °C (12 min hold). The temperature of the injector and temperature of the transfer line was set at 250 °C, respectively and the ion source temperature was kept at 230 °C. The split flow rate was maintained at 50 ml/min with a split ratio of 71.4. The carrier gas in the study was helium with a constant flow rate of 0.7 ml/min. The volume of the sample injected was taken at 20 μL. TG-5MS column (Thermo Scientific, USA) with the dimension of 40 m × 0.15 mm × 0.15 μm was used to separate the components. The mass spectrum was obtained at 1 scan/s and 70 electron volt energy, mass range 45–450 aμ, for MS analysis. Comparing mass spectra with those in the NIST library, version 2.2, 2014, the components were identified.

12.2.4 Microbial Strains and Culture Media

Candida albicans (MTCC277) and *Candida albicans* (ATCC90028) are two fungal strains. *C. albicans* is an opportunistic yeast that can cause disease and is a usual member of the gastrointestinal tract of humans [5]. Additionally, it can live outside of the human body 40–60% of healthy persons have traces of it in their mouths and gastrointestinal tracts. It is a commensal organism most of the time, but it has the potential to become pathogenic in immunocompromised persons under a number of different circumstances. *Bacillus subtilis* (MTCC441) and *Klebsiella pneumonia* (MTCC39) were the two bacterial strains used for the research. *B. subtilis* is considered to be one of the most well-characterized bacteria and is frequently utilized in the study of Gram-positive bacteria as a model species. *B. subtilis* is a rod-shaped bacterium that generates endospores [17]. These endospores allow the bacteria to survive in extreme conditions such as heat and drought. *K. pneumoniae* is a gram-negative, enveloped, non-motile bacterium that is prevalent in the environment [30]. It has been linked to cases of pneumonia in clinical groups that have diabetes mellitus or alcohol use problems. It is common for the bacterium to colonize the skin and mucous membranes of the oropharynx and the gastrointestinal (GI) tract of humans. The microorganisms were received from the Shoolini University Yeast Biology Lab (YBL). The fungi were cultured in YEPD broth (HiMedia, M1363) at 30 °C for 48 h, while the bacteria were cultured in Nutrient broth (HiMedia) at 37°C for 24 h, both of which were then kept on agar plates (HiMedia, GRM666) at 4 °C.

12.2.5 Antimicrobial Activity Using Agar Well Disk Diffusion Method

In order to effectively treat microbial infections and conduct a thorough examination of antimicrobials, it is necessary to conduct an accurate estimate of the degree to which bacteria are susceptible to various antibiotics [25]. The assessment of the MIC in solid media makes numerous uses of the agar diffusion technique [26]. It entails applying antibiotic solutions of varying doses to cups, wells, or paper discs, which are then placed on top of or punched into agar plates that have been inoculated with the test microbial species. Because antibiotics can diffuse from these sources into the agarose medium, they can limit the development of bacteria in the medium's immediate vicinity. This results in the production of clear zones that are devoid of microbial lawns. The agar well disk diffusion method is employed primarily to evaluate the antimicrobial activities of plant extracts. The antimicrobial effect of *C. viminalis* essential oil was determined using the agar well diffusion technique [30]. Vancomycin and Amphotericin B were chosen as positive controls for antibacterial and antifungal activity, respectively. DMSO was utilized as a negative control. Using a HiAntibiotic Zone scale-C, the area of the inhibition zone was determined after a 24–48 h incubation period. The tests were carried out in triplicate, and the findings were represented as the mean standard deviation. Studies have shown the utility of essential oils as a powerful tool to reduce microbial growth [10].

12.2.6 Antimicrobial Activity by Broth Dilution Method

The testing method known as broth dilution can provide both quantitative data known as the minimum inhibitory concentration (MIC) and qualitative responses known as category interpretation. The choice to use an antimicrobial medicine may be profoundly influenced by the results of a MIC test, which can help to establish the level of resistance displayed by a particular microbial species [41]. MIC is the concentration at which the growth of the strain is completely inhibited (as evidenced by the absence of detectable microorganisms).In the process of broth dilution, containers containing equal quantities of broth containing antimicrobial solution in gradually (often geometrically) rising concentrations are infected with a known number of microbes. This is done in order to test the efficacy of the solution. The antimicrobial effect of *C. viminalis* essential oil was determined by the methodology of Hannan [22]. In a 96-well microtiter plate (ThermoFisher Scientific), essential oil (1:10 diluted in 99.9% DMSO (Loba Chemie, LJ1661802) was evaluated. Vancomycin and Amphotericin B were chosen as positive controls for antibacterial and antifungal activity, respectively. DMSO was utilized as a negative control. For fungal strains, microtiter plates were incubated for 48 h at 30 °C and for bacterial strains microtiter plates were incubated for 24 h at 37 °C. The colour shift was then examined visually after the addition of Resazurin dye (HiMedia, RM125). The growth was observable as the colour changed from purple to pink. The MIC value was determined at the minimum concentration at which the colour changed.

12.3 Results and Discussion

12.3.1 Extraction of C. Viminalis Essential Oil and Its Yield

The extraction of essential oil from *C. viminalis* leaves was carried out using a hydrodistillation process for 2–3 h. The extraction of bioactive substances, most commonly in essential oils can be accomplished by the use of hydro distillation. This technique involves the utilization of water/steam. The method is carried out using an apparatus known as the Clevenger apparatus. The extraction temperature is the same as the boiling point of water at normal atmospheric pressure (100 °C). The flask containing leaves mixed with water was heated for 30 min to reach this extraction temperature (100° C) and therefore acquires the distillation of the first essential oil droplet. The essential oil yield from *C. viminalis* leaves was 0.40 ± 0.07% (v/w). The yields are measured in milliliters of essential oil per 200 g of leaves. Similarly, one study found that the yield of oil derived from hydrodistillation of *C. viminalis* leaves was 0.9% (w/w) [32].

12.3.2 Characterization of Various Phytocompounds of C. Viminalis Essential Oil by GC-MS

The GC-MS evaluation of CVEO revealed ten main phytocompounds, each covering area percentage (%) of 32.09, 24.45, 21.29, 5.37, 3.19, 3.92, 2.26, 2.26 and 2.21 which were identified as Eucalyptol; 2-Phenyl-4-methyl-thiazolidine; Amrinone; 4-tert-Octylphenol, TMS derivative; 2,4 (1H,3H)-Quinolinedione, 3-benzoyl-3-(phenylmethyl)-; Decanoic acid, 2-methyl-; 9-Oxononanoic acid; Propanoic acid, 2-(aminooxy)-; Cyclotrisiloxane, hexamethyl- and 1,2,4-Benzenetricarboxylic acid 1,2-dimethyl ester respectively. The spectra from gas chromatography-mass spectroscopy are shown in Fig. 12.2. Plant extracts are an improvement over traditional chemical and physical methods because they are safe for the environment, easy to use, inexpensive, and can be easily scaled up for large-scale synthesis. In addition, it is not necessary to make use of high pressure, energy, temperature, or toxic chemicals when using plant extracts for making pharmaceutical formulations. Further to that, literature from many countries all over the world demonstrated that the plant's various components each have a unique chemical makeup, and these differences were shown to be quite significant (leaves, bark, wood, fruit, and flower).

The corresponding constituents of the GC-MS analysis of CVEO from this work are shown in Table 12.1. Eucalyptol (32.09%) was the major constituent in CVEO, followed by 2-Phenyl-4-methyl-thiazolidine (24.45%) and Amrinone (21.29%). Studies have shown several reports on chemical analyses of *C. viminalis* essential oils acquired from Egypt, Australia, India and Pakistan [8, 16, 18]. The oils mostly composed of 1,8-cineole and with some other major ingredients. These are myrcene, α-pinene, β-pinene, linalool, limonene and methyl acetate. Differences in phytocompounds are attributed to genotype, geographical location, plant sample season, and environmental variables. Due of the existence of these chemical groups, the extracts from C. viminalis have a high level of bioactivity. Additionally, these chemical groups have been demonstrated to have beneficial antimicrobial, antifungal, and antioxidant properties in the past [31].

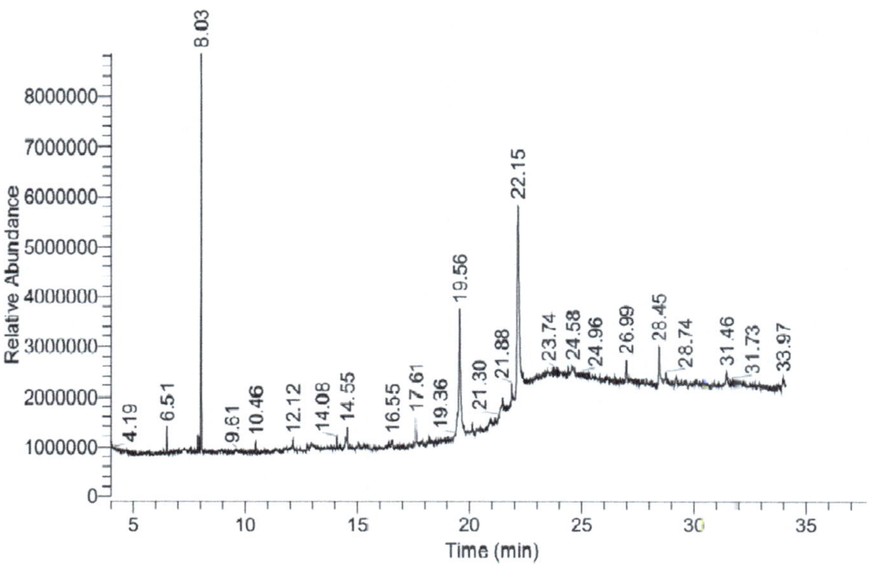

Fig. 12.2 GC-MS profile of *C. viminalis* essential oil (CVEO) in terms of relative abundance alteration with retention time

Table 12.1 A summary of phytocompounds identified from the GC-MS of essential oil of *C. viminalis*

S.no	Compound name	RT	Area %	Molecular formula
1	Eucalyptol	22.15	32.09	$C_{10}H_{18}O$
2	2-Phenyl-4-methyl-thiazolidine	8.03	24.45	$C_{10}H_{13}NS$
3	Amrinone	19.56	21.29	$C_{10}H_9N_3O$
4	4-tert-Octylphenol, TMS derivative	28.45	5.37	$C_{17}H_{30}OSi$
5	2,4(1H,3H)-Quinolinedione, 3-benzoyl-3-(phenylmethyl)-	17.60	3.19	$C_{23}H_{17}NO_3$
6	Decanoic acid, 2-methyl-	21.47	3.92	$C_{11}H_{22}O_2$
7	9-Oxononanoic acid	21.88	2.26	$C_9H_{16}O_3$
8	Propanoic acid, 2-(aminooxy)-	21.88	2.26	$C_3H_7NO_3$
9	Cyclotrisiloxane, hexamethyl-	24.58	2.21	$C_6H_{18}O_3Si_3$
10	1,2,4-Benzenetricarboxylic acid, 1,2-dimethyl ester	26.99	2.56	$C_{11}H_{10}O_6$

12.3.3 Antimicrobial Activity

Using the agar well diffusion method and the broth dilution method, the antimicrobial activity of essential oil isolated from *C. viminalis* leaves was determined. Antimicrobial activity encompasses all active principles (agents) that limit bacterial or fungal development, hinder the formation of microbial colonies, and kill microorganisms [8]. This activity can also be described as the capacity of a material to destroy microorganisms.

Agar well disk diffusion. CVEO was found to be more effective against both microbial strains in this method (Bacterial and fungal). CVEO was found to be antibacterial against *Bacillus subtilis* (18.7 0.5 mm) and *Klebsiella pneumonia* (17.5 0.5 mm). CVEO, like antibacterial activity, demonstrated superior antifungal activity against the studied fungal strains. CVEO inhibited *C. albicans* growth the most (19.5 0.5 mm) for the ATCC90028 strain and 20.15 0.5 mm for the MTCC277 strain. CVEO's zone of inhibition against bacterial and fungal pathogens. The zone of inhibition against bacteria is depicted in Fig. 12.3, and the zone of inhibition is quantified in Table 12.2.

The antibacterial activity was evaluated based on the zone of inhibition (in millimeters) shown CVEO against bacterial strains- *Bacillus subtilis* (A) and *Klebsiella pneumonia* (B) as indicated. Vancomycin was used as a positive control, whereas DMSO was used as a negative control. Antifungal activity was evaluated by measuring the zone of inhibition (in mm) that CVEO exhibited against fungal strains—*C. albicans* (ATCC90028) (A) and *C. albicans* (MTCC277) (B) as indicated. Amphotericin B was taken as a positive control, whereas DMSO was taken as a negative control. The same can be seen in Fig. 12.4 with Table 12.3 showing the zone of inhibition.

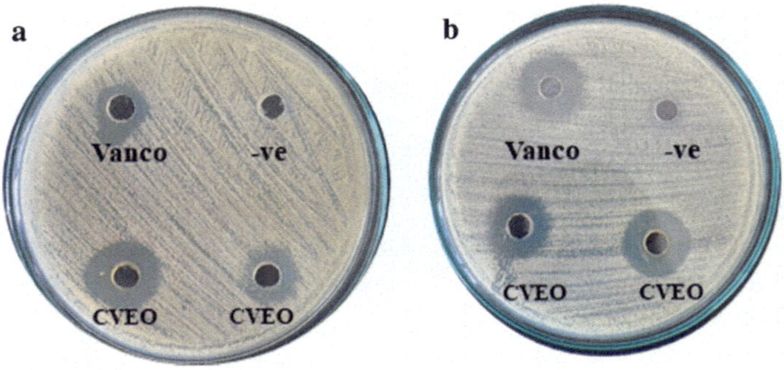

Fig. 12.3 Nutrient agar plates depicting zone of inhibition of *C. viminalis* essential oil and Vanco-Vancomycin against. **a** *Bacillus subtilis* and **b** *Klebsiella pneumonia*

Table 12.2 Inhibition zone data of *C. viminalis* essential oil (CVEO) against bacterial strains

Sample	Zone of inhibition (mm)	
	Bacillus subtilis (mm)	*Klebsiella pneumonia* (mm)
CVEO	18.7 ± 0.5	17.5 ± 0.3
Vancomycin	15 ± 0.7	16.5 ± 0.5

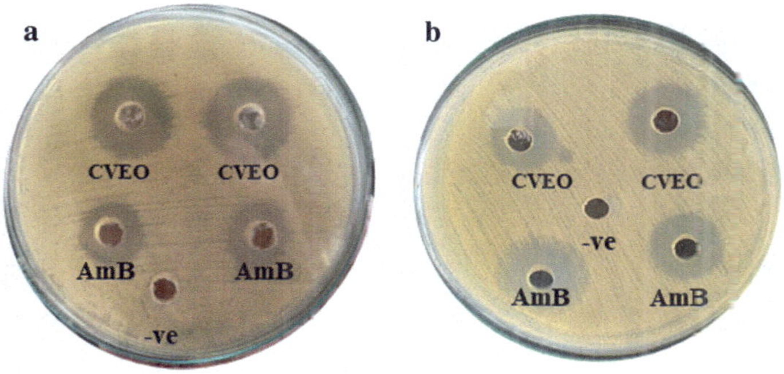

Fig. 12.4 YPD agar plates illustrating zone of inhibition of *C. viminalis* essential oil (CVEO) and AmB-Amphotericin B against *C. albicans*. **a** ATCC90028 and **b** MTCC277

Table 12.3 Inhibition zone data of *C. viminalis* essential oil (CVEO) against fungal strains

Sample	Zone of inhibition (mm)	
	C. albicans (ATCC90028) (mm)	*C. albicans* (MTCC277) (mm)
CVEO	19.5 ± 0.5	20.15 ± 0.5
Amphotericin B	18 ± 0.5	17.5 ± 0.3

Broth dilution method. Furthermore, the broth dilution method was used to investigate the antibacterial potential of CVEO. In Fig. 12.5, the findings of the broth dilution procedure are displayed as the minimum inhibitory concentration (MIC) and its inhibitory activity is presented in Table 12.4. The blue colouration indicates inhibition of growth; pink/red indicates that organisms are active. Labels: CVEO-*C. viminalis* essential oil, Van-Vancomycin, AmB-Amphotericin B, BS-*Bacillus subtilis*, KP-*Klebsiella pneumonia,* ATCC90028 and MTCC277-*Candida albicans* strains.

It has been reported that the MIC ranged from 1.25 to 0.625% against bacterial strains (*Bacillus subtilis* and *Klebsiella pneumonia*) and 0.625–0.312% against fungal strains [*C. albicans* (ATCC90028 and MTCC27)]. In the case of antibacterial activity, vancomycin was used as a positive control, whereas DMSO was used as a negative control, and for antifungal activity, Amphotericin B was used as a positive control, whereas DMSO as a negative control. The oil's potency is attributed to the chemical composition of the plant [36]. MIC values show strong antibacterial and antifungal activity. Given the non-toxic nature of *the C. viminalis* plant, essential oils from this plant can be further developed into anti-microbial formulations. The presence of terpenoids, flavonoids and alkaloids as confirmed by GC-MS is responsible for significant activity against *Bacillus subtilis* and *Klebsiella pneumonia.*

It is worth mentioning that the essential oil of *C. viminalis* is stronger than normal pharmaceutical formulations in inhibiting the development of *Bacillus subtilis, Klebsiella pneumonia,* and *Candida albicans* (ATCC90028 and MTCC277). CVEO has been reported for a variety of medical uses, although its antibacterial potential has

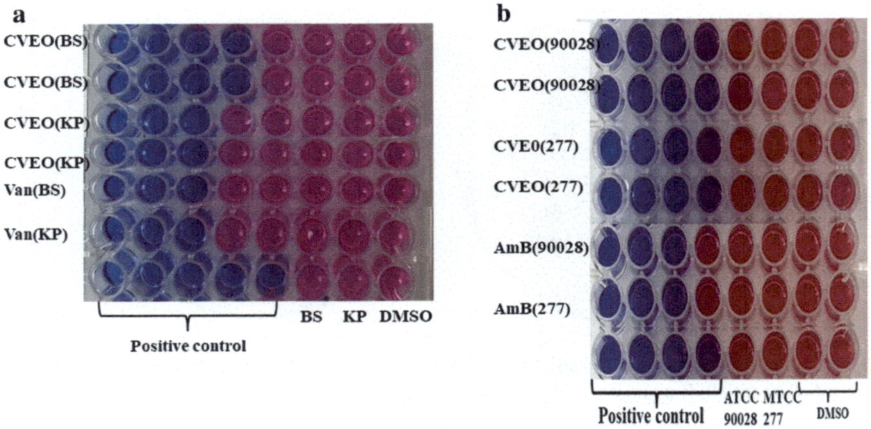

Fig. 12.5 A schematic representation of the 96-well resazurin broth microdilution model. **a** Plate for bacterial strains. **b** Plate for fungal strains

only been studied in small research [33]. These findings reaffirm the superior antibacterial efficacy of CVEO over conventional antibiotics in clinical settings. Moreover, the data indicate that CVEO is more effective than antibiotics against a wider variety of bacterial and fungal species.

Table 12.5 shows the efficiency of essential oils, and some of them, to suppress the growth of various fungal and bacteria species. This is also apparent when the essential oils were used in very small amounts. These investigations show that the bioactive components of EOs have the potential to bind to the cell surface and subsequently pass through the phospholipid bilayer that is part of the cell membrane [14]. Their accumulation impairs the membrane's integrity, which in response can have devastating effects on cellular metabolism and even result in cell death. The composition of the oils, the functional groups that are present in active components, and the interactions between those functional groups and synergistic components are the factors that determine the activity of EOs. The antimicrobial mechanism of action differs depending on the type of EO utilized as well as the strain of the microorganism that is being treated.

Oils extracted from plants, in particular, are more efficient against Gram-positive bacteria than they are against Gram-negative bacteria [9]. Hydrophobic substances have a difficult time penetrating the outer membrane of Gram-negative bacteria because it is more complicated, contains more lipopolysaccharide (LPS), and is more rigid. Gram-positive bacteria, in contrast, have a thin peptidoglycan outer barrier that easily allows tiny antimicrobial compounds to penetrate and kill them [37]. This is because Gram-negative bacteria have a hard outer barrier, making them more difficult to penetrate. Gram-positive bacteria may facilitate the penetration of hydrophobic components of EOs due to the presence of lipophilic ends of lipoteichoic acid in the cell membrane.

Table 12.4 Data summary of minimum inhibitory concentrations of *C. viminalis* essential oil (CVEO) against microbial strains

SAMPLES (Concentration)	MIC against bacteria				MIC against fungus			
	Bacillus subtilis (MTCC441)		*Klebsiella pneumonia* (MTCC39)		*C. albicans* (ATCC90028)		*C. albicans* (MTCC277)	
	MIC	Nature	MIC	Nature	MIC	Nature	MIC	Nature
CVEO 10%(v/v)	1.25–0.625	Bactericidal	1.25–0.625	Bactericidal	0.625–0.312	Fungicidal	0.625–0.312	Fungicidal
Vancomycin 1 mg/ml (w/v)	0.625–0.312	Bacteriostatic	0.625–0.312	Bactericidal	–	–	–	–
Amphotericin B 1 mg/ml (w/v)	–	–	–	–	1.25–0.625	Fungistatic	1.25–0.625	Fungistatic

Table 12.5 A summary of various essential oils and their antimicrobial activity with their mode of action

S.no	Essential oil	Microbes	Mode of action	References
1	*M. longifolia*	*S. typhimurium* *M. luteus*	Targets cell wall of the microbes	Ali et al. [6]
2	*Curcuma longa*	*E. coli* *A. flavus*	Reduced ergosterol biosynthesis	Hua et al. [23]
3	*C. nepalensis*	*Candida* species	Membrane integrity compromised and biosynthesis of ergosterol inhibited	Ahmad et al. [4]
4	*A. sativum*	*E. coli*	Leakage of intracellular materials	Hyldgaard et al. [24]
5	*C. cyminum*	*B. subtilis* *E. coli*	Alteration in cytoplasm	Wongkattiya et al. [41]
6	Cinnamon	*E. coli* *B. cereus*	Distortion of cell membrane	Zhang et al. [42]
7	*O. vulgare*	*P. aeruginosa* *E. coli*	Cell membrane permeabilized	Diao et al. [17]
8	*L. cubeba*	*E. coli*	Cell membrane disruption	Li et al. [29]
9	*F. vulgare*	*S. dysenteriae*	Cell membrane disruption	Akhbari et al. [5]
10	*D. gracilis*	*B. cereus* *E. coli* *P. mirabilis*	Targets cell membrane	Srivastava et al. [39]

Against fungal species such as the Candida isolates, the antifungal activity of EOs is caused by an inhibition in the synthesis of ergosterol and a disruption in the integrity of the membrane. Other research has indicated the usefulness of the mechanism in the process of developing new formulations [4]. This mechanism causes damage to the membrane that surrounds the cytoplasm, which then leads to the leakage of intracellular components like DNA. Natural EOs have been shown to have antifungal properties in some situations, and this has been linked to the EOs' ability to damage the endomembrane system of fungal cells (this includes the cell membrane and the mitochondria) [23]. Particularly, this meant that ergosterol production, mitochondrial ATPase, succinate dehydrogenase, and malate dehydrogenase were all suppressed.

Bioactive compounds that operate slowly as opposed to compounds that act quickly are the two categories of essential oils that can be differentiated depending on the amount of time it takes for the essential oil to generate a meaningful effect. Carvacrol, cinnamaldehyde, and geraniol are three examples of antimicrobials that are regarded to be fast-acting compounds [20]. This is because they can inactivate organisms such as *E. coli* and *Salmonella* in a short amount of time, in some cases within five minutes. In some cases, a time length of 30–60 min was necessary for the compounds that acted slowly in order to display effective antibacterial action [14]. The antifungal activity of carvacrol is equivalent to that of thymol, characterized by the disruption of H^+ homeostasis and Ca^{2+} up-and down-regulation of transcriptional activation, membrane integrity disruption, and impairment of biosynthesis of ergosterol in Candida strains, all of which are also observed in response to Ca^{2+} stress and nutrient deprivation. Despite essential oils' long history of therapeutic usage, relatively little research has been published on many of them. More and more research is being done on essential oils in laboratories all over the world, and this is starting to influence perceptions. In recent years, there has been a rise in interest in the use of plant extracts and essential oil, both of which have been shown to have antimicrobial and other beneficial properties.

12.4 Conclusions

The presence of Eucalyptol, 2-Phenyl-4-methyl-thiazolidine and Amrinone (21.29%) as important phytocompounds in EO isolated from the leaves of *C. viminalis* from the Gangtok district of Sikkim, India has been reported in this study. Agar well diffusion showed a significant zone of inhibitions against bacterial and fungal strains. Antibacterial and antifungal experiments have shown that this product has the potential to reduce the minimum inhibitory concentration necessary to prevent bacterial and fungal development, making it superior to pure essential oil. As a result, with low concentration, activity was significantly increased. Thereby, the findings of this study justify the use of CVEO for effective drug discovery. More research on the dosages to be applied is needed to determine a safe limit for use. Further research on the mechanism of antibacterial and antifungal action of oil can help to develop commercial-scale pharmaceutical formulations.

The high rates of morbidity and mortality are evidence that germs are developing an increasing level of resistance to several drugs at an alarming rate. It is one of the most significant obstacles that physicians and researchers must overcome. The ineffectiveness of the already available medical therapies has made it necessary to look for new pharmaceuticals that are also effecting non-order to solve this issue. Essential oils contain significant volatile chemicals that have a variety of bioactivities, including the capacity to inhibit the growth of microorganisms. Essential oils are garnering fresh interest as an alternative treatment for a variety of health issues, including infections, stress, and other health complications. Essential oils have a pleasant aroma, are effective in treating bacterial and fungal infections, relieve tension, and facilitate restful sleep. They are highly concentrated extracts obtained from various plant sources that can be used for a variety of therapeutic and recreational purposes. Herbal medicines have been employed as the major remedy in traditional medical systems for thousands of years, and they have made a substantial contribution to improving human health. Antimicrobial activity can be investigated using secondary metabolites (potential drug sources) and therapeutically significant essential oils found in medicinal plants. Essential oils have been found to be a promising approach for application in pharmaceuticals, food, and even agricultural commodities. Pollination biology, flora traits in Callistemon viminalis, and its wider application for traditional medicine contribute to bio-diversity conservation. Due to its pharmaceutical properties that are often used as spasmolytic and sedative, this plant also possesses antimicrobial and antioxidant properties. These properties are assumed to support the use of conventional medicine. The biological use of this plant extract from its branches, leaf, flowers, and bark has also found its application as a potential insecticide. Therefore, this plant gains its purpose for bio-diversity conservation with respect to ornamental and forestry plantations.

Acknowledgements The authors acknowledge the support received from Shoolini University for conducting this study in various capacities.

Declaration of competing interest The authors state that they do not have any known competing financial interests or personal ties that could seem to have influenced the work disclosed in this study.

References

1. Abd-ElGawad, A.M., El Gendy, A.E.N.G., Assaeed, A.M., Al-Rowaily, S.L., Alharthi, A.S., Mohamed, T.A., Elshamy, A.I.: Phytotoxic effects of plant essential oils: a systematic review and structure-activity relationship based on chemometric analyses. Plants **10**(1), 36 (2020)
2. Abdelmalek, E.M., Ramadan, M.A., Darwish, F.M., Assaf, M.H., Mohamed, N.M., Ross, S.A.: Callistemon genus-a review on phytochemistry and biological activities. Med. Chem. Res. **30**(5), 1031–1055 (2021)
3. Ahmad, K., Athar, F.: Phytochemistry and pharmacology of Callistemon viminalis (Myrtaceae): a review. Nat. Prod. J. **7**(3), 166–175 (2017)
4. Aijaz Ahmad, A., Khan, A., Kumar, P., Bhatt, R.P., Manzoor, N.: Antifungal activity of Coriarianepalensis essential oil by disrupting ergosterol biosynthesis and membrane integrity against Candida. Yeast **28**, 611–617 (2011)
5. Akhbari, M., Kord, R., Jafari Nodooshan, S., Hamedi, S.: Analysis and evaluation of the antimicrobial and anticancer activities of the essential oil isolated from Foeniculum vulgare from Hamedan, Iran. Nat. Prod. Res. **33**(11), 1629–1632 (2019)
6. Ali, H.M., Elgat, W.A., El-Hefny, M., Salem, M.Z., Taha, A.S., Al Farraj, D.A., Elshikh, M.S., Hatamleh, A.A., Abdel-Salam, E.M.: New approach for using of Mentha longifolia L. and Citrus reticulata L. essential oils as wood-biofungicides: GC-MS, SEM, and MNDO quantum chemical studies. Materials **14**(6), 1361 (2021)
7. Al-Yousef, H.M., Amina, M.: Mechanisms of plant defense against pathogens: phytoalexins induction. In: Phytochemistry, pp. 79–118. Apple Academic Press (2018)
8. Badawy, M.E., Abdelgaleil, S.A.: Composition and antimicrobial activity of essential oils isolated from Egyptian plants against plant pathogenic bacteria and fungi. Ind. Crops Prod. **52**, 776–782 (2014)
9. Balouiri, M., Sadiki, M., Ibnsouda, S.A.: Methods for in vitro evaluating antimicrobial activity: a review. J. Pharm. Biomed. Anal. **6**, 71–79 (2016)
10. Bhattacharya, R., Dev, K., Sourirajan, A.: Antiviral activity of bioactive phytocompounds against coronavirus: an update. J Virol Methods **290**, 114070 (2021)
11. Bhattacharya, R., Rolta, R., Dev, K., Sourirajan, A.: Synergistic potential of essential oils with antibiotics to combat fungal pathogens: present status and future perspectives. Phytother. Res. **35**(11), 6089–6100 (2021)
12. Chachad, D.P., Dias, A., Uniyal, K., Varma, U., Jadhav, P., Satvekar, T., Doshi, N.: Chemical characterization; Antimicrobial and Larvicidal activity of essential oil from Callistemon citrinus (Bottle brush) leaves (2021)

13. Chammartin, F., Ostinelli, C.H., Anastos, K., Jaquet, A., Brazier, E., Brown, S., Dabis, F., Davies, M.A., Duda, S.N., Malateste, K., Nash, D.: International epidemiology databases to evaluate AIDS (IeDEA) in sub-Saharan Africa, 2012–2019. BMJ Open **10**(5), e035246 (2020)
14. Chouhan, S., Sharma, K., Guleria, S.: Antimicrobial activity of some essential oils—Present status and future perspectives. Medicines **4**(3), 58 (2017)
15. Dadgostar, P.: Antimicrobial resistance: implications and costs. Infect. Drug Resist. **12**, 3903 (2019)
16. de Oliveira, C.M., das Graças Cardoso, M., da Silva Figueiredo, A.C., de Carvalho, M.L.M., de Miranda, C.A.S.F., Albuquerque, L.R.M., et al.: Chemical composition and allelopathic activity of the essential oil from Callistemon viminalis (myrtaceae) blossoms on lettuce (Lactuca sativa L.) seedlings. Am. J. Plant Sci. **5**(24), 3551 (2014)
17. Diao, W.R., Hua, Q.P., Zhang, H., Xu, J.G.: Chemical composition, antibacterial activity and mechanism of action of essential oil from seeds of fennel (Foeniculum vulgare Mill). Food Control **35**, 109–116 (2014)
18. Fall, R., Ngom, S., Sall, D., Sembène, M., Samb, A.: Chemical characterization of essential oil from the leaves of Callistemon viminalis (DR) and Melaleuca leucadendron (Linn.). Asian Pacific J. Trop. Biomed. **7**(4), 347–351 (2017)
19. Govaerts, R., Nic Lughadha, E., Black, N., Turner, R., Paton, A.: The world checklist of vascular plants, a continuously updated resource for exploring global plant diversity. Sci. Data **8**(1), 1–10 (2021)
20. Guimarães, A.C., Meireles, L.M., Lemos, M.F., Guimarães, M.C., Endringer, D.C., Fronza, M., Scherer, R.: Antibacterial activity of terpenes and terpenoids present in essential oils. Molecules **24**(13), 2471 (2019)
21. Hamad, Y.K., Abobakr, Y., Salem, M.Z., Ali, H.M., Al-Sarar, A.S., Al-Zabib, A.A.: Activity of plant extracts/essential oils against three plant pathogenic fungi and mosquito larvae: GC/MS analysis of bioactive compounds. BioResources **14**(2), 4489–4511 (2019)
22. Hannan, P.C.: Guidelines and recommendations for antimicrobial minimum inhibitory concentration (MIC) testing against veterinary mycoplasma species. Vet. Res. **31**(4), 373–395 (2000)
23. Hua, Y., Zhang, J., Kong, W., Zhao, G., Yang, M.: Mechanisms of antifungal and anti-aflatoxigenic properties of essential oil derived from turmeric (Curcuma longa L.) on Aspergillus flavus. Food Chem. **220**, 1–8 (2017)
24. Hyldgaard, M., Mygind, T., Meyer, R.L.: Essential oils in food preservation: mode of action, synergies and interactions with food matrix components. Front. Microbiol. **3**, 12 (2012)
25. Iqbal, Z., Ahmed, S., Tabassum, N., Bhattacharya, R., Bose, D.: Role of probiotics in prevention and treatment of enteric infections: a comprehensive review. 3 Biotech **11**(5) 1–26 (2021)
26. Khalil, N., Ashour, M., Fikry, S., Singab, A.N., Salama, O.: Chemical composition and antimicrobial activity of the essential oils of selected Apiaceous fruits. Future J. Pharm. Sci. **4**(1), 88–92 (2018)
27. Khan, W., Khan, N., Jamila, N., Masood, R., Minhaz, A., Amin, F., Nishan, U.: Antioxidant, antibacterial, and catalytic performance of biosynthesized silver nanoparticles of Rhus javanica, Rumex hastatus, and Callistemon viminalis. Saudi J. Biol. Sci. **29**(2), 894–904 (2022)
28. Kheloufi, A., Mansouri, L.M., Telali, I., Slimani, S.: Effect of temperature on seed germination of two Callistemon species (Myrtaceae). Reforesta **9**, 1–8 (2020)
29. Li, W.-R., Shi, Q.-S., Liang, Q., Xie, X.-B., Huang, X.-M., Chen, Y.B.: Antibacterial activity and kinetics of Litseacubeba oil on Escherichia coli. PLoS ONE **9**, e110983 (2014)
30. Martins, L.N., Venceslau, A.F., Brandão, R.M., Braga, M.A., Batista, L.R., Cardoso, M.D.G., Pinto, L.: Antibacterial and antifungal activities and toxicity of the essential oil from Callistemon viminalis complexed with β-cyclodextrin. Curr. Microbiol. **78**(6), 2251–2258 (2021)
31. Moghaddam, M., Mehdizadeh, L.: Chemistry of essential oils and factors influencing their constituents. In: Soft Chemistry and Food Fermentation, pp. 379–419. Academic Press (2017)
32. Oyedeji, O.O., Lawal, O.A., Shode, F.O., Oyedeji, A.O.: Chemical composition and antibacterial activity of the essential oils of Callistemon citrinus and Callistemon viminalis from South Africa. Molecules **14**(6), 1990–1998 (2009)

33. Rathore, R., Rai, N.: Pharmacological action and underlying molecular mechanism of Callistemon: a genus of promising medicinal herbs. Phytomedicine **3**, 154013 (2022)
34. Razzaque, M.S.: Implementation of antimicrobial stewardship to reduce antimicrobial drug resistance. Expert Rev. Anti Infect. Ther. **19**(5), 559–562 (2021)
35. Salem, M.Z., Mervat, E.H., Ali, H.M., Abdel-Megeed, A., El-Settawy, A.A., Böhm, M., Salem, A.Z.: Plants-derived bioactives: novel utilization as antimicrobial, antioxidant and phytoreducing agents for the biosynthesis of metallic nanoparticles. Microb. Pathog. **158**, 105107 (2021)
36. Salem, M.Z., Mervat, E.H., Nasser, R.A., Ali, H.M., El-Shanhorey, N.A., Elansary, H.O.: Medicinal and biological values of Callistemon viminalis extracts: history, current situation and prospects. Asian Pac J Trop Med **10**(3), 229–237 (2017)
37. Semeniuc, C.A., Pop, C.R., Rotar, A.M.: Antibacterial activity and interactions of plant essential oil combinations against gram-positive and gram-negative bacteria. J. Food Drug Anal. **25**(2), 403–408 (2017)
38. Sharma, M., Arora, K., Sachdev, R.K.: Bio-herbicidal potential of essential oil of-a review Callistemon Indian. J. Ecol. **48**(1), 252–260 (2021)
39. Srivastava, M.P., Fatima, A.: Essential oils (Eos) as the advantages of its micro-encapsulation in cosmetic industry. G-J. Environ. Sci. Technol. **8**(2), 28–42 (2021)
40. Subramanian, K., Selvaraj, H., Sampath Renuga, P., Aruni, W.: Anti-quorum sensing in pathogenic microbes using plant-based bioactive phytochemicals. Adv. Mater. Sci. Eng. (2022)
41. Wongkattiya, N., Sanguansermsri, P., Fraser, I.H., Sanguansermsri, D.: Antibacterial activity of cuminaldehyde on food-borne pathogens, the bioactive component of essential oil from Cuminum cyminum L. collected in Thailand. J. Complement. Integr. Med. (2019) **16**(4) (2019)
42. Zhang, Y., Liu, X., Wang, Y., Jiang, P., Quek, S.Y.: Antibacterial activity and mechanism of cinnamon essential oil against Escherichia coli and Staphylococcus aureus. Food Control **59**, 282–289 (2016)
43. Zheljazkov, V.D., Cantrell, C.L., Semerdjieva, I., Radoukova, T., Stoyanova, A., Maneva, V., Salamon, I.: Essential oil composition and bioactivity of two Juniper species from Bulgaria and Slovakia. Molecules **26**(12), 3659 (2021)

Chapter 13
Alpinia nigra and Its Bioactive Compound, Labdane Diterpene: A Review of Their Phytochemical and Biopharmaceutical Potential

Ishani Chakrabartty and Latha Rangan

13.1 Introduction

Nature has always been the best source of medicines since ancient times. Organisms, starting from plants and animals, to microorganisms, have been explored since long in the world of medicine. For the longest time, a predominant source of medicine were crude extracts obtained from natural resources. These extracts have various secondary metabolites as their principal component—most of which are chemicals possessing the ability to cure ailments. With the help of traditional knowledge that has been passed down through the generations, most modern medicines have evolved from naturally occurring compounds [1]. For instance, the description of the medicinal properties of the bark and leaves of willow was discovered in tablets belonging to the old civilizations of Greece. Reverend Edward Stone was the first to perform a scientific study on the benefits of willow to treat patients with fever in 1763. Then, in 1897, acetylsalicylic acid (ACA) was synthesized by Felix Hoffmann, and was soon marketed as 'aspirin' [2]. Thus, it indeed is astonishing on what can be created out of leaves and barks of trees. The Nobel Prize in Physiology or Medicine [2] for discovering artemisinin, a new bioactive compound against malaria was granted to

I. Chakrabartty (✉) · L. Rangan (✉)
Department of Biosciences and Bioengineering, Indian Institute of Technology Guwahati, Guwahati, Assam 781039, India
e-mail: ishani.kaushik@gmail.com

L. Rangan
e-mail: lrangan@iitg.ac.in

I. Chakrabartty
Department of Applied Biology, School of Biological Sciences, University of Science and Technology Meghalaya, Baridua, Meghalaya 793101, India

L. Rangan
School of Agro and Rural Technology, Indian Institute of Technology Guwahati, Guwahati, Assam 781039, India

R. V. S. Uppaluri and L. Rangan (eds.), *Conservation of Biodiversity in the North Eastern States of India*, https://doi.org/10.1007/978-981-99-0945-2_13

Tu Youyou. Man has an age-old relationship with nature that allows him to explore natural sources for medicines.

The plant family, Zingiberaceae is known as the "pharmacy" of the plant kingdom. These plants mostly propagate vegetatively by rhizomes and are rich in essential oils, which impart a characteristic and distinctive odor; some also serve as ornamental elements for decoration due to their beautiful flowers. As the members of Zingiberaceae are extensively used as culinary spice and home remedies for diseases, they have been of great importance for researchers and scientists. Plants of Zingiberaceae have a cosmopolitan distribution but are majorly concentrated in Southeast Asian nations. Twenty-two genera and 178 species of Zingiberaceae are known to be present in India, of which a variety of approximately 88 species are found in North-East (NE) India. Tribal populations of this region have abundant knowledge of the medicinal properties of plants of Zingiberaceae, in addition to them being important condiments. Curcumin is a popular bioactive compound from important plants of the Zingiberaceae family, such as *Curcuma longa* and *Zingiber officinale.* Despite the fact that it has immense pharmacological potential, curcumin has extremely low aqueous solubility and hence, poor bioavailability, as it cannot be formulated into drug leads. *Alpinia,* i.e., the largest genus of this family, is immensely used in folk medicine, and many of its uses have been critically documented [3]. *Alpinia galanga* is the predominantly explored member of this genus and has the highest number of reported bioactive compounds, most of which possess a myriad of biological activities, including analgesic, antimicrobial, and anticancer activities, among others [4–6]. In the scientific community's quest of finding natural medicines, species such as *A. katsumadai, A. purpurata,* and *A. zerumbet*, among others, have been receiving attention in recent decades.

Alpinia nigra (Gaertn. B.L. Burtt) is a plant arising from rhizomes. It has not been able to receive due attention from the scientific community till date [7]. This plant is commonly known as "Tora" or "Tara" in Assamese and "Jangli ada" in Bengali, and is naturally abundant in NE India. The plant is known for its ethnomedical properties locally and is the only member of the Zingiberaceae family to serve as fodder for some animals. Although there is limited research available on bioactive compounds of *A. nigra*, it has emerged as a potential cure for helminthes, as reported by some scientists in Assam, India [7, 8]. However, a proper documentation and extensive literature on the bio-pharmaceutical potential of *A. nigra* and its important bioactive compounds identified till date is lacking.

Effort has been made to collect the works on the different aspects of *A. nigra* and its importance, and to compile all relevant information in the present article. This review consolidates previous reports available on the existing traditional knowledge, along with the phytochemical and pharmaceutical potential of this plant. This article reviews numerous works of experimental research reports on the medicinal importance of the different organic extracts and bioactive compounds obtained from various organs of this plant. In Sect. 13.2, an analysis of the literature on the use of plants from the Zingiberaceae family as medicines in NE India is reported. A major part of the review (Sect. 13.3) consolidates the research work carried out in *A. nigra,* which includes its botanical and chemical description, bioactive compounds, and biological activity, such as antimicrobial, antiparasitic, analgesic, anti-inflammatory, neuroprotective, and so on. Section 13.4 highlights an extensively studied bioactive compound, labdane diterpene—its structure and biological activity in this plant. The subsequent sections i.e., Sects. 13.5 and 13.6, respectively provide directions for future work and conclusions, highlighting the importance of *A. nigra* and the need to conserve this valuable plant in NE India.

13.2 Zingiberaceae: The Scenario in North-East India

The eight states of NE India viz. Arunachal Pradesh, Assam, Meghalaya, Sikkim, Manipur, Nagaland, Mizoram, and Tripura that lie between 21°34′N and 29°50′N latitude and 87°32′E–97°52′E longitude, comprise an area of 262,060 km^2; this area alone is responsible for 50% of the biodiversity hotspots of the country [9]. Out of the 50% angiosperms present in NE India, 40% are endemic, which include members of families such as Poaceae, Fabaceae, Asteraceae, Zingiberaceae, Euphorbiaceae, Rubiaceae, and Orchidaceae. The geographical location, dynamics, climatic conditions, and hilly terrains of this region is believed to contribute to its rich biodiversity, including different varieties of medicinal plants [10]. Owing to the existence of such indispensable medicinal plants, traditional knowledge of healing with plants and extracts is extremely common here—the local healers, also known as bej or kobiraj, use various plants to cure ailments, including bites from poisonous snakes. As such, traditional knowledge among individuals in NE India is an integral part of the ethnic culture of the country, which renders the country with pride. In this region, the use of various plants (among which many are now endangered or threatened) by the tribal communities for ethnomedical purposes has been documented [11]. The knowledge of traditional plants is therefore important and contributes immensely to the Indian folklore and traditional knowledge, something which is exploited by the ethnic groups of this region.

As a house of different products like cosmetics, perfumes, medicines, dyes, and most importantly, spices, the Zingiberaceae family is considered as the "blessing of nature". The presence of large quantities of essential oils, oleoresins, and different volatile aromatic compounds impart the characteristic aroma to the members of this rhizomatous family. India is overflowing with many members of Zingiberaceae, and the NE region of India houses approximately 19 genera and 88 species of this family. The ethnomedical use of the members of the Zingiberaceae family has been critically studied and documented in the past [3]. The application and traditional use of these plants by tribal populations in NE India suggest that the plants synthesize a variety of secondary metabolites and other chemicals that serve as defense mechanisms and help to fight diseases. Highly effective drugs can be developed by extracting bioactive compunds from these plants, together with organic synthesis and combinatorial chemistry. Several countries are now heavily investing in developing the field of "herbal medicine," and India, a country extremely rich in ethnicity and folklore, is also taking an initiative towards developing this field.

Alpinia, the largest and most diverse genus of Zingiberaceae, with around 230 species of monocotyledonous, angiospermic plants growing in tropical and subtropical Asia alone, spreads across Sri Lanka, the Western Ghats of India, China, Japan, southeast Asia, the Pacific region, and Australia. Most of its members grow as herbs on a well-developed pseudostem, indicating the absence of a true stem; in addition, they propagate by thick, underground rhizomes [12]. Owing to the presence of a high quantity of essential oils, these plants are generally aromatic. When different parts of these plants are fractionated, they yield different kinds to extracts that are rich in bioactive compounds and essential oils. Currently, there are in-depth ongoing studies on certain members of this genus viz. *Alpinia galanga, A. purpurata, A. katsumadai, A. officinarum*, etc. However, not all members of genus *Alpinia* have received equal importance till date; a herbaceous plant, *A. nigra* still remains to be scientifically unexplored even though it is widely used as a traditional medicine in NE India.

13.3 *Alpinia nigra*: The Unexplored Goldmine

Alpinia nigra, a medium-sized rhizomatous, herbaceous plant, is endemic to South-East Asia, including China, Bhutan, Burma, Thailand, India, Sri Lanka, and Bangladesh. It is consumed by the indigenous people of NE India; "deragong" and "pullei eromba" are some of the delicacies savored in Tripura and Manipur, respectively. In certain regions around the globe, *A. nigra* is considered to be plant with magico-religious beliefs (Dobur Uie) [13]. This is the only plant belonging to the Zingiberaceae family which serves as a fodder to rhinoceros [14]. It thrives abundantly in the marshy and swampy areas; they propagate by means of stolons in addition to rhizomes. To date, original research and review papers outlining the importance of various extracts obtained from *A. nigra* are limited. Moreover, research has mainly been focused on isolating various bioactive compounds of *A. nigra* and enumerating their mechanism of action towards controlling diseases.

13.3.1 Botanical Description

Alpinia nigra is a herbaceous plant that is found to grow in shady areas, wastelands, and under forest canopies. The plant arises from a rhizome under the soil that develops into a strong pseudostem on the top. The taxonomical classification of this plant is provided in Table 13.1. Its fruits are berries with many seeds in each locule and are green when young (Fig. 13.1); they change from red to black as they mature [15].

Table 13.1 Taxonomical classification of *A. nigra*

Kingdom	Plantae
Class	Angiosperm
Sub-class	Monocotyledonae
Order	Zingiberales
Family	Zingiberaceae
Sub-family	Alpiniodeae
Genus	*Alpinia*
Species	*nigra*

Fig. 13.1 Plant parts of *A. nigra*. **a** Elongated leaf sheath; **b** recemose arrangement of flowers; **c** irregular and pink flowers; **d** unripe, green fruits in cluster

13.3.2 Chemical Composition

Mineral content: Minerals and miniscule elements are not involved in the body weight as they do not have any caloric contributions but are vital for maintaining body growth and metabolism [16]. Although this is not well-known, but small contents of elements like potassium, manganese, etc., have an indirect role in the lifestyle-related disorder, type II diabetes mellitus [17].

The mineral content of *A. nigra* is poor. This plant can be used as a food supplement and developed into medicines as it is rich in some primary elements, such as carbohydrates and iron. This plant is unique because it has an approximately 10% lower concentration of macronutrients, in comparison with *Zingiber officinale* [18].

Oil content: As discussed before, plants from *Alpinia* have a distinctive aroma that can be attributed to the high essential oil content of this genus. An interesting fact is that the presence of different kinds of chemical compounds, instead of the percentage yield, is responsible for the aroma-inducing properties of the plant [19]. As such, *A. nigra* is also no exception to it. It has been observed that the concentration of oil in the leaves is slightly higher in the plants growing in Assam [20], in comparison to those growing in Bangladesh [21]. This can be attributed to the fact that the climate, soil, harvest time, temperature, as well as the method of collection of oil, all have an influence on the quantity of oil obtained [22], and the same may be assumed to be true in the case of *A. nigra*.

13.3.3 Phytochemistry

The vast array of secondary metabolites that are present in different medicinal plants impart different properties to them, such as their characteristic color or aroma; besides, they also play a role in the plant defense. These compounds are mostly composed of metabolite groups, including alkaloids, flavonoids, tannins, lignans, saponins, stilbenes, quinones, and others, that possess a myriad of bio-pharmaceutical activities, such as antimicrobial, antioxidant, insecticidal, hepato-protective, anti-inflammatory activities, and so on [23]. In the same way, the organic and aqueous extracts that are obtained from different parts of *A. nigra* are rich in these secondary metabolites, and are crucial to its therapeutic potential. Researchers have isolated and studied an array of bioactive compounds from the genus *Alpinia*; however, as mentioned above, the available literature on *A. nigra* and its phytochemicals is limited [7]. It is highly surprising that till date only seven bioactive compounds have been isolated from *A. nigra*, some of which are yet to be screened for their biological activity (Table 13.2). In addition, researchers have isolated some hydroxypropyl esters viz. 3 dihydroxypropyl ester (III), 3 dihydroxypropyl ester (IV), and 3 dihydroxypropyl ester (V) from the seed clusters of *A. nigra* [24]. Kaempferol-3-*O*-glucuronide has major concentration in the fruit pulp; this is around six times higher than that present in the seeds (3.043% w/w) [25].

13.3.4 Bio-pharmaceutical Potential

Previous literature has already stressed on the biological potential of the different members of the genus *Alpinia*. Additionally, various spices obtained are utilized in

Table 13.2 Bioactive compounds isolated from *A. nigra* [24, 26–30]

Sr. no	Name of compound	Biological potential	Class of compound	Structure
1	Astraglin	Antibacterial, antiprotozoal, antioxidant, hepato-protective	Flavone glycoside	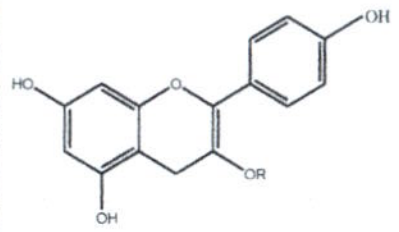 R= Glucose R = Glucose
2	Kaempfero-3-*O*-glucuronide	Antibacterial, antiprotozoal, antioxidant, hepato-protective	Flavone glycoside	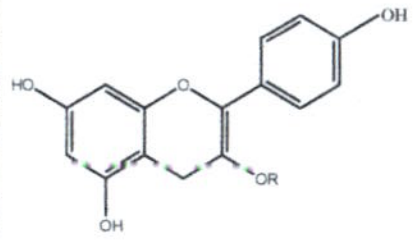 R= Glucouronic acid R = Glucouronic acid
3	(E)-labda-8(17), 12-diene-15,16-dial	Antibacterial, Antidiabetic, Antileishmanial, Anti-*Candida*	Diterpene	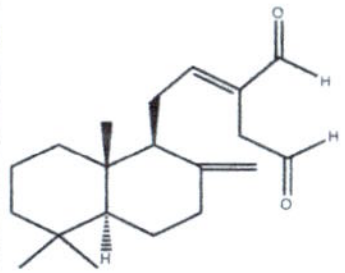
4	(E)-8β,17-Epoxylabd-12-ene-15,16-dial	Antibacterial, Antidiabetic, Antileishmanial	Diterpene (epoxy derivative)	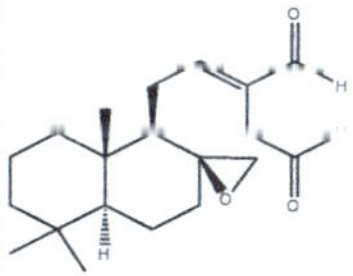

(continued)

Table 13.2 (continued)

Sr. no	Name of compound	Biological potential	Class of compound	Structure
5	Stigmasterol	–	Sterol	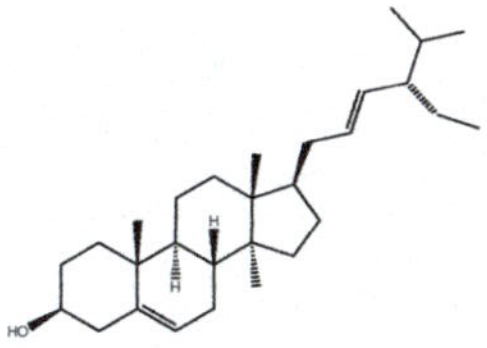
6	γ-Sitosterol	–	Sterol	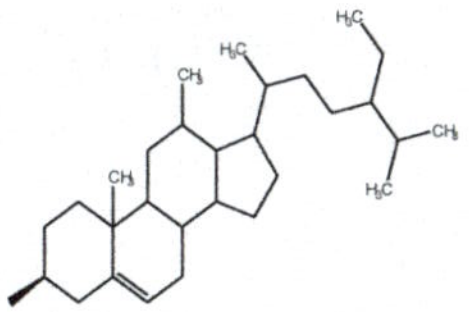
7	3,5-dihydroxy-4,7-dimethoxy flavone	Antioxidant, anti-tyrosinase, anti-inflammatory	Flavone	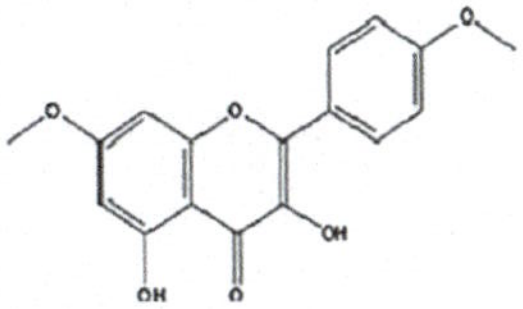

culinary preparations [31]. It has been already stated that *A. nigra* is consumed with rice as part of a vegetarian diet and also as a condiment. Not only that, natives of Tripura consume the aqueous shoot extract of this plant to treat intestinal infections, particularly those caused by flukes [32]; the rhizomes are used to treat gout and colic diseases [18]. The biological efficacy of this plant (Fig. 13.2) is outlined as follows:

Antimicrobial activity: A lot of diseases are caused due to infections by a number of microorganisms. Traditionally, different plants and their parts as well as extracts have been used to treat these infections. Scientific reports are available on the antibacterial activity of the organic extracts obtained from the available portions (whole plant and its parts) of *A. nigra*. Extracts of the rhizome of *A. nigra* with methanol exhibit very mild antibacterial activity against *Staphylococcus aureus* and *Bacillus cereus* [33] and the extract from the same part with water have antibacterial potential against *S. aureus, B. cereus, Salmonella paratyphi* and *Escherichia coli*. The antibacterial potential of the water extracts of flowers and seeds are of moderate intensity [20]. Table 13.3 lists the antibacterial efficacies of the various organic and aqueous extracts of *A. nigra* and its plant parts. It is to be noted that the aqueous leaf extract of this plant shows low antibacterial activity but the intensity is approximately moderate for the organic extract obtained by using methanol as the solvent; this is similar for both Gram positive and Gram negative bacteria [34]. The crude extracts of *A. nigra* have

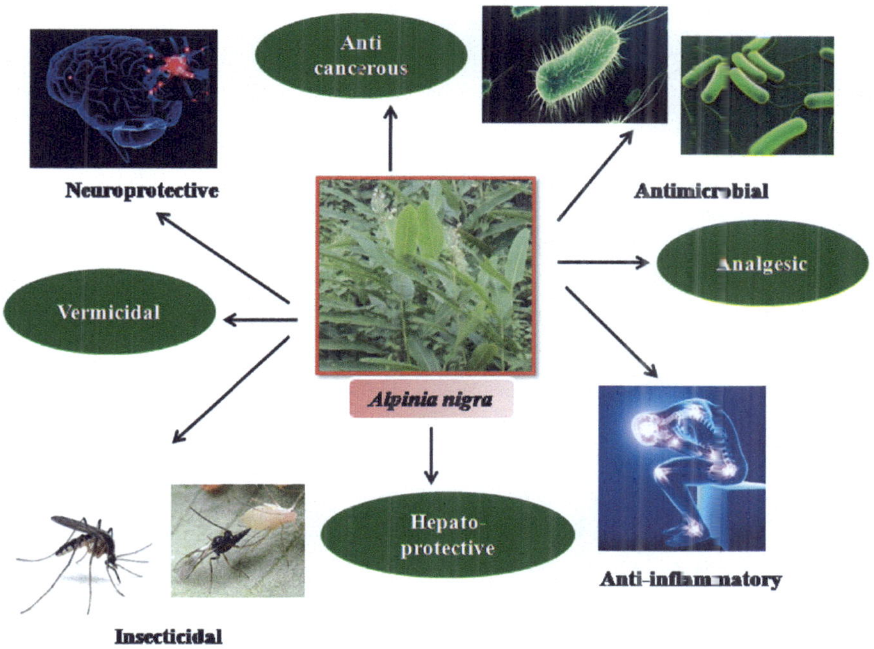

Fig. 13.2 Graphical depiction of the biological activity of *A. nigra*

tremendous antimicrobial potential and can be formulated as potential antimicrobial agents.

Antiparasitic and insecticidal activity: Parsitic organisms, particularly those residing in the human gut, pose a severe threat to human health. It has been reported that the extract obtained from the shoots of *A. nigra* (using alcohol) exhibit vermicidal activity against *Taenie solium*, the intestinal parasitic tapeworm [32]; the extract constituents does not allow digestion and absorption of food by the helminth by disrupting the activity of a digestive enzyme, alkaline phosphatase [8]. The essential oils from different plant parts exhibited 100% mortality against *Aedes aegypti* larvae (125 ppm) as well as prevent from bitting after topical application. Amongst these, the seed essential oil was found to be most effective against the mosquito as opposed to the oil from the flowers [20]. Like most essential oils, oils from *A. nigra* also possess a characteristic and pleasant aroma—a feature that can be explored industrially in the manufacture of soaps, perfumes, other cosmetics, and even sanitizers with high alcohol content.

Table 13.3 Organic solvent extracts of *A. nigra* and their antibacterial potential [33, 34, 45]

Sr. no.	Solvent used	Plant part	Zone of inhibition (mm)	Bacteria
1	Methanol	Rhizome	13	*S. aureus*
	Essential oil (water)		8	
	Methanol	Leaf	9	
	Essential oil (water)		7	
	Essential oil (water)	Flower	6	
		Seed	6	
2	Methanol	Rhizome	12	*B. cereus*
	Essential oil (water)		7	
	Methanol	Leaf	8	
	Essential oil (water)		6	
	Essential oil (water)	Flower	6	
		Seed	6	
3	Essential oil (water)	Rhizome	6	*S. paratyphi*
	Methanol	Leaf	7	
	Essential oil (water)		6	
		Flower	6	
		Seed	6	
4	Essential oil (water)	Rhizome	7	*E. coli*
		Leaf	8	
		Flower	8	
		Seed	7	

Antioxidant and cytototoxic activity: Phytochemicals and secondary metabolites from plants possess a no. of additional activities like antioxidant, anti-inflammatory, analgesic, anti-cancer etc. A large number of foods are rich in natural antioxidants that can scavenge the harmful free radicals and reactive oxygen species (ROS) generated in the body during metabolic activities and stress—this contributes immensely in lowering the risk of life-threatening diseases like hypertension, cardiovascular disease (CVDs) [35]. The high antioxidant activity of the leaves of *A. nigra* is due to the presence of flavonoids and phenolic acids [36]; rhizome extract (with methanol) has the highest antioxidant activity, but has moderate activity with hexane and lowest activity with ethanol [37]. However, cytotoxicity of plant extracts has been addressed. Extracts obtained from rhizomes (using methanol) of *A. nigra* when fractionated using petroleum ether, ethyl acetate and chloroform have been found to be considerably cytotoxic; further assays also confirm pesticidal and anti-tumor activity of these organic fractions of the extract [38]. The LC50 value of the leaf extract is 57.12 μg/ml—this is indicative of the pharmacological property of the extract [34]. Ideally, most bioactive compounds tend to cause toxicity to cells at high dosages. Thus, the crude extracts and different compounds from *A. nigra* have been found to be utilized as a potential tumor-destructive agent.

Analgesic and anti-inflammatory activity: Inflammation is a line of defense initiated by the innate immune system of the body to register injury/disease and further initiate the complex process of healing—this involves recognition and elimination of any harmful stimuli by the defensive immune cells [39]. Steroids have been along used against inflammation, along with a number of non-steroidal drugs; however, these are always accompanied by one or more side effects like nausea, fever, diarrhea, kidney and liver problems besides others [40]. Hence, pain and inflammation can be extensively subdued with plant-based medicines. The genus *Alpinia* houses a large number of bioactive compounds that exhibit analgesic and anti-inflammatory potential [41]. The different biological (and chemical) substances present in sufficient amounts in these extracts (both aqueous and organic) is responsible for these properties as well. A mixture of equal amounts of extracts from the bark of *A. nigra* (fractionation with petroleum ether, ethyl acetate and methanol) is known to exhibit tremendous anti-inflammatory effect [42]. In addition, the ethanolic fruit extract exhibits anti-inflammatory activity in a dose-dependent manner [43] along with the extracts of shoot-bark and leaves of *A. nigra* (using methanol) [44]. However, the anti-inflammatory potential of the methanolic leaf extract of *A. nigra* is extremely weak in comparison to a standard drug, nalbuphine HCl [34]; in contrast, the fruit extracts has a significant analgesic activity [43]. Appropriate bioactive compounds and other responsible agents, if isolated in the pure form, can be developed as antipyretic and analgesic agent. With the currently available analgesics being armed with many side effects, an effective agent from *A. nigra* may be considered to be successful as there are limited chances of any possible side effects.

Neuroprotective activity: Due to the stressful lifestyle of the modern times, psychiatric disorders have become commonplace—most of which are characterized by mild to extreme episodes of anxiety and depression; about 20% of adult population has been reported to be in its grasp in their lifetime [41]. Antidepressants are the most common solution to these disorders but they have high risk to benefit ratio along with the baggage of serious and long-term side effects [46]. Hence, the need of the hour is the search for active drugs that treat psychological disorders like anxiety and depression without any harmful effects on the physical and mental health. Studies have revealed that the extract of leaves of *A. nigra* (using methanol) is known to lower anxiety and fight depression [47]. The plant extract is rich in flavonoids and steroidal components that have the ability to bind to GABAA receptors in central nervous system (CNS); this can have preventive activity on depression as previously described [48]. At a concentration of 200–400 mg/kg of body weight of mice, *A. nigra* extract has sedative or anxiolytic effect, without any serious side effects unlike diazepam (positive control) at the same concentration. If developed appropriately, these extracts can be used as an anesthetic in the near future.

Other activities: In addition, *A. nigra* also exhibits certain other lesser known properties. Extract of the whole plant using methanol has a very strong antipyretic activity [44]. Further, the plant has the ability to breakdown clots and thrombus—this activity is magnified in combination with commonly used thrombolytic agent, streptokinase [49]. As *A. nigra* is consumed as food, it can act as a dietary supplement for diabetic patients; the ethanolic extract has a pronounced hypoglycemic effect as it can significantly decrease digestion and absorption of carbohydrates [50].

In general, this herbaceous plant from Zingiberaceae may have answers or cure for a number of infectious diseases. Hence, it needs to be studied properly and strategically in order to derive maximum benefit from it for human welfare.

13.4 Labdane Diterpene: A Prominent Compound Group in *A. nigra*

Physiological functions of plant secondary metabolites are rare but not unheard of. Such defined functions are possessed mostly by alkaloids—terpenes being the largest class of this group. The basic skeleton of any terpene is composed of a simple hydrocarbon called isoprene (Fig. 13.3a). A family of four isoprene units, represented as C20, constitute the large group of diterpenes. The mevalonate pathway is utilized for their synthesis via geranylgeranyl pyrophosphate (GGPP). Addition of H^+ to GGPP gives coplyl PP, a cyclic compound, and finally yields labdadienyl PP [51]. A very commonly available diterpene is taxane or taxol that is marketed as docetaxel and other brand names. It is used mostly in the treatment of cancer during chemotherapy. There are many other lesser-known plant-derived diterpenes like clerodane, kaurane, gibberellane and so on [52, 53]. Plant-based terpenes, terpenoids and terpene derivatives have various biological activities like antimicrobial against pathogens as well as reducing associated side effects of diabetes, soreness, rheumatism etc. [54–56]. It has been shown that the sesquiterpene, cnicin, together with immunomodulators like cytokines, can cause cell death, particularly of cancerous cells [57]. Terpenoids

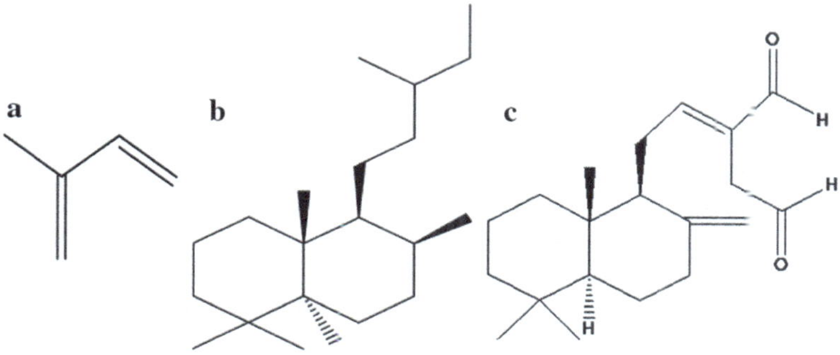

Fig. 13.3 Structure of **a** Isoprene unit, **b** labdane and **c** labdane diterpene

isolated from *A. officinarum* [58] have been reported to be antagonistic to influenza virus.

In all labdane diterpenes, the central core constitutes bicyclic diterpenes (Fig. 13.3b) [59] (https://pubchem.ncbi.mlm.nih.gov/search/compound/labdane#section=2D-Structure). The "Labdane" family of diterpenes gets its name from "Labdanum"—a plant-derived resinous substance obtained from few plants in 1956. Thereby, it is very first compound to be described from the group of diterpenes [60, 61]. A no. of labdane diterpenes and their derivatives have been isolated from different plants belonging to Zingiberaceae family that are endowed with numerous bio-pharmaceutical potentials [62–68]. The early reports available are from González-Burgos et al. [53] who highlighted the anti-oxidative property of a large no. of terpenes including labdane diterpenes. Demetzos et al. in their review have described the therapeutic potential of different types of plant-based labdane diterpenes; such conceptual overview captures information on the antimicrobial, anti-cancer, immunomodulatory, analgesic and many other properties [69].

The labdane diterpene, (E)-labda-8(17), 12-diene-15, 16-dial, the compound highlighted in this review (Fig. 13.3c), was isolated for the very first time from the seeds of *A. galanga* [52]; it also has a very cosmopolitan distribution [70–82]. In addition to *A. galanga*, its presence has been reported in many other *Alpinia* species viz. *A. chinensis* [81], *A. katsumadai* [83, 84], *A. calcarata* [85–87], *A. mutica* [88] and *A. zerumbet* [89, 90] etc. High amounts of labdane diterpene have been reported in few lesser known species of *Alpinia* viz. *A. oxymitra, A. javanica* and *A. speciosa* [71]. The labdane diterpene, (E)-labda-8(17), 12-diene-15, 16-dial, isolated from the germ holders of the plant described here was found to possess good antibacterial, antidiabetic, anti-*Candida* and anti-leishmanial properties of the compound [28, 29, 91]. It was found to have antifungal properties and is cytotoxic against KB cell line as previously reported [52]. The compound also possesses a myriad of activities such as antiviral [68, 92], antiplasmodial [93] cancer chemoprevention [94] and others [95, 96]. In addition, the compound shows huge enhancement in its otherwise feeble Raman spectrum upon utilization with Cu nanoparticles [97].

13.5 Challenges and Future Prospects

It can be deduced from the present review that *A. nigra* has immense importance in terms of its medicinal properties. Researchers have studied and exploited different properties of this plant according to their specific study objectives. The plant part mostly exploited for their biological potential is the rhizome followed by leaf; reports from other plant parts indicated very moderate studies. In the isolation of bioactive compounds, seeds are the mostly exploited plant part (Fig. 13.4). However, a question that remain is that what can be done to explore the full potential of this plant by the entire scientific community? In this context, the following points are suggested:

- In-depth phytochemical investigation of *A. nigra*-derived extracts may pave the way for the screening of different varieties of *A. nigra* of NE India and preparing their pharmaceutical profile
- Structurally, labdane diterpene is a very stable compound—as such, synthetic and combinatorial chemistry can be employed to prepare more reactive and efficient analogues. This can provide a lead for the development of a very effective antimicrobial drug—this compound can become the future 'taxol'!

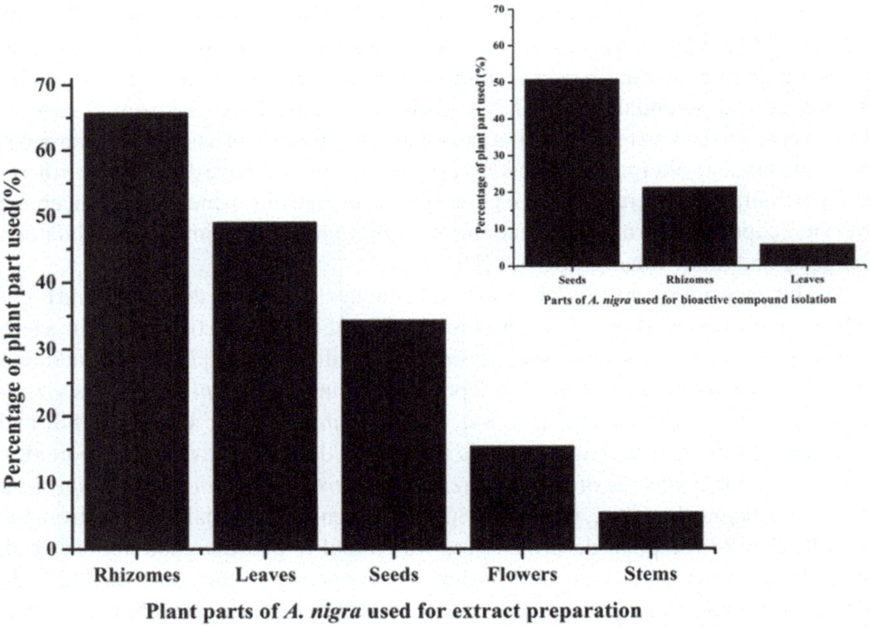

Fig. 13.4 Graphical representation of the different parts of *A. nigra* used for discussed bioactivity studies; inset represents various plant parts exploited for bioactive compound isolation

- By adsorbing labdane diterpene on different nanoparticles from noble metals like Ag, Au, Pd, Pt and others, a comparative SERS profile picture of the compound can be drawn and critically analysed
- Labdane diterpene can be identified easily by its spicy aroma—something that can prove to be a deterrent for mosquitoes and insects. Hence, the compound can be developed into suitable mosquito repellants or conjugated with other available products of the market.
- The mechanism of action of labdane diterpene needs to be thoroughly understood for further studies. Investigations need to be carried out to evaluate whether it would be able to inhibit other differentially stained microbial pathogens.

In addition, labdane diterpene is a hydrophobic compound [98] and it can be suitably used at required concentrations as a nano-composite with metallic nanoparticles. Nanotechnology is armed with a lot of advantages in numerous fields, particularly in medicine and therapeutics; hence, a preparation of a nano-composite may prove to be an excellent mechanism for the delivery and appropriate action of labdane diterpene as a potential therapeutic agent in the long run. Along with experimental research, efforts should be taken to comb through the age-old documents that shed light on the use of *A. nigra* plant parts as medicines; if no such documentation is available then there is an urgent need to preserve this traditional legacy (in the form of written document) for the future generations.

On the other hand, owing to the aromatic nature of the plant extracts, it can be deduced that the plant is rich in essential oils. These essential oils can be extracted and used in perfumery and soap industries. In this respect, a thorough quality control (QC) analysis is required by considering certain markers before establishing the plant oil in the industry. Essential oils can be further screened against various diseases in order to explore their therapeutic potential. Another important question is why the seven bioactive compounds isolated from *A. nigra* till date has not been registered as a drug yet? In this context, studies should be carried out regarding the absorption of these compounds (bioavailability) to develop these compounds as a potent therapeutic drug.

However, due its geographical location, climate and topology, a limitation of the NE India, it is a bitter truth that it did not received enough attention till date from the government as well as the global scientific community. As a result, the huge wealth of medicinal plants, particularly from Zingiberaceae (including *A. nigra*), remained unexplored till date. This review, one of first of its kind for *A. nigra*, is a maiden attempt by the authors to summarize the biopharmaceutical importance of this plant in a nutshell and to draw the attention of the scientific community to promote further rigorous research on *A. nigra* as well as other endemic plants of Zingiberaceae present in NE India.

13.6 Conclusions

A comprehensive review on the biopharmaceutical potential of *A. nigra*, its aqueous and organic extracts, along with the bioactive compound, labdane diterpene has been addressed in this chapter. Just like other species of this genus, *A. nigra* also possesses a huge myriad of bioactive compounds that bring about its bio-pharmaceutical potential. Research can also be directed towards the isolation of bioactive compounds from the rhizomes of *A. nigra* which can serve as chemically interesting and biologically important drug candidates. In addition, a wide range and mixture of solvents are employed in experimental works—methanol has been mostly used for the preparation of crude extracts and isolation of the active ingredients. Crude extracts from this plant can thus, serve as potential therapeutic agents in the future and do have the ability to overcome a wide range of challenges and risks of present day medicines.

In addition, labdane diterpene is one of the major bioactive compounds that is widely distributed among the different plants of Zingiberaceae, particularly *A. nigra*. With the wide range of activities exhibited by this compound, it can serve as one of the most promising drug candidates, that can be utilized immensely, especially in the field of synthesis and design of highly potent synthetic analogues. All in all, *A. nigra* is one of those medicinal plants of NE India that can serve as a good source of a plethora of nutraceuticals, drugs and bioactive agents, if exploited and investigated wisely. A little has been achieved but strategic, scientific and well-designed research needs to go a long way for establishing *A. nigra*, along with the isolated phytochemical labdane diterpene, as a hope for futuristic medical challenges.

However, owing to different developmental activities in terms of industrial setup, large areas of land are being cleared including places that housed dense population of medicinal plant like *A. nigra* (Fig. 13.5). As a result, this valuable plant is getting depleted without being explored to its full potential. As this plant is valuable not only to plants but also animals, it is necessary to take appropriate measures (on individual as well as govt. level) to conserve the natural diversity of *A. nigra*—this plant has not yet received its due importance in the scientific world.

2014-2019 **2022**

Fig. 13.5 Depletion of biodiversity, including *A. nigra*, as observed near IIT Guwahati

Acknowledgements The authors also thank the Ministry of Human Resource Development, Government of India, for providing student fellowship to IC. No other funding was provided for this work.

References

1. Kumar, A.: Nature has been a source of medicinal agents for thousands of years. Science 2.0 (2009). http://www.science20.com/humboldt_fellow_and_science/blog/nature_has_been_source_medicinal_agents_thousands_years-62053
2. Chamberlain, R.: From bark to bacteria: the natural sources of modern medicine. Bichemist 42–44 (2015)
3. Tushar, Basak, S., Sarma, G.C., Rangan, L.: Ethnomedical uses of Zingiberaceous plants of Northeast India. J. Ethnopharmacol. **132**, 286–296 (2010)
4. Janssen, A.M., Scheffer, J.J.: Acetoxychavicol acetate, an antifungal component of *Alpinia galanga*1. Planta Med. **51**, 507–511 (1985)
5. Oonmetta-aree, J., Suzuki, T., Gasaluck, P., Eumkeb, G.: Antimicrobial properties and action of galangal (*Alpinia galanga* Linn.) on *Staphylococcus aureus*. LWT—Food Sci. Technol. **39**, 1214–1220 (2006)
6. Weerakkody, N.S., Caffin, N., Lambert, L.K., Turner, M.S., Dykes, G.A.: Synergistic antimicrobial activity of galangal (*Alpinia galanga*), rosemary (*Rosmarinus officinalis*) and lemon iron bark (*Eucalyptus staigerana*) extracts. J. Sci. Food Agric. **91**, 461–468 (2011)
7. Roy, B., Swargiary, A., Ranjan Giri, B.: *Alpinia nigra* (Family Zingiberaceae): an anthelmintic medicinal plant of North-East India. Adv. Life Sci. **2**, 39–51 (2012)
8. Swargiary, A., Roy, B., Ronghang, B.: Partial characterisation of alkaline phosphatase in *Fasciolopsis buski*—An intestinal fluke treated with crude extract of *Alpinia nigra* (Zingiberaceae) J. Pharm. Technol. Drug Res. **2**(5) (2013)
9. Mao, A., Hynniewta, T., Sanjappa, M.: Plant wealth of Northeast India with reference to ethnobotany. Indian J. Tradit. Knowl. **8**, 96–103 (2009)

10. Upadhaya, K., Choudhury, H., Odyuo, N.: Floristic diversity of Northeast India and its conservation. In: Mandal, R. (ed.), International Environmental Economics: biodiversity and Ecology, pp. 61–71. Discovery Publishing House Pvt. Ltd., (2012)
11. Sajem, A.L., Gosai, K.: Ethnobotanical investigations among the Lushai tribes in North Cachar Hills district of Assam, Northeast India. Indian J. Tradit. Knowl. **9**, 108–113 (2010)
12. Wu, D., Larsen, K.: Zingiberaceae. In: Flora of China, pp. 322–377 (2000)
13. Sharma, U.K., Pegu, S.: Ethnobotany of religious and supernatural beliefs of the Mising tribes of Assam with special reference to the "Dobur Uie". J. Ethnobiol. Ethnomed. **7**(16) (2011)
14. Ghosh, C, Das, A.P.: Rhino-Fodders in Jaldapara wildlife sanctuary in duars of West Bengal, India. Nature 14–20 (2007)
15. Roy, B., Swargiary, A.: Anthelmintic efficacy of ethanolic shoot extract of *Alpinia nigra* on tegumental enzymes of *Fasciolopsis buski*, a giant intestinal parasite. J. Parasit. Dis. **33**, 1–6 (2009)
16. Ismail, F., Anjum, M.R., Mamon, A.N., Kazi, T.G.: Trace metal contents of vegetables and fruits of Hyderabad retail market. Pak. J. Nutr. **10**, 365–372 (2011)
17. Kar, A., Choudhary, B.K., Bandyopadhyay, N.G.: Preliminary studies on the inorganic constituents of some indigenous hypoglycaemic herbs on oral glucose tolerance test. J. Ethnopharmacol. **64**, 179–184 (1999)
18. Shahid, M., Hussain, F.: Chemical composition and mineral contents of *Zingiber officinale* and *Alpinia allughas* (Zingiberaceae) rhizomes. Int. J. Chem. Biochem. Sci. **2**, 101–104 (2012)
19. Roy, J.: Karyophotometrical analysis and exploration of major oil constituents of Zingiberaceae. Mahatma Gandhi University, Kottayam (1998)
20. Ghosh, S., Ozek, T., Tabanca, N., Ali, A, ur Rehman, J., Khan, I.A., Rangan, L..: Chemical composition and bioactivity studies of *Alpinia nigra* essential oils. Ind. Crops Prod. **53**, 111–119 (2014)
21. Islam, F., Islam S., Shahjahan, M., Nandi, N.C., Satter, M.A.: Chemical constituents of essential oil from the leaf of *Alpinia nigra* of Bangladesh. Int. Food Res. J. **21**, 161–164 (2014)
22. Mejdoub, R., Katsiotis, S.T.: Factors influencing the yield and the quality of the obtaining essential oil from the leaves of *Eucalyptus citriodora* Hook. Growing in crete. Sci. Pharm. **66**, 93–105 (1998)
23. Huang, W.Y., Cai, Y.Z., Zhang, Y.: Natural phenolic compounds from medicinal herbs and dietary plants: potential use for cancer prevention. Nutr. Cancer **62**, 1–20 (2010)
24. Qiao, C., Wang, Z., Dong, H., Xu, L., Hao, X.: The chemical constituents of Blackfruit Galangal (*Alpinia nigra*). Chinese Tradit. Herbs Med. **31**, 404–405 (2000)
25. Qiao, C., Han, Q., Song, J., Wang, Z., Xu, L., Xu, H.: HPLC determination of two bioactive flavone glycosides and GC-MS analysis of volatile oil constituents in *Alpinia nigra*. Asian J. Tradit. Med. **2**, 85–91 (2007)
26. Ghosh, S., Rangan, L.: Molecular docking and inhibition kinetics of α-glucosidase activity by labdane diterpenes isolated from Tora Seeds (*Alpinia nigra* B.L. Burtt.). Appl. Biochem. Biotechnol. **175**, 1477–1489 (2014)
27. Sethi, S., Prakash, O., Pant, A.K.: Phytochemical analysis and isolation of stigmasteroal and sitosterol from the rhizomes of *Alpinia allughas* Rosc. Int. J. Curr. Res. Biosci. Plant Biol. **2**, 124–129 (2015)
28. Ghosh, S., Singh, R.K., Dubey, V.K., Rangan, L.: Antileishmanial activity of labdane diterpenes isolated from *Alpinia nigra* seeds. Lett. Drug Des. Discov. **14**, 119–124 (2017)
29. Chakrabartty, I., Vijayasekhar, A., Rangan, L.: Therapeutic potential of labdane diterpene isolated from *Alpinia nigra* : Detailed hemato-compatibility and antimicrobial studies. Nat. Prod. Res. **35**(6), 1000–1004 (2019)
30. Gupta, M.K., Senthilkumar, S., Chiranjivi, A.K., Banik, K., Girisa, S., Kunnumakkara, A.B., Dubey, V.K., Rangan, L.: Antioxidant, anti-tyrosinase and anti-inflammatory activities of 3,5-dihydroxy-4′,7-dimethoxyflavone isolated from the leaves of *Alpinia nigra*. Phytomed. Plus **1** (2021)
31. Parthasarathy, V.A., Kandiannan, K.: Spices and condiments. In: Horticulture (Vegetables, Tubers, Spice and Crops) (2007)

32. Roy, B., Dasgupta, S., Tandon, V.: Ultrastructural observations on *Fasciolopsis buski* and its alterations caused by shoot extract of *Alpinia nigra*. Microsc. Res. Tech. **72**, 61–66 (2009)
33. Bhunia, D., Mondal, A.K.: Antibacterial activity of *Alpinia* L. (Zingiberaceae) from Santal and Lodha tribal areas of Paschim Medinipur district in Eastern India. Adv. Bioresour. **3**, 54–63 (2012)
34. Abu Ahmed, A.M., Sharmen, F., Mannan, A., Rahman, M.A.: Phytochemical, analgesic, antibacterial, and cytotoxic effects of *Alpinia nigra* (Gaertn.) Burtt leaf extract. J. Tradit. Complement. Med. **5**, 248–252 (2015)
35. Chakrabartty, I., Mohanta, Y.K., Nongbet, A., Mohanta, T.K., Mahanta, S., Das, N., Sarvanan, M., Sharma, N.: Exploration of Lamiaceae in cardio vascular diseases and functional foods: medicine as food and food as medicine. Front. Pharmacol. (2022). https://www.frontiersin.org/articles/10.3389/fphar.2022.894814/abstract
36. Sahoo, S., Ghosh, G., Das, D., Nayak, S.: Phytochemical investigation and in vitro antioxidant activity of an indigenous medicinal plant *Alpinia nigra* B.L. Burtt, Asian Pacific J. Trop. Biomed. **3**, 871–876 (2013)
37. Sattar, N., Hussain, F., Iqbal, T.: Antioxidant activities of *Z. officinale* Roscoe and *A. allughas* Roscoe (Zingiberaceae) rhizomes, Bangladesh. J. Sci. Ind. Res. **48**, 115–118 (2013)
38. Paul, B., Sen, C.K., Saha, M.R.: Cytotoxic effect of *Alpinia nigra* crude extracts obtained from its rhizomes. Int. J. Pharmacogn. **2**, 49–54 (2015)
39. Ferrero-Miliani, L., Nielsen, O.H., Andersen, P.S., Girardin, S.E.: Chronic inflammation: Importance of NOD2 and NALP3 in interleukin-1β generation. Clin. Exp. Immunol. **147**, 227–235 (2007)
40. Urhausen, A., Torsten, A., Wilfried, K.: Reversibility of the effects on blood cells, lipids, liver function and hormones in former anabolic-androgenic steroid abusers. J. Steroid Biochem. Mol. Biol. **84**, 369–375 (2003)
41. Ghosh, S., Rangan, L.: *Alpinia*: the gold mine of future therapeutics. 3 Biotech **3**, 173–185 (2013)
42. Das, B.N., Biswas, B.K.: Anti-inflammatory activity of the rhizome extract of *Alpinia nigra*. Int. Res. J. Pharm. **02**, 73–76 (2012)
43. Das, B.K., Fatema, U.K., Hossain, M.S., Rahman, R., Fatema, K.: Anti-inflammatory and analgesic activities of Alpinia nigra fruit extract in laboratory animals. Eur J. Intern. Med. **4**, 1158–1166 (2014)
44. Ameen, S.S., Jahan, S.R., Ali, M.Y.: Evaluation of the anti-inflammatory and antipyretic activities of the plant *Alpinia nigra* (Family: Zingiberaceae). World J. Pharm. Res. **3**, 164–170 (2014)
45. Ghosh, S., Padilla-González, G.F., Rangan, L.: *Alpinia nigra* seeds: a potential source of free radical scavenger and antibacterial agent. Ind. Crops Product **49**, 348–356 (2013)
46. Newman, M.F., Lhuillier, A., Poulsen, A.D.: Checklist of the Zingiberaceae of Malesia. Universiteit Leiden branch No, Netherland, Nationaal Herbarium Nederland (2004)
47. Sharmen, F., Mannan, A., Rahman, M.M., Chowdhury, M.A.U., Uddin, M.E., Abu Ahmed, A.M.: Investigation of in vivo neuropharmacological effect of *Alpinia nigra* leaf extract. Asian Pacific J. Trop. Biomed. **4**, 137–142 (2014)
48. Jäger, A.K., Saaby, L.: Flavonoids and the CNS. Molecules **16**, 1471–1485 (2011)
49. Mannan, A., Ahmed, A.M.A., Sharmin, F., Fatema, T., Omar, M., Sikder F.: In vitro thrombolytic assay of *Alpinia zerumbet*, *Alpinia nigra* and *Urena sinuata*. Int. J. Res. Phytochem. Pharm. Sci. **1**, 187–191 (2011)
50. Ghosh, S., Indukuri, K., Bondalapati, S., Saikia, A.K., Rangan, L.: Unveiling the mode of action of antibacterial labdane diterpenes from *Alpinia nigra* (Gaertn.) B L. Burtt seeds. Eur. J. Med. Chem. **66**, 101–105 (2013)
51. Dewick, P.M.: Medicinal Natural Products: a Biosynthetic Approach, 2nd edn. Wiley, England, UK (2002)
52. Morita, H., Itokawa, H.: Cytotoxic and antifungal diterpenes from seeds of *Alpinia galanga*. Planta Med. **54**, 117–120 (1988)

53. Gonzalez-Burgos, E., Gomez-Serranillos, M.: Terpene compounds in nature: a review of their potential antioxidant activity. Curr. Med. Chem. **19**, 5319–5341 (2012)
54. Steinbach, A., Scheidig, A.J., Klein, C.D.: The unusual binding mode of cnicin to the antibacterial target enzyme MurA revealed by X-ray crystallography. J. Med. Chem. **51**, 5143–5147 (2008)
55. Fortuna, A.M., de Riscala, E.C., Catalan, C.A.N., Gedris, T.E., Herz, W.: Sesquiterpene lactones from *Centaurea tweediei*. Biochem. Syst. Ecol. **29**, 967–971 (2001)
56. Cabral, C., Goncalves, M.J., Cavaleiro, C., Salgueiro, L., Antunes, T., Sevinate-Pinto, I., Sales, F.: *Vitex ferruginea* Schumach. Et. Thonn. subsp *amboniensis* (Gurke) Verdc.: Glandular trichomes micromorphology, composition and antifungal activity of the essential oils. J. Essent. Oil Res. **20**, 86–90 (2008)
57. Jöhrer, K., Obkircher, M., Neureiter, D., Parteli, J., Zelle-Rieser, C., Maizner, E., Kern, J., Hermann, M., Hamacher, F., Merkel, O., Wacht, N., Zidorn, C., Scheideler, M., Greil, R.: Antimyeloma activity of the sesquiterpene lactone cnicin: impact on Pim-2 kinase as a novel therapeutic target. J. Mol. Med. **90**, 681–693 (2012)
58. Sawamura, R., Shimizu, T., Sun, Y., Yasukawa, K., Miura, M., Toriyama, M., Motohashi, S., Watanabe, W., Konno, K., Kurokawa, M.: In vitro and in vivo anti-influenza virus activity of diarylheptanoids isolated from *Alpinia officinarum*. Antivir. Chem. Chemother. **21**, 33–41 (2010)
59. PubChem, Labdane|C20H38, PubMed. 2012. https://pubchem.ncbi.nlm.nih.gov/compound/labdane#section=2D-Structure. Accessed 16 Feb. 2018
60. Cocker, J.D., Halsal, T.G., Bowers, A.: The chemistry of gum labdanu. Part I. Some acidic constituents. J. Chem. Soc. **819**, 4259–4262 (1956)
61. Cocker, J.D., Halsal, T.G.: The chemistry of gum labdanum. Part II. The structure of labdanolic acid. J. Chem. Soc. **820**, 4262–4271 (1956)
62. Kunnumakkara, A.B., Ichikawa, H., Anand, P., Mohankumar, C.J., Hema, P.S., Nair, M.S., Aggarwal, B.B.: Coronarin D, a labdane diterpene, inhibits both constitutive and inducible nuclear factor-kappa B pathway activation, leading to potentiation of apoptosis, inhibition of invasion, and suppression of osteoclastogenesis. Mol. Cancer Ther. **7**, 3306–3317 (2008)
63. Prabhakar, R.P., Ranga, R.R., Shashidhar, J., Sastry, B.S., Madhusudana, R.J., Suresh Babu, K.: Phytochemical investigation of labdane diterpenes from the rhizomes of *Hedychium spicatum* and their cytotoxic activity. Bioorg. Med. Chem. Lett. **19**, 6078–6081 (2009)
64. Suresh, G., Prabhakar, R.P., Suresh Babu, K., Shaik, T.B., Kalivendi, S.V.: Two new cytotoxic labdane diterpenes from the rhizomes of *Hedychium coronarium*. Bioorg. Med. Chem. Lett. **20**, 7544–7548 (2010)
65. van Wyk, A.W.W., Davies-Coleman, M.T.: Semisynthesis of labdane diterpene metabolites from the nudibranch *Pleurobranchaea meckelii*. Tetrahedron **63**, 12179–12184 (2007)
66. Sob, S.V.T., Tane, P., Ngadjui, B.T., Connolly, J.D., Ma, D.: Trypanocidal labdane diterpenoids from the seeds of *Aframomum aulacocarpos* (Zingiberaceae) Tetrahedron **63**, 8993–8998 (2007)
67. Tomla, C., Kamnaing, P., Ayimele, G.A., Tanifum, E.A., Tsopmo, A., Tane, P., Ayafor, J.F., Connolly, J.D.: Three labdane diterpenoids from *Aframomum sceptrum* (Zingiberaceae). Phytochemistry **60**, 197–200 (2002)
68. González, M.A., Mancebo-Aracil, J., Tangarife-Castaño, V., Agudelo-Goméz, L., Zapata, B., Mesa-Arango, A., Betancur-Galvis, L.: Synthesis and biological evaluation of (+)-labdadienedial, derivatives and precursors from (+)-sclareolide. Eur. J. Med. Chem. **45**, 4403–4408 (2010)
69. Demetzos, C., Dimas, K.S.: Labdane-type diterpenes: chemistry and biological activity. Stud. Nat. Prod. Chem. **25**, 235–292 (2001)
70. Sirat, H.M., Hadi, A.H.A.: A labdane diterpene from *Enicosanthum congregatum*. J. Sci. Technol. Pertanika **1**, 189–193 (1993)
71. Sirat, H.M., Masri, D., Rahman, A.A.: The distribution of labdane diterpenes in the Zingiberaceae of Malaysia. Phytochemistry **36**, 699–701 (1994)

72. Kenmogne, M., Prost, E., Harakat, D., Jacquier, M.J., Frédérich, M., Sondengam, L.B., Zèches, M., Waffo-Téguo, P.: Five labdane diterpenoids from the seeds of *Aframomum zambesiacum*. Phytochemistry **67**, 433–438 (2007)
73. Sheeja, A.D.B., Nair, M.S.: Facile isolation of (E)-labda-8 (17), 12-diene-15, 16-dial from Curcuma amada and its conversion to other biologically active compounds. Indian J. Chem. **53**, 319–324 (2014)
74. Singh, S., Kumar, J.K., Saikia, D., Shanker, K., Thakur, J.P., Negi, A.S., Banerjee, S.: A bioactive labdane diterpenoid from *Curcuma amada* and its semisynthetic analogues as antitubercular agents. Eur. J. Med. Chem. **45**, 4379–4382 (2010)
75. Abas, F., Lajis, N.H., Shaari, K., Israf, D.A., Stanslas, J., Yusuf, U.K., Raof, S.M.: A labdane diterpene glucoside from the rhizomes of *Curcuma mangga*. J. Nat. Prod. **68**, 1090–1093 (2005)
76. Abe, M., Ozawa, Y., Uda, Y., Yamada, Y., Morimitsu, Y., Nakamura, Y., Osawa, T.: Labdane-type diterpene dialdehyde, pungent principle of Myoga, *Zingiber mioga* Roscoe. Biosci. Biotechnol. Biochem. **66**, 2698–2700 (2002)
77. Firman, K., Kinoshita, T., Itai, A., Sankawa, U.: Terpenoids from *Curcuma heyneana*. Phytochemistry **27**, 3887–3891 (1988)
78. Kimbu, S.F., Ngadjui, B., Sondengam, L.B., Njimi, T.K., Connolly, J.D., Fakunle, C.O.: A new labdane diterpenoid from the seeds of *Aframomum daniellii*. J. Nat. Prod. **50**, 230–231 (1987)
79. Kimbu, S.F., Nijmi, T.K., Sondengam, B.L., Akkinnmyi, J.A., Connolly, J.D.: The structure of labdane dialdehyde from *Afromomum danielli* (Zingiberaceae). J. Chem. Soc. Perkin Trans. **1**, 1303–1304 (1979)
80. Abreu, P.M., Noronha, R.G.: Volatile constituents of the rhizomes of *Aframomum alboviolaceum* (Ridley) K. Schum. from Guinea-Bissau. Flavour Fragr. J. **12**, 79–83 (1997)
81. Sy, L.K., Brown, G.D.: Labdane diterpenoids from *Alpinia chinensis*. J. Nat. Prod. **60**, 904–908 (1997)
82. Tatsimo, S.J.N., Tane, P., Melissa, J., Sondengam, B.L., Okunji, C.O., Schuster, B.M., Iwu, M.M., Khan, I.A.: Antimicrobial principle from *Aframomum longifolius*. Planta Med. **72**, 132–135 (2006)
83. Hua, S.Z., Luo, J.G., Wang, X.B., Wang, J.S., Kong, L.Y.: Two novel monoterpene-chalcone conjugates isolated from the seeds of *Alpinia katsumadai*. Bioorg. Med. Chem. Lett. **19**, 2728–2730 (2009)
84. Ngo, K.S., Brown, G.D.: Stilbenes, monoterpenes, diarylheptanoids, labdanes and chalcones from *Alpinia katsumadai*. Phytochemistry **47**, 1117–1123 (1998)
85. Kong, L.Y., Qin, M.J., Masatake, N.: Two new bis-labdanic diterpenoids from *Alpinia calcarata*. Acta Bot. Sinica **46**, 159–164 (2004)
86. Kong, L.Y., Qin, M.J., Niwa, M.: Diterpenoids from the rhizomes of *Alpinia calcarata*. J. Nat. Prod. **63**, 939–942 (2000)
87. Hema, P.S., Nair, M.S.: Flavonoids and other constituents from the rhizomes of *Alpinia calcarata*. Biochem. Syst. Ecol. **37**, 52–54 (2009)
88. Sirat, H.M., Jani, N.A.: Chemical constituents of the leaf of *Alpinia mutica* Roxb. Nat. Prod. Res. **27**, 1468–1470 (2013)
89. Chompoo, J., Upadhyay, A., Kishimoto, W., Makise, T., Tawata, S.: Advanced glycation end products inhibitors from *Alpinia zerumbet* rhizomes. Food Chem. **129**, 709–715 (2011)
90. Upadhyay, A., Chompoo, J., Kishimoto, W., Makise, T., Tawata, S.: HIV-1 integrase and neuraminidase inhibitors from *Alpinia zerumbet*. J. Agric. Food Chem. **59**, 2857–2862 (2011)
91. Ghosh, S., Rangan, L.: Molecular docking and inhibition studies of α-amylase activity by labdane diterpenes from *Alpinia nigra* seeds. Med. Chem. Res. **23**, 4836–4852 (2014)
92. Grienke, U., Schmidtke, M., Kirchmair, K., Pfarr, J., Wutzler, P., Dürrwald, R., Wolber, G., Liedl, K.R., Stuppner, H., Rollinger, H.M.: Antiviral potential and molecular insight into neuraminidase inhibiting diarylheptanoids from *Alpinia katsumadai*. J. Med. Chem. **53**, 778–786 (2010)
93. Duker-Eshun, G., Jaroszewski, J.W., Asomaning, W.A., Oppong-Boachie, F., Olsen, C.E., Christensen, S.B.: Antiplasmodial activity of labdanes from *Aframomum latifolium* and *Aframomum sceptrum*. Planta Med. **68**, 642–644 (2002)

94. Endringer, D.C., Taveira, F.S.N., Kondratyuk, T.P., Pezzuto, J.M., Bragaa, F.C.: Cancer chemoprevention activity of labdane diterpenes from rhizomes of *Hedychium coronarium*. Braz. J. Pharmacogn. **24**, 408–412 (2014)
95. Martins, C.P., Gomes, O.A., Martins, M.L., de Carvalho, L.D., de Souza, J.G., Da Fonseca, F.G., dos Santos, R.G., Andrade, M.S., Zani, C.L., de Souza-Fagundes, E.M., Barbosa-Stancioli, E.F.: A reduction of viral mRNA, proteins and induction of altered morphogenesis reveals the anti-HTLV-1 activity of the labdane-diterpene myriadenolide in vitro. BMC Microbiol. **14**(331) (2014)
96. Awang, K., Abdullah, N.H., Hadi, A.H.A., Su Fong, Y.: Cardiovascular activity of labdane diterpenes from *Andrographis paniculata* in isolated rat hearts. J. Biomed. Biotechnol. (2012). https://doi.org/10.1155/2012/876458
97. Chakrabartty, I., Baruah, P.K., Panda, A.N., Khare, A., Rangan, L.: Hybrid formulation of Cu nanoparticles and labdane diterpene from *Alpinia nigra*: a vibrational spectroscopic study. J. Appl. Spectrosc. **85**, 983–990 (2018)
98. Chakrabartty, I., Kalita, N.K., Boruah, P., Hakeem, K.R., Katiyar, V., Rangan, L.: Physico-rheological characterization of organically derived seed samples from *Alpinia nigra* (Gaertn.) B.L. Burtt, an ethnic medicinal plant of Northeast India. Ind. Crops Prod. **152** (2020)

Chapter 14
Enhancing Bio-Diversity Conservation through Integrated Horticultural Development Strategies: A Case Study of Kiwi Fruits in Arunachal Pradesh of North-East India

Sanjib Kumar Dutta, Papiya Dutta, Saikat Majumdar, Gayatri Saikia, Lilika K. Zhimomi, and TechiYajoTara

14.1 Introduction

The horticulture industry is one of the most profitable and practical options for the diversified farming. Through such ventures, bio-diversity conservation can be ensured in the north-east. Biodiversity loss has become a global concern, owing in partly to agricultural intensification. In such a scenario, the fruit growers wish to adopt environmentally friendly production techniques. In order to develop an environment favourable for beneficial living beings, they strive to restrict or reduce intense pesticide and chemical fertilizer usage and thereby improve organic management of the orchards. Horticulture is critical to biodiversity conservation and sustainable development region [1]. Consequently, Kiwi farming is a sustainable method of land management as it is a good alternative for sloped land management that accustoms soil erosion reduction. It guarantees more effective usage of land, lowers downstream flooding, makes pollen for bees easier to access, and promotes mixed farming, which subsequently enhances biodiversity. Also, the high-value crop itself promotes biodiversity. Also, it enhances livelihoods by offering a source of monetary income [2]. Furthermore, horticulture holds the key to crop and food diversification, and for the assurance of nutritional security. Additionally, natural pollination for horticulture crops is crucial to maintain an array of farmers' varieties that are adapted to the local environment and to climate change. A number of best practices can be used to maintain and improve pollination services. Synchronization between pollinators and horticulture crops is essential to guarantee optimality of pollination strategies and practices.

S. K. Dutta (✉) · P. Dutta · S. Majumdar · G. Saikia · L. K. Zhimomi · TechiYajoTara
Department of Rural Development, University of Science and Technology Meghalaya, Baridua, India
e-mail: rohanbaba1998@yahoo.co.in

R. V. S. Uppaluri and L. Rangan (eds.), *Conservation of Biodiversity in the North Eastern States of India*, https://doi.org/10.1007/978-981-99-0945-2_14

Sustainable and cost competitive agro-business management holds the key for the flourishing of the horticulture industry. It provides a broad range of crops with the potential for various cropping systems that are suited for production under various agro-climatic conditions and topography, increasing yields per unit of area, creating employment, and ensuring nutritional and food security [3]. The demand for vegetarian cuisine enabled dietary behaviors to change. This inturn enhanced attention upon the horticulture crops. Any crop-based enterprise has to venture into profitable engagement from both production and marketing perspectives. Thereby, disorganized engagement of horticulture production and marketing enterprises demand upon the definite need for an integrated horticulture development plan. True benefits from such an approach will be achieved with improved availability of quality fruits for marketing, given the perishable nature of produce such as the Kiwifruit and its seasonal production. Accordingly, needs are to be matured and emphasized. Departmental inventions are mandatory to address the problems and constraints in the prominent areas such as production and marketing [4]. The proposed integrated horticultural development plan aims to address various issues in order to ultimately assist in the restructuring of the post-harvest management and marketing of horticultural commodities. Such holistic approaches will involve interventions in three major engagements. These are production sustainability and enhancement, post-harvest management and value addition.

The state of Arunachal Pradesh lies between the North latitude of 26,028–29,030 and East longitude of 90,030–97,030. The chapter devotes to the case study conducted in the north-east district of Arunachal Pradesh. This state is a typical example with vast agricultural land scape and practices with its very remote location. The state's major produces demand an effective post-harvest management like spices, vegetables and fruits which are still under-developed and especially for the perishable product sectors such as fruits and vegetables. Both lower density, higher return crops are often used to augment the ribbon-style growth in the interior of the state and thereby support village economy in less communicated and isolated areas. The availability of plain, wide, sloppy terrain at higher elevations with steeper slopes and small valleys is highly challenging for effective post-harvest management. In these farm regions, agriculture and horticulture take precedence over food cultivation and farming. The state government promotes horticulture as a means to reduce poverty by providing regular earnings and newer job possibilities [3]. Horticultural farming is becoming the primary regional economic policy in the government services. Thereby, the inclusion of sale in the governance activity is gradually considered by the State Government for the effective post-harvest administration.

Considering few critical aspects, this book chapter attempts to identify various problems and difficulties faced by the Kiwi growers in the Arunachal Pradesh and thereby, provide useful information to the stakeholders and administrators of the government. Further, the chapter aims towards needful awareness programme on financial status of kiwi growers and suggest suitable measures that assist in the overall well-being of the farmers.

14.2 Overview of Horticultural Produces of Arunachal Pradesh

Based on geography and current agro-climatic considerations, the state can be divided among four horticulture sections. These have been presented in Table 14.1. The table conveys that abundant produce of perishable fruits and vegetables has been achieved due to potential area of farming.

Depending on the potential of different areas and micro-climatic conditions, few selected fruit crops and spices can be identified. These have been presented in Table 14.2.

In its subtropical and temperate environment, the state grows a large variety of fruits, spices, vegetables, orchids, and medicative herbs with a reported area of 69,500 ha as per 2020 data. Fruits occupy an area of 45,091 hectare among the horticulture crops and generate a production of 1,17,950 MT (2020). Spices represent the next category of horticultural crops and occupy 11,612 ha that are crucial

Table 14.1 A summary of horticultural zones, cultivated crops and potential area in the state of Arunachal Pradesh, North-east India

S. no	Horticultural zones	Crops suitable for cultivation	Potential area (ha)
1	Foot Hills & Valley (170–915 M altitude)	Litchi, Citrus, Papaya, Pineapple, Banana, Seasonal Vegetables	6,05,147
2	Mid Hills (915–1830 M altitude)	Peach, Plum, Apricot, Almond, Low chilling apple, Pomegranate, Flowers, Seasonal Vegetables	5 84,280
3	High Hills	Cherry, Apple, Pomegranate, Walnut, Kiwi, Off seasonal Vegetables, seed production of temperate vegetables	4 79, 944
4	Rain shadow area with rainfall (with less than 40″ rainfall)	Plum, Walnut, Apple, Peach, Areca nut, Almond, Pear, Off seasonal Vegetables	1,30,629
	Total		18,00,000

Data source Directorate of Horticulture, Government of Arunachal Pradesh

Table 14.2 A summary of areas-climate-crops synergy in the state of Arunachal Pradesh

S. no	Area	Climate	Crops
1	Tawang, West Kameng, Lower Subansiri, Mechuka, Anini, Hawai, walong	Temperate	Apple, Walnut, Kiwi, Pear, Plum
2	Middle foot hill areas of all the districts	Sub-tropical	Citrus, Pineapple, Guava, Litchi

Table 14.3 A summary of production (MT) of various fruits and species in the year 2020 in the state of Arunachal Pradesh

S. no	Name of the fruits	Area (ha)	Production (MT)
A. Temperate Fruits			
1	Apple	4664	6557
2	Walnut	1000	630
3	Pear	1164	2956
4	Kiwi	320	450
	Sub-total	**6,901**	**10,901**
B. Tropical & Sub-Tropical Fruits			
1	Mandarin	32,730	69,740
2	Pine apple	3090	23,110
3	Banana	2210	14,080
4	Lime/Lemon	160	111
	Sub-total	**38,190**	**1,07,041**
C. Spices			
1	Large Cardamom	2342	512
2	Black Pepper	820	118
3	Ginger	7650	56,580
	Turmeric	800	3840
	Sub-total	**11,612**	**61,050**
	Total	**56,703**	**1,79,002**

Data Source Directorate of Horticulture, Arunachal Pradesh

to generate income based on a production rate of 61,050 MT per year. Table 14.3 provides information on the production and area for various spices and fruits. Besides the crops mentioned in the Table 14.3, the other major fruit produced are Papaya, Guava, Jack Fruit, Plum, litchi, and Peach. Similarly, under spices, Large Cardamom and Ginger fall in this category.

14.3 Salient Features of Horticultural Marketing Scenario in Arunachal Pradesh

Horticultural marketing in Arunachal Pradesh is highly unorganized and underdeveloped. Few salient features of the horticultural marketing scenario in Arunachal Pradesh are as follows:

(a) Poorly developed market infrastructure: Most markets do not have adequate facility to display, auction and facilitate seller-buyer interactions. While storage facility for non-perishable commodities are available to certain extent, they have been nil for such temperate fruits.

(b) Non-established Market Yards: A major portion of the commodities do not flow through the market yards. Farmer's participation in market operation is seldom found even at the primary rural markets. Traders have their own established channel for procuring and distributing the commodities in the trading network.
(c) Middlemen dominating the Linkages: Backward and forward linkage is formed through a long chain of middlemen that percolate right up to the village level. Middlemen at backward link very often change their role in market management, as agricultural trading is not the primary occupation of about 79% of surveyed such middlemen. Often, they disassociate themselves from the marketing chain and especially when the markets are down.
(d) Lack of Institutional Market Functionaries: The entire marketing system has been created and managed by private traders and middlemen Institutional market functionaries do not exist at any stage of marketing.
(e) Buyer Dominance: Buyer dominance at the primary stage of marketing is very evident. This is due to the existence of area and commodity-specific monopoly of large traders. They operate through their own established channels. Farmers receive a throw-away price which is 16–28% of the actual whole sale price. The price depends upon the area and commodity.
(f) Non-existence of Credit-linked Marketing for Fruits: Pre-harvest sales and credit-linked marketing have been witnessed with respect to cash crops like ginger, areca nut etc. However, such options are scarce for the fruits.
(g) Higher Post-harvest Loss: The extent of post-harvest loss of horticultural produces is very high. Depending upon the commodity and their production sites it is estimated that the post-harvest loss is about 18–32% during marketing and transport. Nevertheless, around 30–40% of the perishable produces especially vegetables and certain types of fruits perish at the field itself and even prior to their harvest.

14.4 Kiwi Production Status in World and Arunachal Pradesh

The Scientific name of kiwi is *Actinidiadeliciosa.* This fruit originated from China. The first commercial plantings were established in New Zealand in 1937. In 1960s, the expansion of kiwi fruit began throughout the world. By 1980s, other countries around the world began to produce and export kiwi. Presently, the highest commercial growth of kiwi fruit is in the countries of China, followed by New Zealand and Italy. Figure 14.1 presents a comparative kiwi fruit produce in the world. India is not in the graph due to its sub-tropical and non-chilling climate.

In India, Kiwi was 1st planted in 1960 and in Lal Bagh Garden at Bangalore. However, the plant did not come into bearing due to the lack of chilled environment. Later in 1963, it was again reintroduced by National Bureau of Plant Genetic Resource (NBPGR), Regional Station, Shimla and the plant came into bearing in 1969 [5].

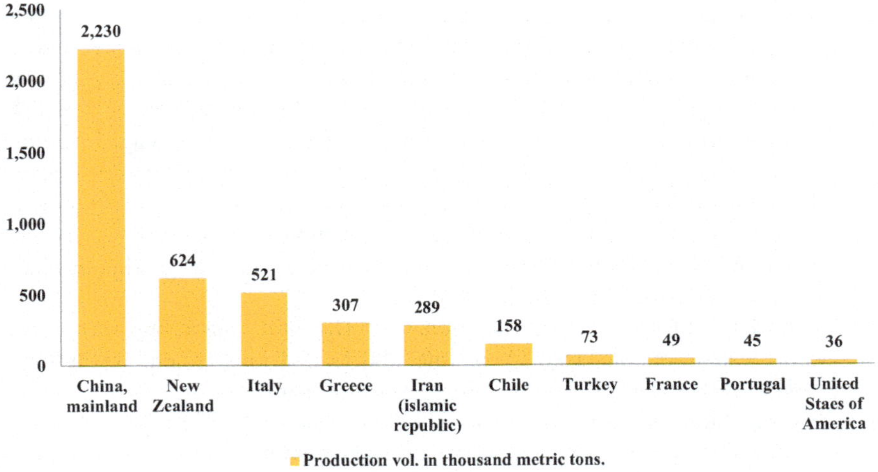

Fig. 14.1 Bar chart depicting the Kiwi fruit production volume (in thousands of MT) in various countries. *Data source* Food and Agriculture Organization Corporate Statistical Data base

In Arunachal Pradesh, twenty years ago, the kiwifruit was called "**Anteri**" and it grew as a wild species in the hills of Arunachal Pradesh's Ziro valley. The fruit was then used to feed the animals and was also consumed by the local people and barely caught anyone's attention. A domesticated variety of kiwi was introduced as a commercial fruit only in the recent year of 2000. Only when the local market of Arunachal Pradesh started selling kiwis, the farmers of Ziro slowly and steady recognized the commercial value of the fruit and started establishing kiwi orchards.

The percentage share of major states of North-East India for kiwi production has been depicted in Fig. 14.2 [6]. Being the largest kiwi producing state, the overall productivity of kiwi in Arunachal Pradesh is very low (3.08 MT/ha). Approximately, 56% of the total 14,000 tons of the kiwi in 2020 is produced from the state of Arunachal Pradesh. The state has a tremendous scope for cultivation owing to its natural agro-climate condition. Thus, there is a huge scope for the kiwi production in the country and especially in the northeastern part of India and in states such as Arunachal Pradesh, Nagaland, Manipur, Sikkim and Meghalaya. Arunachal Pradesh became the first state to obtain organic certification of kiwi under Mission organic value chain Development for North East Region (MOVCD-NER). Such certification in India can be obtained after strict scientific assessment being conducted by the Agricultural and Processed Food Products Export Development Authority (APEDA).

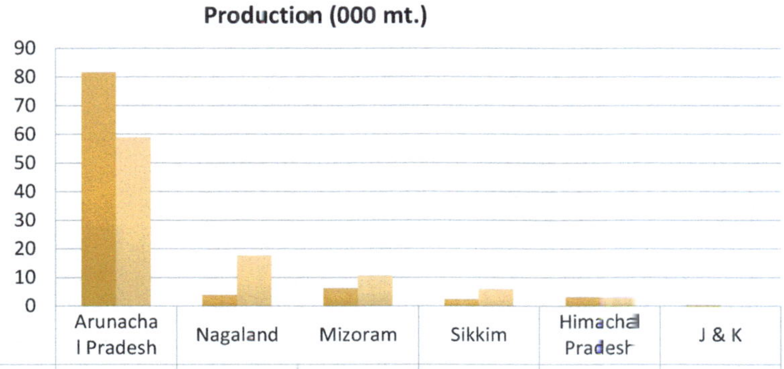

	Arunachal Pradesh	Nagaland	Mizoram	Sikkim	Himachal Pradesh	J & K
Area	81.76	4.05	6.32	2.53	3.26	0.47
Production Share	59.11	17.78	10.78	6.03	3.27	0.06

Fig. 14.2 Percentage share of area and production volume of kiwi fruit production in Indian states during 2020–21. *Data source* Agricultural and Processed Food Products Export Development Authority (APEDA)

14.5 Critical Issues for Kiwi Fruit Production and Marketing in Lower Subansiri District of Arunachal Pradesh

Lower Subansiri district of Arunachal Pradesh is the only district to produce kiwi fruit in the cluster. The district is divided into four blocks and among these, only block Ziro-I produced the cluster of Kiwi in the district. It is located in and around the district headquarters i.e. within 10–15 km from Ziro. According to APEDA (2020) report, Arunachal Pradesh produced the highest Kiwi fruit in the India. Also, according to District Horticulture Office, Lower Subansiri, the Kiwi fruit produced was 2575 MT. during 2020–21.Incidentally, the production and area has been increasing in the district from 2018 till 2021 (Fig. 14.3). In the year 2017, "NAARA-AABA" brand brewed by M/s LambuSubu Food & Beverages claimed to become the first Indian pure organic kiwi winery. "NaaraAaba" Kiwi winery unit was set by the Mrs. Tage Rita in Hong village and has a capacity of 60,000 L per batch. Apart from kiwi wine, the unit is now producing wine from fruits such as Pears, Peaches & Plums which are abundantly grown in Lower Subansiri district.

The lower Subansiri district of Arunachal Pradesh has a very favorable climate condition as well as fertile soil for Agriculture and Horticulture development. The climate condition and soil type of Ziro is the most favorable for the Apple and Kiwi cultivation. Cool summer and cooler winter is the main characteristic feature of the Ziro's climate. Tremendous commercial production in Ziro has an impact on the life of rural people. It increased the living-standard and socio-economic parts of the district. In this chapter, studies have been devoted to resolve few problems being faced by kiwi crop which can be listed as follows:

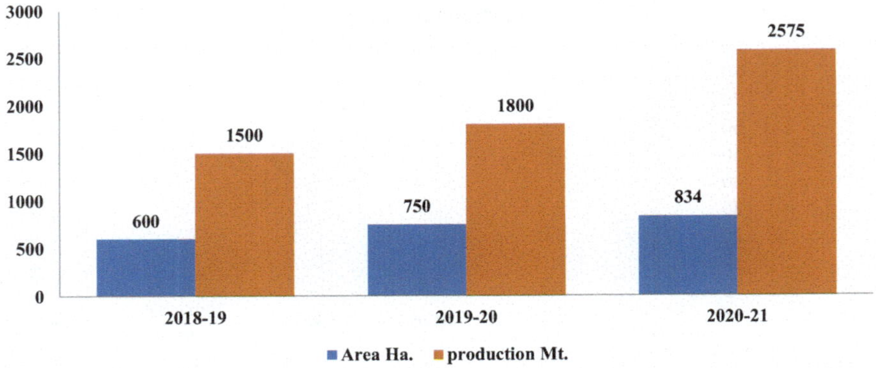

Fig. 14.3 Bar chart depicting cultivation area and production volumes of Kiwi fruit in lower Subansiri district of Arunachal Pradesh during the year range of 2018–2021. *Data source* District Horticulture Office, Lower Subansiri

- There is very limited awareness among the kiwi growers on the agriculture loans, insurance and other financial products.
- There is a lack of awareness in the distant market with respect to kiwi cultivation in Arunachal Pradesh. Farmers also do not know where to market, at what time to market and at what price.
- There is no availability of permanent traders due to remote location of kiwi and apple orchards.
- There are various market channels for kiwi fruit. However, only few farmers have direct contact with the distant market.
- No subsidy is provided by the Government for the transportation of the produce. Thus, the farmers bear the transportation of annual produces from the remote located orchards to the markets without proper connectivity.
- Sorting, grading and packing are done through manual process. No mechanical system exists for these processes. Due to improper mechanisation, the farmer's reveal that the traders many a times complain that the farmers are not following the standard grading practices of the fruit produces.
- No cold storage exists in the market cluster. Thus, perishable goods can't be stored.
- Soil testing is often not conducted by the farmers. This could lead to nutritional deficiency in the plants and thereby foster issues such as lower production, fruit dropping and small fruit size. All these are not good options for bio-diversity conservation in the state.
- Natural calamities such as heavy rainfall and wind during flowering season in the month of April foster greater losses for farmers and up to 5 tons. Also, due to heavy rainfall in the month of flowering, the annual yield will be lower for a new orchard that aims to establish the produce.
- Pollination also plays a very important role in realising maximum annual yield. This is due to the fact that both male and female flower are in different but not same plants.

- Commercial production involves production after 6 years of DAP (Day after plantation). Hence farmers have to wait for 5 years to generate their cost input.
- Processing units do not exist in the district after harvesting kiwifruit. Presently only one kiwi processor exists as a kiwi winery with a capacity 60,000 L per batch.

From research perspectives, two questions can be framed. These are:

(a) What is the present socio-economic status of kiwi growers in Arunachal Pradesh and lower Subansiri district for needful analysis and resolution?
(b) What kind of infrastructure is needed to bring a positive change in kiwi cultivation and marketing channel?

Considering these questions, the following objectives have been set for the book chapter:

(a) Study issues related to sustainable Kiwi fruit production in Arunachal Pradesh.
(b) Study the Post-Harvest Management of kiwi fruit in Arunachal Pradesh.
(c) Analyze the Supply-Chain Management and associated bottlenecks.

14.6 Overall Methodology

To study the status of Kiwi production in Arunachal Pradesh and Post-Harvest Management of the fruit in the stage, the overall methodology involved a study area in which relevant sampling methods have been applied for primary data collection. This was supplemented by the secondary data. All these have been briefly presented as below:

14.6.1 Study Area

The research was carried out in Arunachal Pradesh as well as in Lower Subansiri district, which is one of the 25th administrative district of the state of Arunachal Pradesh in Northeastern India.

There are 7 administrative circles in the district, namely- Ziro (Sadar), old Ziro, Yachuli, Pistana, Deed, Yazali and ParamPutu (Loth).The district is also divided into 4 blocks—Ziro-I, Ziro, Pistana and Hong-hari. Notably Ziro-I Block is the only production cluster for kiwi fruits in the district. It is located in and around the district headquarter i.e. 10–15 km from Ziro.

14.6.2 Sampling Methods

The data were collected by using multi stage random sampling procedure. Interviews and discussion were conducted with Kiwi growers, processors and local retailers.

14.6.3 Data Collection

The conducted investigations generated both primary and secondary data. A structured questionnaire, interview and personal observation have been used to collect the primary data. The secondary data on the number of regulated market, local market, and agricultural institution were collected from various government sources, publication and journal.

Data were collected from 50 respondents (kiwi Growers) to represent the entire district. Five villages were selected and from each village, 10 respondents were considered. Kiwi growers of Zirodo not maintain any records. Information with respect to kiwi prices, market channel, problems faced by the kiwi growers and facility were retrieved from the Department of Horticulture, Ziro. These were eventually collected from the farmers.

14.7 Sustenance and Enhancement of Kiwi Fruit Production

14.7.1 Optimality of Agro-Climatic Conditions

Kiwi plants can be grown in high altitude ranging from 800 to 1500 m above the mean sea level. The study area is located at an altitude of about 1500 and henceforth provides sufficient and appropriate climatic condition for the kiwi cultivation. Kiwi fruit also requires about 700–800 chilling hours (number. of hours during which temperature remains at or below 7 °C during the winter season). For good fruit formation and also for well ripening of the fruit, a well distributed rainfall of 150 cm in the year is sufficient. The average rainfall in the district is recorded at about 190 cm which is appropriate for kiwi cultivation.

14.7.2 Soil Requirements

For kiwi cultivation, a deep brown loamy soil being characterized as well-drained and fertile soil is the ideal soil for kiwi farming. Most land in the study areas have deep, rich, well drained and sandy loamy soil, and hence ideal for kiwi cultivation.

14.7.3 Land Preparation

Steep land on the hills is contoured into terraces and thereby ensured to become ideal for kiwi cultivation. Since kiwi plants require maximum possible sunlight for good flower formation, the plants have been planted in north-south direction and thereby received good amount of sun light. During field survey, it was found that the kiwi farming required 4 m spacing between rows and 6 m between plants. Therefore, on an average, 400–416 plants exist per hectare. To establish new kiwi orchard gardens, proper soil preparation and input of farm yard manure is required. Land preparation for kiwi cultivation takes place in the month of January. For successful farming, pit digging and filling are carried out to provide proper nutrients.

14.7.4 Planting

Kiwi plants are mostly propagated by seed/seedling/grafting/cutting mechanisms. Commercially grown important pistillate and staminate cultivars are Abbott, Allison, Bruno Monty and Allison. Plantation is usually done in the month of January. Arbor or pergola system framework is used with 6 m distance between plants and 4 m distance between rows. For pollination in kiwi plants, male to female plants ratio was maintained as 1:5. During plantation, iron T-bars are also laid down with new plants. T-bars are required to withstand and provide protection from winds, hailstorms, heavy rainfall etc. Due to the high cost of iron T-bars, many farmers used bamboo as an alternative source. However, bamboos have short life span and need replacement after every 2–3 years. On an average, kiwi planting material costs about Rs. 80–150/per plant.

14.7.5 Fertilizer and Nutrition Requirement

For a residence of 400 kiwi plants per hectare, 1600 kg farm yard manure/compost is required. Thus in general, 4–6 kg of manure is required per plant. As per sampled respondents, the farmers do not need or use chemical fertilizer. This is due to the perception that the soil is already rich in nutrients and other fertilizers are hence forth not required.

14.7.6 Training and Pruning

Since kiwi vines grow in large quantities, pruning is required to avoid over loading. This practice is also important to divert the flow of nutrients towards fruiting vines and thereby ensure more fruit produce. During the first year of plantation, the primary focus is on the straight growth of kiwi plants. Hence, kiwi plants are tied loosely to T bars for straight growth. To support kiwi laterals, wire tension is installed from one T-bar to another T-bar. The wire extension should not be over-strained, otherwise the wire can break at knot due to crop load. Later, the arising kiwi laterals from main branch are trained onto a canopy of 2–3 wires. Since this system is costly, most farmers during survey affirmed bamboo usage for both T-bars and wires. Winter (January) is the best season for pruning and training activity.

Training is required to establish and maintain a well-formed framework of main branches and fruiting arms. During pruning, farmers cut off the diseased, weaker and over lapping branches. Thereby, regrowth of strong branches can be ensured. If the kiwi vines do not create healthy side branches at the top, trimming is done at the main branch by approximately 2 feet. Thereby, it is allowed to grow and form proper side branches at the top portion of the vine and in the following year. The kiwi vines not being subjected to timely pruning would become over crowded.

14.7.7 Thinning

Thinning encourages the growth of high-quality fruit. To help the leftover fruits bloom more quickly and to avoid the establishment of such a massive crop that the plant is unable to blossom and produce a commercial harvest in the next year, flowers or early fruit are to be removed. This is called thinning. Reducing the fruit numbers always increases the yield and size of kiwi fruits.

Excepting Hayward cultivator, all other Kiwi fruit cultivars bear heavily on an yearly basis. In cultivars bearing more than 5–6 fruits per flowering shoots, the thinning is beneficial for the size of the crop. The sampled farmers informed that thinning is done in the month of June or July. Generally, 8–10 fruits are kept in a bunch after the thinning process.

14.7.8 Pests and Diseases

In kiwi plants, root rots can be attacked by *Phytophthora* soil fungi. This especially occurs on poorly drained soil or over moist soil. Bootlace fungus *Armillarianova ezelandiae* spreads to kiwi fruit from infected dead tree stumps or buried wood and causes fatal infection. In humid climate, grey mold *Botrytis cinerea*rot infects flowers and young fruit. The cluster respondents conveyed that any such type of serious pests and diseases have not been observed. However, very mild infection of Aphids has been evident in the orchards which are also cut off during pruning season. Farmers do not use any kind of pesticide or insecticide to get rid of it. *Phytophthora*type of soil fungi did not infest the orchards due to the kiwi orchards being located in the hilly and slope area. Thereby, excessive water is leached out soon after rainfall and the conditions do not favor the growth of such soil fungi.

14.7.9 Pollination and Fruiting

Flowers are pollinated and thinned in the early summer. Kiwi vine is a not self-pollinating plant. Hence, a part each orchard must be devoted to male vines unless artificial pollination is allowed with the brought pollen. Unlike most other fruits, kiwi fruit needs high levels of pollination (about 13,000 pollen grains per stigma in comparison to only 12 grains for apple flowers). Growers either place numerous honey-bee hives in the orchards temporarily or artificially pollinate the flowers. Honey-bee and others birds play a very important role in the pollination of kiwi flowers.

14.7.10 Harvesting and Yield

Kiwi plants start bearing fruits at the age of 3–4 years. Thereafter, commercial production starts at the age of 6–7 years. Due to the temperature difference, kiwi fruits mature faster at low altitudes than at high elevations. Large fruits are harvested first, and smaller fruits are allowed to grow for a longer period of time. Kiwi fruits ripen between October and December and this depends on climate and variety. Kiwi fruit's maturity is ensured by rubbing of the upper and outer surface of fruit. Thereby, while rubbing the outer surface, stiff hairs drop out easily to indicate the fruit's maturity. Alternatively, a technical procedure can be followed in terms of the sugar content of the fruit being determined with a sugar reader. A fruit with about 6.5 gm of sugar (determined with the sugar reader) is considered to be with maturity for harvesting. Usually kiwi fruits are harvested well before their ripening and maturity stage. This is due to the reason that they can be transported to a longer distance prior to the serious damage to the fruits.

14.7.11 Seasonal & Cultivation Activity of Kiwi Fruit

The seasonal and cultivation activity of the kiwi fruit is presented in Table 14.4. As shown, specific activities need to be addressed in specific months to achieve best harvest conditions and maximum fruit yield.

Table 14.4 A tabulated summary of monthly and seasonal cultivation activities of kiwi fruit in lower Subansiri district of Arunachal Pradesh

Sl. No.	Activities	Month											
		Jan.	Feb.	Mar	April	May	June	July	Aug.	Sept.	Oct.	Nov.	Dec.
1.	Land Preparation	■											■
2.	Planting	■	■										
3.	Manuring	■	■										
4.	Pruning	■						■					
5.	Flowering				■	■	■						
6.	Thinning					■	■						
7.	Fruiting							■	■	■			
8.	Weeding				■		■		■				
9.	Harvesting										■	■	■
10.	Lean Season							■	■				

14.8 Supply Chain Management of Kiwi Fruit in Lower Subansiri District

The supply chain of kiwi fruit in Lower Subansiri district starts with farmers (kiwi growers) who are also responsible for the harvesting, cleaning, grading & sorting and packing. On an average, the fruit yield varies about 25–90 kg per vine in a well-managed orchard. Fresh fruits are directly transported from the farmer's field to the retailers of different states. Futher, some quantity (especially grade C&D) of the fruit also goes to the processing unit.

The functions of various entities of the supply chain of kiwi fruit are as follows:

A. **Growers**: The kiwi growers are responsible for all primary activities such as land preparation, maintenance of orchard, planting of planting materials, nutrient application in the form of farm yard manure or compost, culture practice, harvesting, post-harvest management, sorting & grading and packing of the produce etc. Farmers holding small scale orchard also have to transport their produce to the traders at the nearest local market for sale. Otherwise, they can sale their produce to large framers. The large farmer at times acts like a secondary trader for the small farmers. The uncertainty in trading is of serious concern in this highly disorganized sector. However, it can't be engineered or controlled.
B. **Growers' organization**: From last 2 decades in lower Subansiri district, the number of kiwi farmers swelled. More and more farmers are establishing new kiwi orchard within 10–15 km of the district headquarter. In order to promote and enhance awareness, a growers' organization with the name of "Kiwi growers cooperative society Ltd." was established in 2016. Their main aim is to promote cooperation and help one other during difficult times, as no such help exists from

the state or district Horticulture department. Salient features of kiwi growers' cooperative society are as follows:

i. Farmers having kiwi orchards with more 100 plants are eligible to become a member after paying Rs. 3000 as membership fees.
ii. Only those farmers who practice organic farming are allowed to become a member. Any farmer practicing inorganic cultivation is excluded from the society.
iii. The collected funds are available as a loan (interest rate of 3% per month). For this, land allotment document has to be submitted as a collateral.
iv. The executive body is selected from the registered members. They reside in the Ziro-I area.
v. The society mandates the members to maintain a uniform rate of selling price of kiwi fruit. The price is fixed in the general body meeting of the society in every new season.

C. **Traders and retailers**: The traders and retailers pick up kiwi fruit mostly at the farm gate from kiwi growers. The traders and retailers do come from local markets such as Ziro, Naharlagun, Itanagar and also from distant markets such as Guwahati, Kolkata, Delhi etc. Very few farmers get directly connected with distance market traders. Hence, a huge challenge exists for the kiwi farmers as no permanent traders exist for their commodity. This is due to its remote location of kiwi orchards and poor connectivity through the road network. At times, due to all these detrimental issues, the farmers have to bear the losses in the price of their commodity. Due to perishable nature of the fruits and dynamic supply–demand constraint, minimum price cannot be assured.

D. **Pre-harvest contractors**: During field survey, majority revealed that they very often prefer to sell their commodity to the traders and retailers in local market and in the distant market. Some farmers also informed that very few farmers are connected with pre-harvest contractors. However, such a process is not always a permanent feature of the trading network. Pre-harvest contractors come either from within the state or outside the state. They usually come during flowering stage or fruit formation stage. Prior to the harvesting stage, they inspect the orchard field and thereby forecast and estimate the orchard production volume. Eventually, they negotiate the price of kiwi fruit with the farmers. The price is based on the distance of orchards from road to destination market, condition of road, size and grade of fruit, expected price during the season and market demand etc. Thereafter, the price is decided between farmer and contractor. In such a situation, the contractor takes responsible for all risk and expenses related to the undertaking of the remaining farm operation such as plant protection, harvesting, sorting and grading, packing, transportation and marketing of produce. Till date, very few farmers have been able to establish the pre-harvest contractor arrangement.

E. **Processors**: The processing unit for kiwi, presently in Block-I, Ziro refers to the only processing unit for kiwi wine. Kiwi fruit are mostly acquired from local farmers' orchard and promoter's own orchard. During field survey it also came to the knowledge that many farmers prepare homemade kiwi wine and on a small scale for their own consumption. This kiwi wine is made out of low grade fruit (C & D) and with especially damaged kiwi fruits.

14.9 Supply-Chain Management for the Marketing of Kiwi Fruit

Figure 14.4 illustrates the kiwi fruit commodity flow structure. It involves all possible channels. A brief account of these are as follow:

Channel 1 (Local Traders): The channel is the most prevalent supply channel for traders. Here, most traders approach the farmers and buy kiwi fruit at the farm gate.

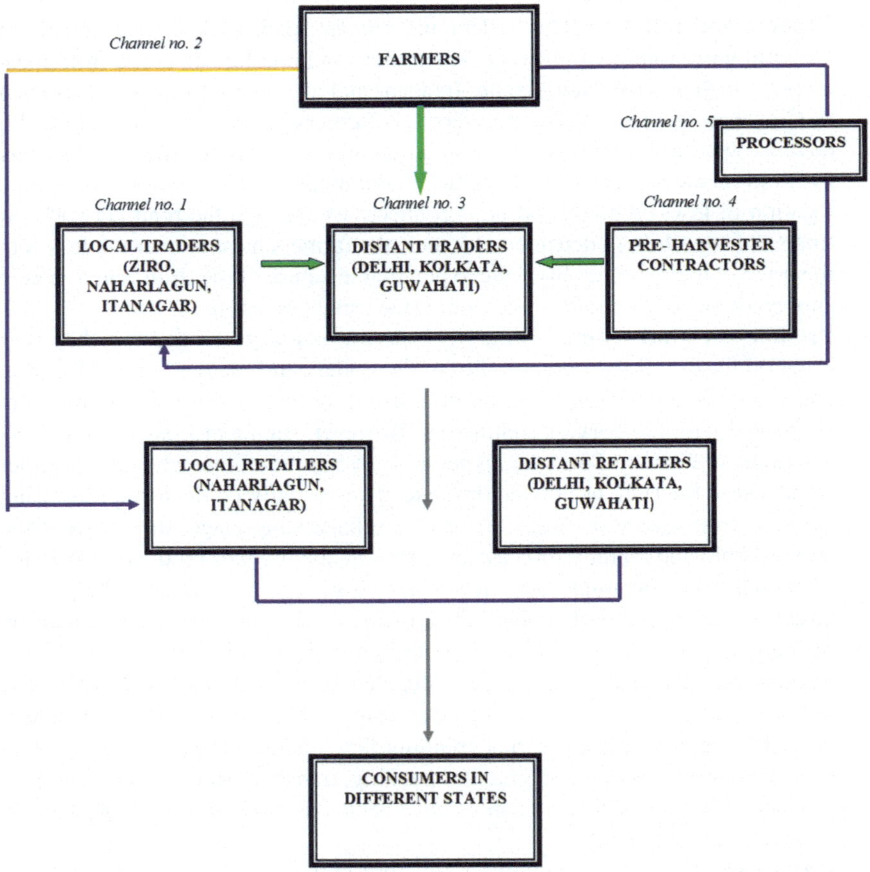

Fig. 14.4 Kiwi fruit product commodity flow diagram. *Data source* District Horticulture Office

Table 14.5 Grade and price of Kiwi fruit

Grade	A+	A	B	C	D
Weight per fruit (gm)	>100	80–100	60–80	50–60	<50
Price (per kg)	Rs. 120	Rs. 100	Rs. 80	Rs. 60	<Rs. 50

Source Primary data

The farmers harvest the kiwi fruit. Thereby, based on size, sorting & grading, the fruits are classified. Thereafter, the produce are packed in the cardboard boxes by farmers. Immediately the payment is made to the farmers by the trader and mostly this is done in cash. The selling price of kiwi fruit is decided by the Kiwi cooperative society limited. Traders are supposed to pay the decided price and as per the grades. However, during the peak season, farmers sell their produce by about 10–15% lower price. This is due to oversupply in the peak season, lack of storage facilities in the district and perishable nature of produce. A brief account of wholesale price of various grades of kiwi fruit has been summarized in Table 14.5.

Thereafter, traders sell the produce to the local and distant retailers, and finally the fruit reaches the consumer. At times, a local trader acts as an intermediate between farmers and distant market traders. For such mechanisms, the local traders buy the produce from farmers and then resell it to the distant market traders.

In certain cases, the farmers themselves bring the produce to the traders in the nearby market. Such a situation arises only when cash is needed on an immediate basis. According to the requirement, the farmers harvest the kiwi fruit and bring them to the local market. In such case, expenses of all transportation are borne by the farmers and a price is negotiated through telephone. Thereafter, based on required quantity, the produce is brought to the market. These modes of marketing are done by the farmers holding large size of the orchard.

Channel 2 (Local retailers i.e., Naharlagun, Itanagar): The channel is another prevalent supply channel for farmers. Here, farmers directly sell the produce to the local retailers in the market of Ziro, Naharlagun, Itanagar etc. Such type of marketing channel is preferred by those farmers having a small-scale production. In such mode of marketing, the selling price of produce is decided on the spot and immediate payment is done in the form of cash. In this marketing mode as well, the farmers have limited bargaining power.

Channels 3 and 4 (Distant Traders & Pre-harvest contractor): In channel-3, the farmers directly sell the produce to the distant market of Guwahati, Kolkata and Delhi. In channel 4, farmers have an agreement with the pre-harvest contractors. Only few farmers are connected with distant market traders and through channel no. 3. This is due to high transportation cost and losses in transit. Further, distant market traders have high bargaining power. Hence, farmers with large scale production are in trade terms with distance tracers.

Channel 5 Processors: Very few quantities of kiwi fruit go to the processing unit for kiwi wine production. Usually under grade kiwi fruit reaches into the processing unit. Since wine is often homemade, very few quantities reach to the only processing unit.

14.10 Post-harvest Management of Kiwi Fruit

Post-harvest management plays a very important role in deciding the economic value of the kiwi fruit in the market. This is due to high value of the crop. The harvested kiwi fruits are sorted, graded and packed manually at the farm level. As mentioned earlier, the graded produce based on weigh is designated as A+, A, B, C and D. Due to lack of sorting and grading machines, farmers are often complained for inability to standardize. For kiwi packaging, egg like trays, corrugated cardboard boxes and cartons are used. Few farmers also use plastic crates, (20–25 kg kiwi per karate) for packing. For long distance transportation, plastic crates are the best. The packing capacity are 3 and 10 kg respectively. Tray packing system is used basically for A+ grade which costs about Rs. 360/- with kiwi fruit. Such a tray system is used only during long distance transportation. Hence, very few farmers use such packing materials.

According to respondents, about 10–15% of produce volume is damaged. Such damaged fruits are very small in size. At times, these are feed to domesticated animals as the farmers believe that the practice improved the yield of animals and make them to grow very fast. With damaged and small sized kiwi fruits, some farmers prepare the homemade kiwi wine at a small scale for self-consumption. At times, it is for sale at a rate of Rs 400/l. During kiwi cultivation, damage is very less to the fruits as most of the kiwi fruit are harvested at the maturity stage prior to the ripening of fruit. This is also a main reason to convey that Lower Subansiri district do not have much losses. The other reason for lower loss is the climatic condition of Ziro. Here, kiwi fruits are harvested during winter season that gives additional shelf life to the kiwi fruit. During field survey, it was found that no cold storage exists in the cluster. For this reason, kiwi fruits are not available during off season. Also, due to no cold storage facility in the district, farmers have to sell out all their produce prior to the damaged or spoiling of the fruit.

According to sampled farmers, the District Horticulture department at lower Subansiri provided few subsidies. Also, there was a limitation in the budget for such subsidies. As a subsidy, few boxes were distributed on a first come first serve basis. Very few farmers received such free boxes for packing from the District Horticulture Department. The actual gains of a good quality produce is only realized upon marketing of the produce at a rewarding price. This would be possible only when due attention is paid at each step of post-production and until the produce reach the consumers. The following steps have been suggested to relevant good for post-harvest management of kiwi fruits:

a. **Maturity standard**: Most of the farmers are unaware of the current maturity standard of kiwi. Very often they harvest the fruit at different stages of maturity.

Most farmers in order to exploit the market advantage often harvest ripened fruits. Thereby, quality assurance and trader satisfaction are jeopardized. This can be easily addressed through science & technology intervention.

b. **Grading**: Discussion with the farmer respondents as well as with the market functionaries in kiwi growing belts indicate that all fruits are always packed for marketing without any reference or grade specification to suit a particular market. Producers generally grade marketing grade as A(Good) and B (poor). These are non-scientific classification procedures. Hence, scientific intervention is required.

c. **Packaging**: A related aspect to grading is the appropriate packing of the produce. Since most produce from the Arunachal Pradesh is sold within the state, no such attention or investment is made on proper packaging. Farmers themselves or the marketing agents, as the case may be, to bring the produce to the market in plastic creates or gunny bags. Once they are being sent to Guwahati/Tezpur or other parts of neighbouring state of Assam, they are being packed into old cartoon boxes which are mostly old egg boxes. Thus, research can be as well directed for the cause. The potential locations for setting up of post-harvest management facilities including cold storage/cold chambers in the state of Arunachal Pradesh are Rupa and Dirang.

d. **Market Analysis**: Several discussions with growers, fruit merchants, horticulture development agencies and other players confirmed on the limited production that necessitated local sales. For this reason, no or little investment was targeted on the post-harvest management. As per information provided by state Directorate of Horticulture, Naharlagun, about 40% of the total produce is consumed locally and the balance (60%) goes outside the state. The local sales refer to sale within the state. The outside market refers to the adjoining state of Assam. The produce has been going to Guwahati and other parts of Assam through Tezpur. Precisely, no marketing system exists for kiwi in the state. Various channels for marketing are in use but they function in a disorganized way. In most cases, growers take their produce to the nearby markets and sell them on their own. In some places, a few agents collect the produce from the growers and organize their sale in the local or nearby markets. Most of the agents from Assam usually negotiate pre-harvest sales. There is no operational public sector undertaking for marketing and distribution of kiwi and their by-products. As such the entire marketing and distribution system is un-intervened and managed by the private traders.

e. **Wastage**: Needless to mention, a sizeable percentage of wastage of kiwi fruits is among the farmers, retailers and wholesalers. It is observed that the percentage of wastage is much higher in the retailers but not the wholesale traders. The estimated loss of produce is around 30%. The loss in farmer's field mainly confines to pre-harvest drop, poor quality in view of under developed color, reduced size etc. These fruits have little or no market as fresh fruit produce either within the state or outside the state. With the present level production of 450 MT of fruit, the marketed surplus would be 426 MT.

14.11 Summary and Recommendation

The horticulture industry is one of the most profitable and practical options for the diversification of farming. Through such engagements, bio-diversity conservation can be ensured in the north-east region of India. Sustainable and cost competitive agro-business management holds the key for the flourishing of the horticulture industry. It provides a broad range of crops with the potential for various cropping systems that are suited for crop production under various agro-climatic conditions and topography, increasing yields per unit of area etc., Thereby, the engagements create employment, and ensure nutritional and food security. The relative position of horticulture compared to other agricultural activities is very poor. Although in hill states it is comparatively better, it is still far below the expected level. A wide range of products with small inventories makes it difficult for the appropriate planning of post-harvest infrastructure. The agricultural marketing system is highly dis-organized and has underdeveloped infrastructure. The marketing system is fully dominated by private traders and a large chain of middlemen. Crop specific establishment of primary producer company (PPC) is an utmost necessity in the production centers.

Kiwi fruit is among several crops being introduced in the north-east region. It became highly renowned in India's Himalayan and sub-Himalayan regions due to its vast economic prospects and nutritious significance. Thereby, huge scope exists in the Kiwi production and especially in the North Eastern part of India. Among all North-eastern states, Arunachal Pradesh produced the highest production of kiwi and accounts to about 56% of the total production in 2020. The State has a tremendous scope for Kiwi cultivation owing to its natural Agro-climate condition. Arunachal Pradesh became 1st State to obtain organic certification of kiwi fruit under Mission Organic Value Chain Development for the North east region. Such Certification in India can be obtained only after strict scientific assessment by the regulatory body (Agricultural and Processed Food Products Export Development Authority). Kiwi fruits have bright prospects and can be an important future commercial crops due to their high demand share in the country's kiwi fruits production scenario. Further, organic certification of kiwi fruits will help to boost the Economy of the state and country.

In this book chapter, a critical summary of Kiwi fruit production in Arunachal Pradesh state of India has been addressed. These include conservation based farming strategies, post-harvest management, supply chain management, etc., Thereby, relevant bottlenecks have been identified that hold the key to further enhance sustenance of the Kiwi farming in Arunachal Pradesh.

Horticultural crops have bright prospects and can be an important future commercial crop due to their high demand. India is currently importing 73% of kiwi fruits from New Zealand. There is a huge potential for kiwi cultivation in the region. Area under cultivation has been increasing over the years. However, till date, it did not pick greater momentum. Few recommendations for the sustenance of Kiwi fruit higher production and efficient marketing approaches can be summarized as follows:

(a) The state government should take initiatives to provide incentives to the farmers for the promotion of kiwi fruit production in the region which includes the development of certified nurseries for the production of quality planting material for area expansion.
(b) Processing units should be set up near the production areas for utilizing the inferior quality fruit. This will surely give impetus to the production of the fruit in the region as well as in the state.
(c) Controlled atmosphere storage facilities should be created for better profitability. Also, good marketing network can be ensured by such technological support to the farmers.
(d) To a certain extent, there must be subsidies for the farmers as orchards are located at remote location and have poor connection to the road network.
(e) Besides improved production, practices feasible for the region should be developed to increase productivity and quality.

In summary, the findings reported in this work convey that diversified cash crop farming such as that of kiwi fruit can stabilize and enhance bio-diversity conservation in North-East India. This is due to the criticality of the region in terms of the micro-climatic parameter optimality for horticultural produces. A similar research work can be undertaken for other regions and for other produces.

References

1. Kumar, A.: Role of horticulture in biodiversity conservation. In: Nandwani, D (ed.), Sustainable Horticulture System. Sustainable Development and Biodiversity, vol. 2. Springer (2014)
2. Raja, W.H., Noomphy, M.: Potential and prospects of Kiwifruit production in the state of Arunachal Pradesh. Pharma Innov. J. SP **10**(12) (2021)
3. Hema Yadav, S.: Kiwi value chain analysis and market. Assessment for Lower Subansiri District, Arunachal Pradesh. CCS National Institute of Agricultural Marketing (A Govt. of India organization under Ministry of Agriculture and Farmers welfare) Kota Road, Bambala, Pratap Nagar, (Rajasthan) (2018)
4. Gyanendra Mani, A.K.: Kiwi value chain in Arunachal Pradesh: issues and prospects. Agric. Econ. Res. Rev. **30**(1), 123–130 (2018)
5. Pandey, G., Tripathi, A.: Kiwifruit. A boon for Arunachal Pradesh. Indian Council of Agricultural Research (2014)
6. *Annual Report 2020–21 Agricultural and Processed Food Products*. Ministry of Commerce & Industry, Government of India (2021)

Part IV
Conservation of Bio-Diversity Through Critical Environmental Factors and Its Sensitivity Towards Urbanization and Holistic Methodologies

Chapter 15
Mapping and Quantifying the Impact of Urbanization and Influence of COVID-19 Induced Lockdown on Eco-Sensitive Deepor Beel Wetland

Mustak Ali, Amlan Saikia, Monisha Das, Kunal Kumar Borah, Dipankar Sarma, Runjun Baruah, Utpal Sarma, and P. L. N. Raju

15.1 Introduction

Statistics from various international agencies and organizations suggest that the world population has increased at an exponential rate and has grown to 7 billion in 2021 [1]. As of 2020, 56% of the world population lives in urban areas as compared to 47% in 2000 [2]. In a developing country, this shift in urban population led to the cities to undergo faster urban extension [3]. With this substantial amount of growth, green cover belts have been shrinking throughout the world. Among all ecosystems within this belt, wetlands can be termed as most important for their highly contributing nature. Freshwater ecosystems are not mere sources of water but they are treasure troves of biodiversity [4]. However, because of the unregulated land conversion and waste management system, wetlands in most urban and suburban areas are in danger [5]. North-East India has several wetlands. Bio-diversity conservation in the region requires their time bound monitoring through appropriate mechanisms. However, such monitoring may not provide a microscopic assessment of the health of the wetland. Nonetheless, such research directions can be considered to support firmly the noble cause of bio-diversity conservation in the region.

The launch of landsat 1 Multispectral scanner imagery and remote sensing methods has been continuously used for regional level mapping of wetlands since 1972. Then onwards, sensors like landsat series data sets, MODIS, SPOT, ASTER images of satellite images were used to map catchment area of the wetland [6]. In the last two decades, remote-sensing images (multi-spectral), drone and lidar have been widely used to monitor the water surface. In a study previously carried out for a lake, wherein the objective was to calculate the water volume of the lake, water surface area and underwater terrain data with the utilization of a triangulated irregular

M. Ali (✉) · A. Saikia · M. Das · K. K. Borah · D. Sarma · R. Baruah · U. Sarma · P. L. N. Raju
Assam State Space Application Centre, Guwahati, India
e-mail: mustakali25@gmail.com

R. V. S. Uppaluri and L. Rangan (eds.), *Conservation of Biodiversity in the North Eastern States of India*, https://doi.org/10.1007/978-981-99-0945-2_15

network, analyzed that the storage capacity of the lake reduced in the last 20 years along with a gradual decline in the surface area of the wetland. Several research articles demonstrate that the water body differences due to result of land cover and land use change and urbanization [7]. Prasad et al. [8] stated that mapping, identifying and delineating wetlands sequentially will pave its way for various changes considered significant for the management of natural resource and activity planning [8]. For the conservation and management of various resources of wetland, it is important to screen wetlands along with their uplands. Thereby, remote sensing provides a cost-effective measure for outlining of wetlands and their characteristics along with useful statistics at different points over a large area [9]. Deepa and Ramachandra [10] reported alterations changes and wetlands interconnectivity for longer time in Bangalore (rural and urban) [10]. Manju et al. [11] presented a report that define wetlands in east Champaran region using the GIS and remote sensing as a major tool [11].

In our preliminary field studies, it was noticed that with time, the ambient environment around the periphery of Deepor Beel has been under constant stress due the rapid urban sprawl of Guwahati city in all directions. In addition, the solid waste dumping ground, brick kiln industry located around the periphery of the wetland at Boragaon has led to the rise of aquatic pollution to alarming levels. The direct and indirect impacts of civilian expansion along with industrial setup around the wetland have been detrimental to the ecosystem at large. Diminishing area of the wetland is a major concern which is again supplemented by the transformation of land use pattern of the immediate surroundings. In this study, we've tried to utilize remote sensing techniques that have been utilized in similar lines to map land cover land use, and map the Deepor beel region. Thereby, the land cover transformation has been quantified from 2005 to 2021.

The hydrosphere has been polluted because of rapid industrialization, exploitation and urbanization for decades [12]. However, the impact of pollution in the hydrosphere such as oceans, rivers, wetlands, lakes and groundwater tanks has not been explored in correlation with the quality of air and atmosphere. The emphasis of this book chapter was to examine the impact of COVID-19 on the environmental challenges in the short period and quantify its pollution status during the entire lockdown.

Based on previous analysis carried out by the European space agency using specific satellite-based data collected from Tropospheric Monitoring Instrument (TROPOMI), the Ozone Monitoring Instrument (OMI) and Sentinel-5 satellite on NASA's Aura satellite found a negative trend for the presence of nitrogen dioxide (NO_2) in the atmosphere during March 25, 2020. This lasted until May 31, 2020 [13]. The lockdown period was further extended into many phases, until June 8, 2020 with a total lockdown period of 75 days with strict regulations and minimal mobilizations. Almost 99% of the world's polluting industrial areas was shut down and nearly six billion people globally stayed indoors. This resulted in the reduction of environmental interference by people [14]. The emission of greenhouse gases namely CO_2, CO, N_2O, SO_2, CH_4 etc. reduced. Lockdowns minimized the degree of gaseous substances being emitted by 10% in comparison to the previous year [15].

Many researchers studied the relationship between PM10 and PM2.5 precipitation and concentrations in different areas and observed that the precipitation removal effect on aerosol particles was closely associated to precipitation intensity, diameter of raindrop and particle size of aerosol [16]. In the tropical Ramsar site, South India, analysis and assessment for water quality was targeted by Aswathy et al. [17] using time series analysis of Landsat 8 images. The authors reported a reduction in overall PM2.5 during the Covid-19 period in comparison to previous years. Another article narrated the causes for the SPM improvement in water quality (lockdown, April–May 2020) and emphasized upon the strategies and anthropogenic activities for the sustainable water resources and its management in the coastal ecosystems (Ashtamudi Lake) [17]. In this context of numerous other related studies, no article targeted to relate the ambient air quality, rainfall, and its immediate impact upon the wetland with relation to Covid-19 induced lockdown. Such studies will provide a newer direction for bio-diversity conservation due to natural effects enforcing non-urbanization for a limited period Therefore, the book chapter attempts to explore whether the global scenario is prevailing in the micro level for the Deepor beel area during COVID lockdown.

The state of Assam is gifted with a multitude of lakes being termed as locally beel. In 2002, Deepor beel, a large natural wetland [18] is endowed with rich biodiversity and was known as a Ramsar Site [19]. Deepor beel is a freshwater lake formed by the abandoned channel of the Brahmaputra river. It resides between 26°05′–26°11′N and 91°35′–91°43′ E that lies on the Southern Brahmaputra bank. Deepor beel lies in the Rani Garbhanga forest in the southern part of region. In addition, five notified Elephant corridors across the forest to the Deepor beel exist i.e., Watch Tower point, Banbandha, Mikirpara, Beltol, and 274 railway gates [20]. These five corridors are situated along the PWD Highway through Deepor Beel and 274 railway gates. The corridors start from the Rani forest in Meghalaya and Assam border and pass through the deep forest down to the Deepor Beel. The railway track cuts the corridor in the tail end along with a VIP road that connects NH-37 to the LGBI airport in Guwahati. The beel covers an area of about 40 km^2 and is also considered to originate off the river Brahmaputra, which lies on the Southwestern periphery of Guwahati City.

Due to various anthropogenic factors, the lake area has undergone huge degradation in the context of biology and marine life [21]. A 2008 report from the Ministry of Environment and Forests (MoEF) has very well highlighted a number of issues that pose a severe risk to the wetland area. One major listing was the municipal garbage dumping by the Guwahati Municipal Corporation (GMC) [22, 23]. Moreover, a study carried out by Mozumder et al. [24] conveyed a loss of approximately 57% of the total wetland area [24]. The catchment water area of the Deepor beel comprises the surrounding highland lying immediately to the north and south. This hydrologic regime brings a unique framework to its origin. Geomorphologically, the origin of the beels and their process of formation is related with the tectonic and geologic background of the region. The beel accumulate their water mainly from the Basistha-Bahini, Bharalu, and Kalmani rivers. And during the monsoon (May–Sep), the depth is about four meters and it drops to one meter during the dry season. The Khoujan channel drains the water out of the beel to the Brahmaputra River, which is 5 km to its north. During the winter months, when the water of the beel is dried up, the exposed catchment area is used for the cultivation of rice. The periphery of Deepor beel has a humid tropical monsoon season starting from May–October. The winter season is relatively cool, and dry. The water level variations of the wetland provide an ecosystem for a variety of habitats. When the water level is high and during break monsoon, one can find hyacinths, water lilies, aquatic grasses and various submerged and emergent floating vegetation. During the winter emergent vegetation, marshy lands, mud flat, grassland patches, paddy field provide livelihood for terrestrial avifauna, migratory birds and residential waterfowl. The railway line diagonally bisects the southern part of the water catchment area of the beel and the national highway passes through the western side of the beel.

During the summer and winter, the availability of a food web abundance and its diversity is largely affected by the fluctuation of the water regime. The lowest level of producer in the beel is the phytoplankton. One can find a total of 18 genres of phytoplankton in the deepor beel area. During the winter season, the phytoplankton population density is high and remains fairly low during the peak rainfall season. Few migratory bird species have been frequent visitors to the beel areas namely Lesser Whistling Teal (Dendrocygna javanica), Common Teal (Anas crecca), Baer's Pochard (A. baeri), Northern Pintail (Anas acuta) and Little Cormorant (P. niger). Figure 15.1 depicts the location of study area.

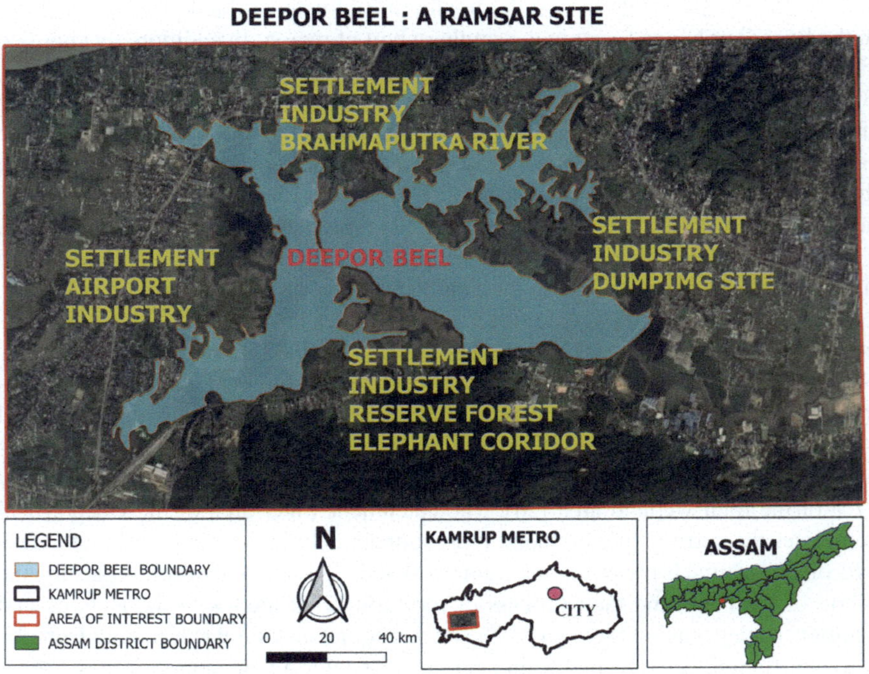

Fig. 15.1 Study area location

In the North-East India, Deepor beel is a famous wetland which that contains large numbers of wetland biota. These includes 11 amphibian species, 32 reptiles, 232 birds, 61 fish and 24 mammals despite heavy natural habitat destruction since decades. Deepor beel, a renowned Ramsar Site [25] has good numbers of IUCN wildlife fauna such as critically endangered, several threatened and endangered species in the ecosystem that indicates the wetland's potential for the conservation of biodiversity. The wetland supports 35 endangered vertebrate and 18 threatened species (Indian Wildlife Protection Act, 1972) along with 2 critically endangered vulture species. The beel is known for its abundance Asiatic wild Elephants and whistling Teal. A large number of inhabitat herbivorous species in the reserve forest are found. From a recent investigation, it was found that the Asian Openbill stork population have been enhancing in comparison to previous 10 year records with occasional sighting records (personal observations and Census report for mid-Winter waterfowl). The migratory bird populations have been reducing due to human conflicts and have changed the avain species diversity. During the survey, the wader population reduced amongst the migratory waterfowl due to human conflicts in the shoreline regions and shallow areas of the wetland.

In summary, the article aims to apply remote sensing techniques for a macroscopic quantification of wetland degradation indices. Thereby, the article provides useful directions for the correlation of these indices with the microscopic quantifications in terms of biota and other vegetative indices.

15.2 Methodology

15.2.1 Data

Remote sensing and geographical data for the study area have been acquired from different sources and geospatial platforms. The study required administrative boundary, road, and railway layers along with satellite imagery for spatio-temporal analysis. For the given area of Interest (AOI), the administrative layers were downloaded from the North Eastern Spatial Data Repository (NeSDR) data portal. Through online registration, NeSDR enables users to geo-process, publish, visualize and retrieve geospatial layers of North-East India. The satellite imagery used was WorldView-3 and Quick Bird 2 Satellite Sensor. The worldView 3 data was used as Quickbird-2 was not available for the period of 2021. The resolution of the two sensors is 1.24 m and 2.62 m respectively. The data were downloaded using google earth. The temporal data for 2005, 2012, and 2021 data were then manually digitized. The digitization of the thematic map for Land Cover and Land Use (LCLU) as well as the Natural Color Composite (NCC) Image from Google Earth (Quick Bird) were classified. The boundary map of Deepor beel was collected from Guwahati Municipality Development Authority. The air quality data was acquired from ENVIS (Centre on Control of Pollution Water, Air, and Noise). ENVIS provides Carbon Monoxide

Table 15.1 Data type & data source

S. no.	Data	Data type	Source	Websites
1	Administrative boundary, road & railway	SHP	NESDR	https://www.nesdr.gov.in/assam/
2	Google earth (Quick bird) Imagery (2005, 2012 and 2021)	JPG	Google Earth	https://www.google.com/earth/
3	Air quality (2005–2022) PM2.5, PM10, SO_2, NO_2	API data and CSV files	ENVIS (Centre on control of pollution water, air and noise and OpenWeather API)	http://cpcbenvis.nic.in/air_quality_data.html
4	Rainfall (2005–2021)	NetCDF files (0.25 * 0.25 km)	IMD Pune	https://www.imdpune.gov.in/

(CO), Sulphur Dioxide (SO_2), Suspended Particulate Matter (RSPM), Ozone (O_3), Nitrogen Dioxide (NO_2) and Suspended Particulate Matter (SPM). Among these, Sulphur Dioxide (SO_2), Nitrogen Dioxide (NO_2) were only taken from the site. The air quality parameters related to PM2.5 and PM10 were taken from Open Weather API for the year starting 2005–2022. Rainfall data was acquired for the same time frame as mentioned from the India Meteorological Department (IMD), Pune web portal. The following table highlights the information related to data types and data sources (Table 15.1).

15.2.2 Data Pre-processing

The data acquisition, analysis, and processing along with the outcome of the study has been shown in a flowchart Fig. 15.2. Thereby, an insight into the overall work process can be achieved. Remote sensing and GIS platforms have been integrated to create Land use land cover (LULC) maps. Remote sensing is a process for monitoring and detecting a location's physical characteristics by remotely computing reflected and emitted waves (satellite Images). Also, a set of information about the earth's remote location could be acquired. The acquired satellite image is processed and viewed in the GIS platform, ArcGIS in the study was utilized for the same. A GIS (Geographic Information System) is a framework that works on the basis of data evaluation, collection and management. It is a field that relates geography and combines data structures. Also, the assembled layers and spatial position of informative data sets are evaluated into images. LULC is the visual representation of various structures of land pattern distribution over a given area. It provides information to help users to understand the current landscape. It shows the area covered by different features such as forests, wetlands, and impervious surfaces over time that has significant role in the study of regional eco-system changes.

The satellite image after downloading is georeferenced and registered in UTM projection using WGS 1984 datum. The LULC map is digitized in ArcMap Version 10.8.2 through visual interpretation. Three LULC maps have been generated for the years 2005, 2012, and 2021 and are shown in Figs. 15.3, 15.4 and 15.5 respectively. The LULC classes have been categorized into six major classes and eighteen sub-classes. The major feature classes include Agriculture, Built-up, Forest, Scrubland, Grazing, and Water Bodies (Table 15.2).

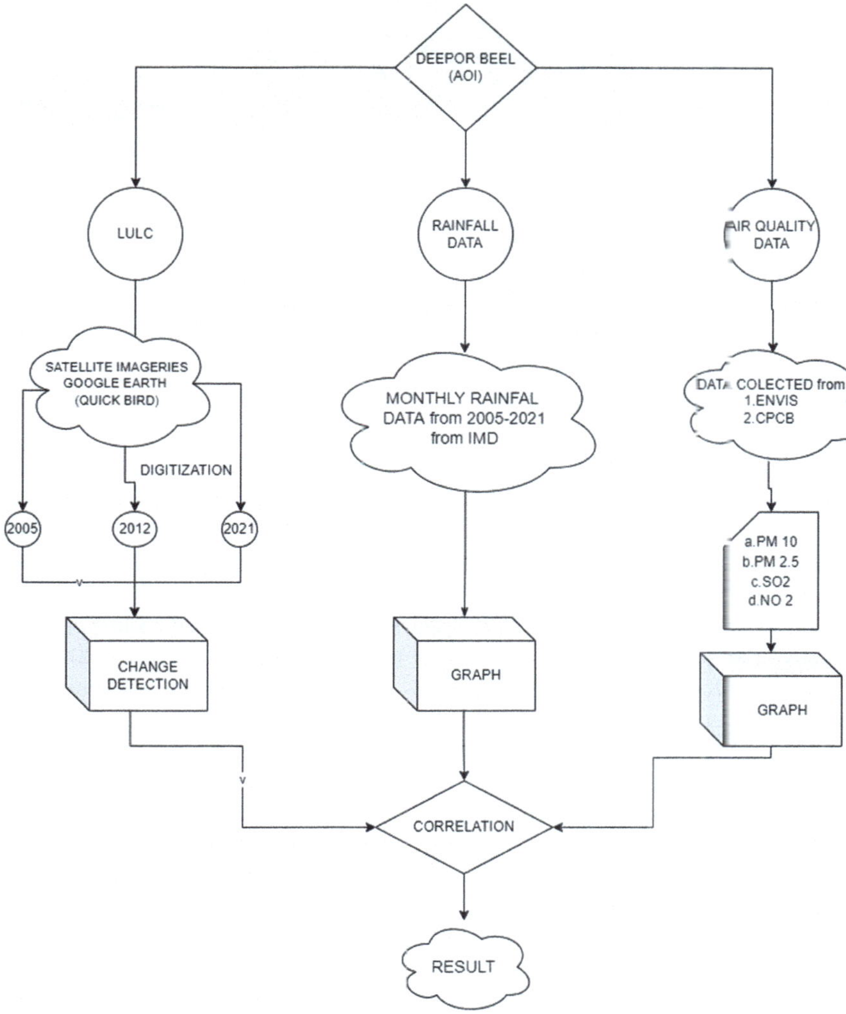

Fig. 15.2 Flowchart of the working model

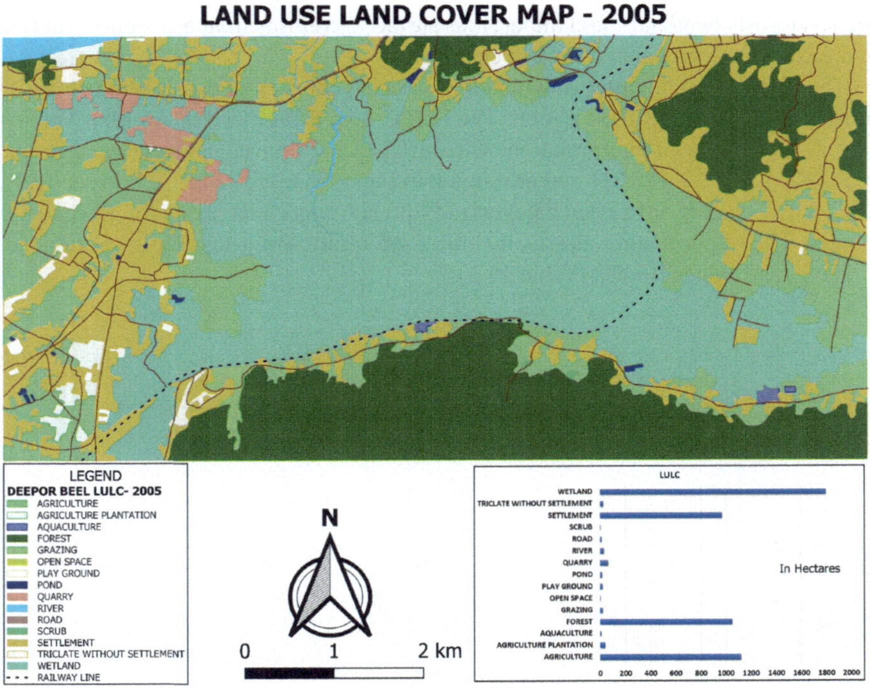

Fig. 15.3 Land cover land use in the year 2005

The process of extraction, clipping, georeferencing, digitization, and the intersection of the study area was carried out in ArcMap and the final map composition was done in the open-source software in QGIS Version 3.03. The comparative study to understand the changes in the LULC maps is done by the geoprocessing analysis known as the intersection process. The intersection tool is used to calculate the geometric intersection of any feature class or feature layer of the same class. In the study area, the intersection was done in two parts, first for the years 2005 and 2012 and the second part for 2012 and 2021. After the analysis, the statistics of the feature class area have been calculated and exported as a .xls file. The .xls file is then visualized in Microsoft Excel and a Pivot table is generated to study the comparative change of different feature classes. The statistical alteration in the area for various feature areas has been shown in the form of a Bar Graph (Fig. 15.6).

The Rainfall data was collected from IMD Pune through the official website [26]. The data is available in NetCDF file (0.25×0.25 km) format and the time period is considered as per the LULC map i.e. for a time period from 2005 to 2021. The NetCDF data which is available on a daily basis is converted into monthly and yearly scales by converting the daily average to monthly and monthly to yearly format. After sorting the data, the NetCDF file was converted to Raster format in ArcGIS. The raster image was converted to a vector point layer and was attributed with precipitation data

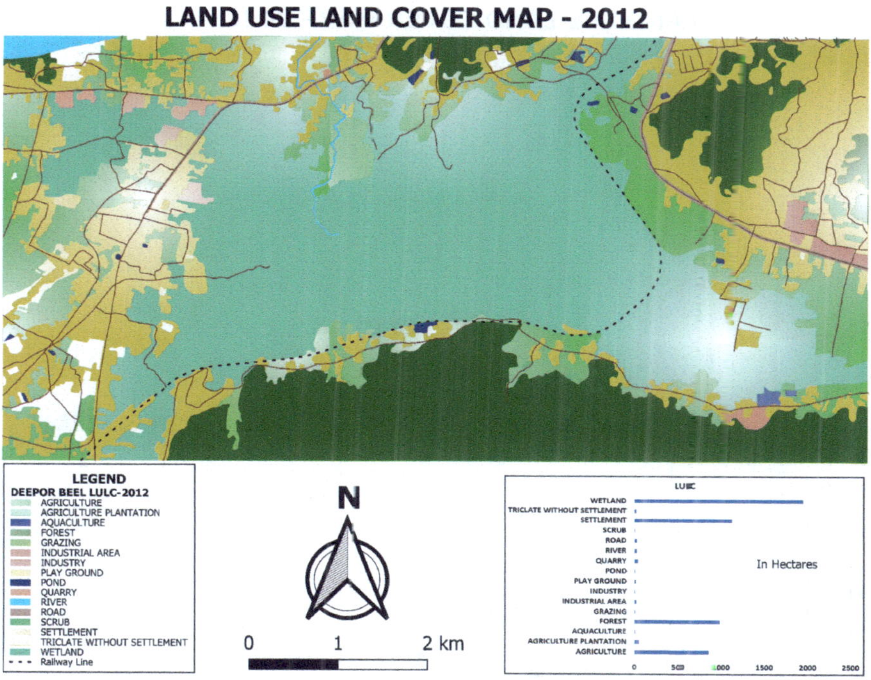

Fig. 15.4 Land cover land use in the year 2012

that was clipped to the AOI. This AOI raster data is converted to tabular data. The tabular monthly and yearly data is converted to data in .xls which was plotted in a line graph for representation as in Figs. 15.7 and 15.8.

In addition to the above two data categories, we also considered air quality to correlate and showcase the changes with respect to the atmosphere over the AOI. The parameters such as SO_2 and NO_2 have been acquired from ENVIS Centre web portal [27]. The data was available in tabular format. In addition, data pertaining to PM2.5 and PM10 from Open Weather API [28] was acquired in JSON format which was later transformed into a tabular format in .xls which was represented as a Bar graph as shown in Fig. 15.9.

15.3 Results and Analysis

From the mapping and analysis of the LULC map for the years 2005, 2012 & 2021, it is evident that the catchment area of the wetland area has diminished. In 2005, the wetland area was 1950 ha that eventually decreased to 1795 ha in 2012 i.e. a negative change of 155 ha. Further, in 2021, the wetland area shrinks to 1678 ha which is 117 ha lower than that in 2012. It has been observed that the total wetland area from 1950

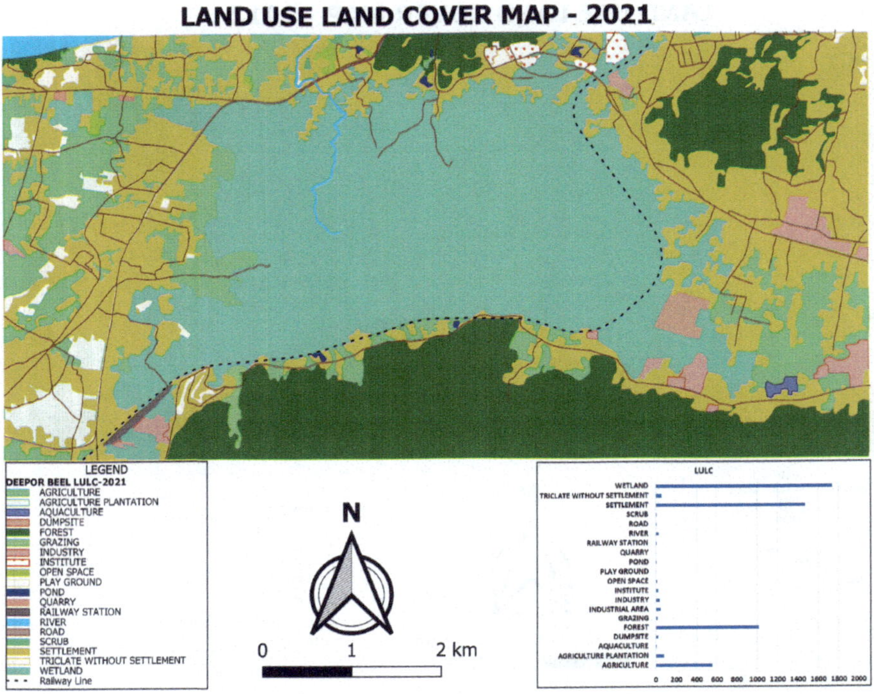

Fig. 15.5 Land cover land use in the year 2021

ha in 2005 has diminished to 1678 ha in 2021 which clearly highlights a negative shift of total wetland area by 272 ha in the last one and half decade i.e. within a period of 15 years (Note: Here the wetland area denotes the sum total of wetland areas in and around the AOI including Deepor Beel). The change in area is subject to the visual interpretation of the satellite imageries in different time frames. Hence, 90–93% accuracy is to be considered for the reported data.

Other classes with major changes in land use can be seen in agriculture and built-up. The agriculture and wetland area shows a negative change over the years (diminishing area) whereas the built-up area has a positive rate of change which in a real sense has an unfavorable influence on the land cover and its use and also the wetland ecosystem as a whole.

The total area under agriculture has diminished to 655 ha in 2021 from 1173 ha in 2005 i.e. a negative change of 518 ha of area, whereas the built-up area escalated to 1761 ha in 2021 from a meager 1085 ha in 2005 which implies a rapid growth of urbanization. This growth in the built-up area accounted for 676 ha of land within 16 years. From the study, it can be observed that the predominant class in the LULC is the built-up area and its unregulated growth has threatened the wetland area to a great extent (shown in Fig. 15.3, 15.4 and 15.5 respectively). The LCLU change for the above-mentioned years is also represented in the bar graph (shown in Fig. 15.6).

Table 15.2 Major classes and its sub-classes of land cover land use (LULC)

S.no.	Major class	Sub-class	S. no.	Major class	Sub-class	S. no.	Major class	Sub-class
1	Agriculture	1. Agriculture 2. Aquaculture 3. Plantation	3	Forest	13. Forest	5	Grazing	15. Grazing
2	Built-up	4. Settlement 5. Tree clad 6. Play ground 7. Open space 8. Railway Station 9. Industry 10. Institute 11. Dump site 12. Quarry (Brick Kiln)	4	Scrubland	14. Scrubland	6	Waterbodies	16. Pond 17. River 18. Wetland

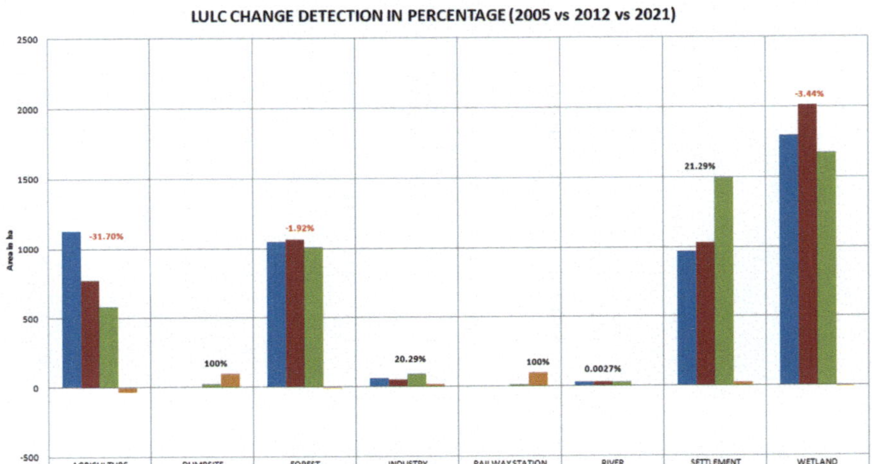

	AGRICULTURE	DUMPSITE	FOREST	INDUSTRY	RAILWAY STATION	RIVER	SETTLEMENT	WETLAND
2005	1126.78445	0	1048.952692	61.965209	0	29.35005	969.775272	1795.661029
2012	770.98151	0	1063.625951	50.600819	0	29.351664	1034.075682	2012.904169
2021	584.344988	23.643488	1009.303363	93.51522	12.391729	29.351664	1494.681846	1676.176858
%age	-31.70066799	100	-1.926355514	20.29195006	100	0.002749494	21.29907517	-3.441525062

Fig. 15.6 Bar chart depicting the land cover land use of major classes and sub-classes

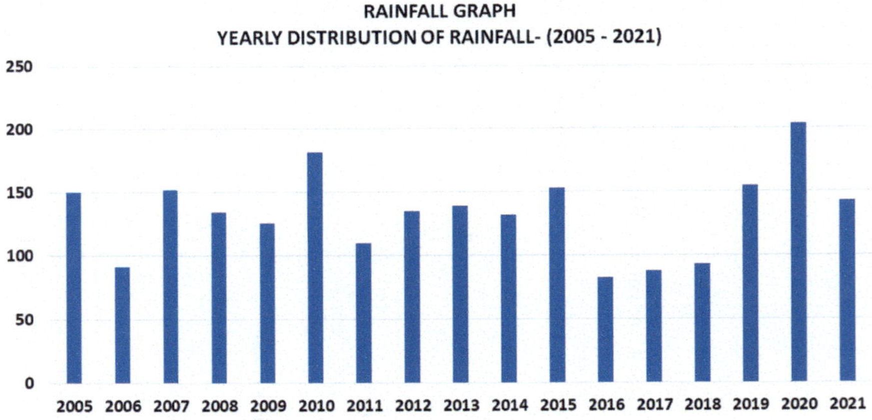

Fig. 15.7 Bar chart demonstrating annual rainfall during the year range of 2005–2021

It is evident from the analysis that the built-up area over time has increased mainly in terms of the settlement, industries, institutions, and infrastructures, and also through the roads as well as Azara Railway Station, a recent development in the southwest part of the Deepor Beel.

The field survey and revalidation of lab generated results on the ground ensured that the encroachment led to a reduction in the wetland area. Moreover, the inlets bring deposits that reduce capacity of water retention. These deposited land masses are used for agricultural practices. The periodic survey also results in changes in the

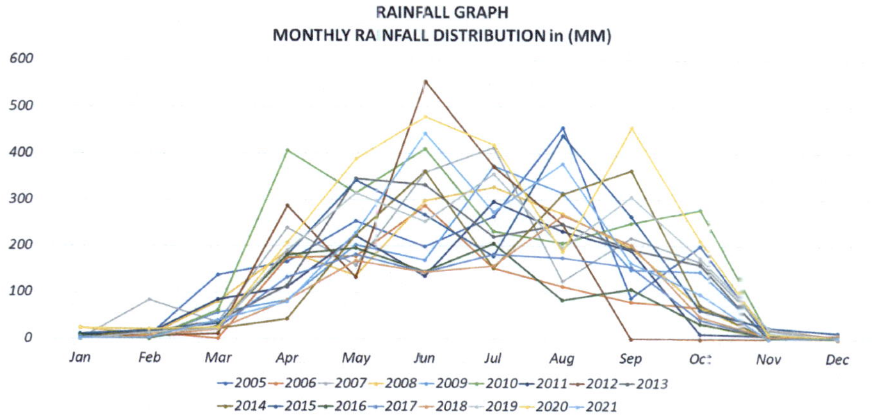

Fig. 15.8 Graph depicting monthly rainfall distribution during the year range of 2005–2021

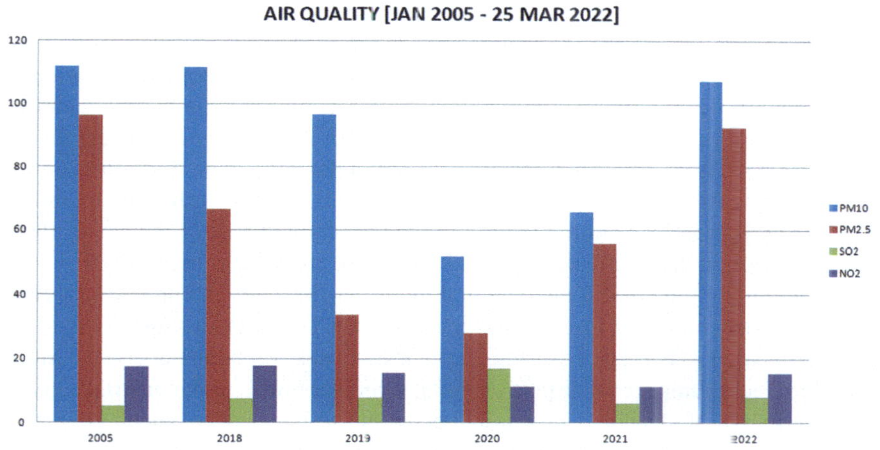

Fig. 15.9 Bar chart depicting air quality data during the period of January 2005–25th March 2022

water quality during summer and winter. It was also seen that when the catchment area dries out during the dry season, agriculture is practiced within the area. Thus, from our study, we try to derive the changes in the recent year from 2005 to 2020. Thus, comparing our study to the previous research it can be seen that with the change in LULC pattern (especially the periphery of the wetland is diminishing), the biodiversity loss and imbalance of the ecosystem has been to a large and significant extent. Upon field study for this wetland, it was noticed that the railway lines construction in the eastern region is a major reason of fragmentation. The threats associated have clearly depicted the shrinking of the lake regions along with the sound of trains and air that severely destroyed the natural eco-system within the wetland for the existence of various flora and fauna. A microscopic quantification of these indices is beyond the scope of the book chapter and relevant research works can be undertaken vis-à-vis literature reported findings.

Upon aerial survey using satellite imageries of temporal data and digital elevation model, it can be mentioned that the Beel is a storage basin and catchment area for Guwahati exceeding 40,000 ha total area. It provides habitats to the natural fisheries that are considered to be a major contribution towards the public (5000 fisher folk and regional custom). Also, it has better bio-diversity in terms of 400 numbers of vertebrate. Enriched floral and faunal diversity of the Beel consists of mammals (24), diatoms (65), reptiles (33), aquatic macrophytes (58 species), molluscs (15), amphibians (11), crabs (2), zooplankton (171), prawns (3), finfish (68), aquatic insects (55), birds (234), 11 bird species occurring in the wetland (Threskiornis melanocephalus, Ephippiorhynchus asiaticus, Leptoptilos dubius, Aythya baeri, A. nyroca, Gyps bengalensis, Haliaeetus leucogaster, L. javanicus, G. tenuirostris, Sarcogyps calvus and Pelecanus philippensis) and bryozoans (5) was found in the IUCN. Also, this wetland contain finfishes that reduced from 67 to 64 numbers in 1999 to 2016–2018 [29]. In the present study, spatiotemporal data was used to calculate the change in the LULC pattern. Thereby, the associated change had an impact on the wetland ecosystem. The statements from mentioned literature suggested on the linkage of measured parameters with the bio-diversity loss and non-conservation.

Correlation of rainfall data reveals that the highest rainfall is in the month of June–August for the considered time period i.e. 2005–2021. The plotted data for the period 2005–2019, conveys that the annual rainfall bar chart altered post lockdown. After lockdown (Mar 2020), a peak exists in rainfall activity during the month of July 2020 followed by a similar pattern in the month of July 2021. However, there is an exception in the month of July, 2012. All these are evident in Fig. 15.8.

Observation for the LULC analysis highlights that due to brick kiln industries and other industrial emissions, PM2.5 was higher in 2005. However, during the Covid19-induced lockdown (Fall 2019–Spring 2021), PM2.5 was comparatively on the lower side. With the resumption of airport services, vehicular traffic, and industrial activity in the periphery of Deepor Beel, particulate matter levels have enhanced gradually. The bar plot clearly indicates the rise of pollution levels which is again inversely related to the precipitation around the study area (Fig. 15.9).

15.4 Conclusions

In recent times, remote sensing and GIS technology have been widely used in different Planning and Development activities. With LULC change detection, an overall picture of the transition of the landscape in the real ground is obtained in both time and space. GIS data not just gives an overview of the changes but also provides statistical information. It clearly depicts the extent of the loss of the wetland area and its surroundings to agriculture, settlement, industries, etc. Thus GIS technique is able to generate accurate results and provides real-time monitoring, cost and time effective. The study highlighted on the extent to which the wetland area gets converted to agriculture, dumpsite, industry and settlement.

The correlation between rainfall and air quality data implies an inverse relationship between the same. One such observation is evident from the plots for (2018–2020). In the article titled "Analysis of the Influence of Precipitation and Wind on Particulate Matter (PM2.5 and PM10) in the Atmosphere" [30], authors have derived a negative correlation between particulate matter and rainfall i.e. with rising precipitation, the rate of PM2.5 and PM10 declines. This is evident in the relevant graph.

Brick industries, airport, and other industries are major causes of disturbance in the Eco Sensitive Zone. Thereby, they threaten the biodiversity. The Boragaon Dumping Site located in the south-eastern periphery of the Deepor Beel has been creating an environmental hazard to the wetland and its surrounding since 2007. Presently, it is under consideration for relocation to a new site by the concerned authority as of 2022. Otherwise, it has enormously contributed to the degradation of the water quality and has resulted in the micro as well a macro aquatic biodiversity loss of the wetland ecosystem. It was also mentioned in the report that no reliable data exists with respect to the biodiversity of the Beel and with a comparative analysis of the past data. Hence, it is recommended for Assam Science Technology and Environment Council (ASTEC) to consider it as a concerted program and thereby bring out a biodiversity atlas of the Deepor Beel [31].

In our study, it was observed that due to Covid19 induced lockdown all the industries, vehicular and air traffic movement were brought to a standstill. Thus, correlating land cover and land use change with rainfall and air quality data, it can be inferred that, for a balanced ecosystem in Deepor beel, the concerned authority will have to formulate an action plan so as to restrict upon the particulate matter levels and thereby monitor the rapid growth of built-up settlement and other industries in and around the wetland area.

In summary, using remote sensing techniques and relevant software, the wetland loss has been quantified in this work for the Deepor beel. However, its correlation with bio-diversity along with climate change factors shall be addressed for the biodiversity conservation of wetlands in the north-eastern states of India. Thereby, a comprehensive and general research framework can be ensured for synchronized future research in the chosen field of study.

References

1. Population Bulletin, UN Homepage. https://unctad.org/statistics. Accessed 18 May 2022
2. World Bank Data Homepage. https://data.worldbank.org/. Accessed 18 May 2022
3. Urban sprawl and land transformation in Rohtak city: a study using remote sensing and GIS. http://hdl.handle.net/10603/9792. Accessed 18 May 2022
4. IndiaWaterportal. https://www.indiawaterportal.org/articles/taking-people-along-conserve-freshwater-ecosystems. Accessed 18 May 2022

5. Mozumder, C., Tripathi, N.K.: Geospatial scenario based modeling of urban and agricultural intrusions in Ramsar wetland Deepor Beel in Northeast India using a multi-layer perceptron neural network. Int. J. Appl. Earth Observ. Geoinf 92–104 (Elsevier) (2014)
6. Sivanpillai, R., Miller, S.: Improvements in mapping water bodies using ASTER data. Eco. Inform. **5**(1), 73–78 (2010)
7. Ahmed, I.A., Baig, M.R.I., Talukdar, S., Asgher, M.S., Usmani, T.M., Ahmed, S., Rahman, A.: Lake water volume calculation using time series LANDSAT satellite data: a geospatial analysis of Deepor Beel Lake, Guwahati. Front. Eng. Built Environ. **ahead-of-print** (2021). https://doi.org/10.1108/FEBE-02-2021-0009
8. Prasad, S.N., Ramachandra, T.V., Ahalya, N., Sengupta, T., Kumar, A., Tiwari, A.K., Vijayan, V.S., Vijayan, L.: Conservation of wetlands of India—A review. Trop. Ecol. **43**(1), 173–186 (2002)
9. Ozesmi, S.L., Bauer, ME.: Satellite remote sensing of wetlands. Wetl. Ecol. Manag. **10**(5), 381–402-9 (2002)
10. Deepa, R.S., Ramachandra, T.V.: Impact of Urbanization in the interconnectivity of wetlands. In: Paper presented at the National Symposium on Remote Sensing Applications for Natural Resources: retrospective and Perspective (XIX–XXI 1999), Indian Society of Remote Sensing, Banglore, India (1999)
11. Manju, G., Chowdary, V.M., Srivastava, Y.K., Selvamani, S., Jeyaram, A., Adiga, S.: Mapping and characterization of inland wetlands using remote sensing and GIS. J. Indian Soc. Remote Sens. **33**(1), 51–66 (2005)
12. Häder, D.-P., Banaszak, A.T., Villafañe, V.E., Narvarte, M.A., González, R.A., Helbling, E.W.: Anthropogenic pollution of aquatic ecosystems: emerging problems with global implications. Sci. Total Environ. **713** (2020), Article 136586-12
13. COVID-19: nitrogen dioxide over China (2020). https://www.esa.int/Applications/Observing_the_Earth/Copernicus/Sentinel-5P/COVID-19_nitrogen_dioxide_over_China. Accessed 17th Apr. 2020
14. COVID-19 and surface water quality: Improved lake water quality during the lockdown. https://www.sciencedirect.com/science/article/pii/S0048969720325298. Accessed 18 Aug. 2022
15. Datuk Seri Dr. Zaini Ujang—January 14, 2021. Impact of Covid-19 on the Environment. https://www.nst.com.my/opinion/columnists/2021/01/657271/impact-covid-19-environment. Accessed 18 Aug. 2022
16. Prancsha, T.S., Kamra, A.K.: Scavenging of aerosol particles by large water drops:3. Washout coefficients, half-lives, and rainfall depths. J. Geophys. Res.: Atmosp. (1984–2012) **102**(D20), 23947–23953 (1997)
17. Aswathy, T.S., Achu, A.L., Francis, S., Gopinath, G., Joseph, S., Surendran, U., Sunil, P.S.: Assessment of water quality in a tropical ramsar wetland of southern India in the wake of COVID-19. Remote Sens. Appl.: Soc. Environ. **23**(2352–9385) (2021)
18. Baruah, P., Deka, C.R., Goswami, D.C.: Study of seasonal variation in water spread and land use pattern in the Deepor Beel area using digital image processing and GIS techniques. In: Proceedings of the Zoological Society of Assam, Spl. edn., pp. 34–48 (1992)
19. Deb, S., Saikia, J., Kalamdhad, A.S.: Ecology of Deepor Beel Wetland, A Ramsar site of Guwahati, Assam with special reference to algal community (2019)
20. Kalita, A., Das, A.N.: Status of elephant corridors in Garbhanga-Rani reserve forest in and around Deepor beel. Int. J. Emerg. Technol. Innov. Res. **7**(3), 329–337 (2020)
21. RIS.: Information sheet on Ramsar Wetlands (RIS). World Wildlife Fund for Nature, Guwahati, India (2002). https://www.wetlands.org/reports/ris/2IN012en.pdf. Accessed 20 Mar. 2013
22. MoEF.: Report on visit to Deepor beel in Assam: a wetland included under national wetland conservation management programme of the Ministry of Environment and Forests, Guwahati (2008). https://www.niti.gov.in/planningcommission.gov.in/docs/reports/E_F/DeeporBeel.pdf. Accessed 07 May 2022
23. Sharma, N., Janauer, G., Mondal, S., Bakimchandra, O., Garg, R.D.: Assessing wetland landscape dynamics in the Deepor Beel of Brahmaputra Basin using geospatial tools. Asian J. Geoinf. **12**(1), 1–8 (2012)

24. Mozumder, C., Tripathi, N.K.: Climatic factors on wetland dynamics of Deepor Beel in India: a remote sensing and GIS based approach. In: Asian Conference on Remote Sensing. Asian Association of Remote Sensing, Taipei, Taiwan (2011)
25. Dr. Saikia, P.K.: Qualitative and quantitative study of lower and higher organisms and their functional role in the Deepor Beel ecosystem (2005)
26. IMD portal. https://www.imdpune.gov.in/. Accessed 30 Sep. 2022
27. ENVIS Centre Web Portal. http://cpcbenvis.nic.in/air_quality_data.html Accessed 30 Sep. 2022
28. Open Weather API. https://openweathermap.org/api. Accessed 30 Sep. 2022
29. Bhattacharjya, B.K., Saud, B.J., Borah, S., Saikia, P.K., Das, B.K.: Status of biodiversity and limno-chemistry of Deepor Beel, a Ramsar site of international importance: conservation needs and the way forward (2022)
30. Liu, Z., Shen, L., Yan, C., Du, J., Li, Y., Zhao, H.: Analysis of the influence of precipitation and wind on PM2. 5 and PM10 in the atmosphere. Adv Meteorol **2020**, 1–13 (2020)
31. Report on Visit to Deepor Beel in Assam—A wetland included under National Wetland Conservation and Management Programme of the Ministry of Environment & Forests (2008)

Chapter 16
A Review on Structure, Floristic Diversity and Functions of Homegardens

Biplov Chandra Sarkar, Gopal Shukla, C. P. Suresh, and Sumit Chakravarty

16.1 Introduction

Homegardens are also known as urban food gardens, kitchen garden, home food gardens, and domestic food gardens [46]. These are believed to be old tree-based land utilization systems in which a multi-stratum vegetation structure comprising both woody perennials annual crops are maintained around homestead. At times, they exist in association with livestock for the primary daily subsistence needs without any external energy inputs [40]. These homegardens have been maintained by the small land owners or forest frontier community as a strategy to achieve food self-sufficiency, nutritional demands and income security for subsistence living. Thereby, they are often targeted to deliberately maintain high plant diversity in a very small unit area of land around the homestead [38].

The cultivation of homegardens must enable a "learning to live with change and uncertainty". Homegardens mimic a natural forest due to its multi-strata heterogeneous plant composition of wild and cultivated plant species. These species are dynamic with regions and ownership are associated with diverse ecosystem services [26]. These gardens are distinct ecological and cultural units and an integrated part of overall agriculture landscapes in moist tropical Asia, Africa and South America conserving plant biodiversity, socio-cultural dignity of the owners, local cultural values and indigenous ecological knowledge [15].

B. C. Sarkar (✉) · C. P. Suresh
Department of Forestry, North Eastern Hill University, NEHU Campus, Tura, Meghalaya, India
e-mail: biplovubkvfor13@gmail.com

G. Shukla · S. Chakravarty
Department of Forestry, Uttar Banga Krishi Viswavidyalaya, Pundibari, West Bengal, India

R. V. S. Uppaluri and L. Rangan (eds.), *Conservation of Biodiversity in the North Eastern States of India*, https://doi.org/10.1007/978-981-99-0945-2_16

16.2 Homegardens—A Diminishing Tradition

Homegardens are dynamic and resilient micro-farming systems that respond effectively to rural transformation and livelihood diversification driven socio-economic changes. However, now they are forced to transform into commercially oriented systems [49]. Unfortunately, from the past decades, driven by socio-economic changes in terms of urbanization, better access and communication, tourism, loss of knowledge and traditional crops, health issues of local public, crop pests and diseases, land shortage, drought and unemployment, the communities and especially the younger generation shy away from their traditional knowledge. Thereby, ongoing trend conveys the non-adoption of traditional farming and abandoning of the homegardens [44]. Population growth increased the land fragmentation and thereby decreased the size of homegardens. Thus with reduced production and income, the farmer is forced to either adopt commercial high value crop or abandon home gardening [20].

In a study, it was found that presently traditional plants were less preferred by farmers. They now prefer commercial species which are easily marketable. Hence, these varieties are cultivated in comparison with traditional crops [18]. Such unfortunate engagement has been linked to the loss of nutritional deficiencies, local diversity, food insecurity and traditional knowledge as indigenous varieties and species that provide diverse ecosystem services, and particularly cultural engagements linked with traditional knowledge [5]. Commercially driven home gardeners highly adopt cash crop production and monoculture species with external synthetic inputs such as fertilizers and pesticides. Such engagements disrupt the heterogeneity of the homegardens and convert it into high input, high yield, and high-risk farming system. Thereby, such engagements hinder the social services offered by the homegardens and as well tear into the social fabric of the community [1].

Industrialization driven alterations in consumption and demand–supply relationship also influence rural farming activities such as homegardens. Thereby, the heterogeneous farming unit transformed into enhanced homogenous cash crop farming unit [31]. The transformation of traditional farming activities such as home gardening with commercially driven modern activities drives social and economic inequalities. Invasive and/or exotic plants being introduced in the homegardens are other potential threats to the traditional farming system [20]. In addition, climate change, over exploitation and pollution are significant threats to homegardens in terms of the disruptions to the ecosystem services [35].

In an agreement to these statements, the disturbing updates in many countries such as Uganda, convey the national governments development plan for rural urbanization and industrialized agriculture in rural settings. Thereby, they threaten traditional farming systems such as homegardens [48]. Many countries are now expressing their concern on the decline of homegarden farming. Losing the tradition of homegarden farming has been feared to reduce the resiliency of the small holder producers to climate change [6]. Thus, immediate policy and institutional interventions are needed evaluation of homegardens which include biodiversity and genetic resource status on a priority basis along with legal support [9]. Fortunately, the significance of traditional technology and knowledge is gaining recognition through international enterprises such as the Convention on Biological Diversity (CBD), Satoyama Initiatives, The United Nations Framework Convention on Climate Change (UNFCCC) and The Globally Important Agriculture Heritage System (GIAHS) introduced by Food and Agriculture Organization of the United Nations [28].

16.3 Structure and Composition of Homegardens

Homegardens are man-managed highly energy subsidized, structurally and functionally heterogeneous micro-environments attached to the homestead within larger farming systems in the regions of higher human population density. Thereby, they satisfy diverse livelihood needs of its owner and demonstrate need-based domestication at ecosystems and landscape levels [23]. The urge of the owners for self-sufficiency in subsistence living with food and livelihood security throughout the year translate into the maintenance of high plant diversity in the smaller areas of homegardens and incessant of newer plant species that protect the naturally growing species. Thereby, homegardens significantly introduction conserve and improve native and exotic biodiversity [41].

Different types of homegardens serving particular functions have been reported such as vegetable gardens (oldest, fruits and vegetables), herb gardens (culinary and medicinal), rose gardens (for flowers and ornamentals), knot gardens (squares of flora or paving encased by dwarf box-hedging for aesthetics), oriental gardens (incorporates water, paths, rocks, sculptured bushes and trees to create tranquillity) and wild gardens that encourages growth of natural plants and wild grasses [2]. Homegardens mimic natural forests due to their canopy structural ability that resemble multi-storey forest-like shape [4]. Such a structure is often resultant either from a lacking apparent outline or deliberate planning to mimic the forest. The dynamic architecture of homegardens adjust diverse plant species temporally and spatially with efficient utilization of ground and aerial space. Thereby, such homegardens ensure upon the realization of sustainable and resilient systems [12].

Homegardens are an integrated system in which energy transfer from the sun is efficiently transmitted to the animals via plants and man. Hence, matter is effectively recycled. Homegardens are often separated from their surroundings by fences, hedges, or other obstructions with specialized edaphic, microclimate and biotic conditions created by the repeated specific management of the household to make them remarkably different from their surrounding landscape. Homegardens are complex vertical and horizontal arrangement of plants. Such resultant characteristics are due to the variation in plant height with 3–5 canopy strata of more than 15 m height and intermixing of the species and individuals [37].

Homegardens have been reported to be dynamic in terms of their structure and composition as they always tend to alter in response to a variation in the socio-economic features [33]. Food provisioning species are the primary component of homegardens across the world [19]. Trees, especially fruit trees are common in homegardens of all landscapes. However, they are lesser numbers due to their higher space requirement [11]. Medicinal plants [45] and bamboos [29] have been reported to be important components of both cultivated and wild traditional homegardens. Exotic plant species have also been reported as an important component of homegardens and on global basis. This is due to the predominant trade offs among medicinal, food, ornamentation and other purposes [50].

Six different types of homegardens have been reported by Peyre et al. [33] Palaghat district, Kerala. They were created on the basis of cluster analysis of shrub species or tree density along with homegarden size. Traditional gardens have been also reported to transform into modern homegardens through four gradual developmental stages. Based on this transition, it was reported that 50% of the Palghat homegardens were traditional and 33% of them incorporated modern farming practices. These practices include reduced woody species diversity, increased yet limited number of cash crops, ornamentals and external inputs i.e., towards homogenization. Similar structural changes of homegardens have been also reported for homegardens along a rural-peri urban gradient in Campeche, México. It was reported that the size of a homegarden was related to the size of the general farm in an agricultural landscape. However, it is not so in an urban landscape that contributes fragmented resource due to scarcity of arable land in urban landscape. Urban and peri urban homegardens differ with species composition and use pattern harbouring with mostly diverse ornamental species in comparison to the diverse range of fruits, vegetables, medicine, shade, fence, wood and energy species in rural homegardens. Plant species composition of central Yucatán homegardens located in the outskirts of cities and remote villages have been reported to vary and even with the variations in the socio-economic condition of the homegarden owners [34].

Overall, low correlation between the size of the homegardens and species diversity has been reported [1]. Generally, small sized homegardens have been reported with higher floristic diversity. Higher number of species (17–51) were reported from Palakkad District in Kerala homegardens with a size of 0.14–1.01 ha [33]. However, homegardens located in hilly landscapes were reported with higher diversity of perennial plants in contrast to higher diversity of annuals in plain landscapes. The number of species and homegarden size encountered at diverse altitudes exhibited a fragile increasing trend but were substantial at the high-altitude gardens [39]. A study conducted at four altitudes with homegardens classified them into three sizes: small (< 0.003 ha), medium (0.004–0.006 ha) and large (> 0.007 ha). Thereby, the study stated 111 plant species with the domination of trees (1200 individual ha^{-1}) at mid-altitude. On the contrary, diversity was maximum (2.42) in large homegardens at very low altitude. In Kumaun Himalaya, both tree and herb density have been found to be highest in large homegardens (825 and 479 plants ha^{-1}, respectively), followed by medium (750 and 317 plants ha^{-1}, respectively) and lowest in small sized homegardens with 317 and 249 plants ha^{-1}, respectively [47]. Table 16.1 summarizes the chief characteristics of a typical homegarden.

Table 16.1 Chief Characteristics of a typical Homegarden [30]

Characteristics	General practice
Species density	High
Species type	Staples, vegetables, fruits, medicinal plants
Production objective	Home consumption
Labor source	Family (woman, elderly, children)
Labor requirements	Part-time
Harvest frequency	Daily, seasonal
Space utilization	Horizontal and vertical
Technology	Simple hand tools
Input-cost	Low
Distribution	Rural and urban areas

16.4 Floristic Diversity

Abundant studies across the globe on homegardens reported extraordinary species diversity and varieties. Twenty-three woody species were grown in homegardens of eastern Tigray, Ethiopia. Further, 55 of them were grown in the homegardens of South-eastern Ethiopia. The Wolayta homegardens were diverse with overall 159 plant species including 112 useful species and 60 crop plants. Homegardens in western Ethiopia were reported with 52 plant species; Basketo and Kafa zone of southern Ethiopia with 213 plant species; Dwaro zone of southern Ethiopia with 224 plant species along with 78 cultivated crop species; north-western Ethiopia with 69 plant species [25] and Dilla Zuria district of Ethiopia with 52–75 plant species. Homegardens of Arba Minch town in southern Ethiopia was documented with 138 plant species. Among these, 12.36% of edible plants, 11.90% were income generating plants and 8.92% were shade plants. Species richness in these homegardens has been reported to be 10–45 with an average of 27.5 species.

Households in Yayo Biosphere Reserve, Southwest Ethiopia harboured 76 plant species and represented 38 families in the homegardens with a size range of 200–1250 m^2. The diversity referred to 13 species of fruits and 8 species of vegetables [24]. Homegardens in Jimma City of Ethiopia were found to mitigate climate change by conserving native biodiversity of 36 woody species with substantial relationship between species richness and diversity. The Shannon Wiener diversity index has been estimated for these urban gardens as 3.15. Similarly, homegardens of Cameroon were also highly diverse with Shannon Wiener diversity of 3.30. In Kiboguwa, the eastern slopes of Uluguru Mountains, Tanzania homegardens size varied as 0.15–1.84 acres along with an average size of 0.66 acres. For the system, plant species richness documented was 285 and involved only trees and herbs. In 360 studied homegardens of Benin, 323 plant species were signified by 226 genera and 81 families were observed. These included 185 wild and 138 crops and plant species. Further, the species referred to 153 native and 170 exotic species. Most ownership was with women. However, all male and female homegardens owners primarily preferred food and medicinal plants. Most plant species (90.82%) in the study of Gbedomon et al. [13] have been documented from less than 10% of the studied homegardens. Also, none of the documented species were found in more than 75% of the studied homegardens. Ugandan studies reported highly diverse homegardens. Especially, the western region being dominated with cooking banana varieties constitute the diversity. This is due to banana being a chief staple food of several Ugandans. Homegardens of southwestern highlands in the western region of Uganda has been documented with a richness of 209 crop species. Many of these were food species.

South American homegardens have been reported to be diverse with studies reporting species richness of 161 in Amazonia; 324 in Nicaragua; 232–280 in Mexico (Poot-Pool et al., 2015); 309 species in Peru and 484 in Ecuador [4]. The bio diversity has been believed to be not an effect of varying sample size in the studies due to the heterogeneity and unpredictability of studied homegarden diversity. Homegardens of Amazonian, Ecuador have been inventoried with 484 plant species that represented 96 families with an average species diversity of 32 [4]. Major life forms in these Ecuadorian gardens were herbs followed by trees, shrubs, epiphytes, vines, lianas and hemi-epiphytes.

Homegardens located along the middle Madeira River in the municipality of Manicoré, Amazonas, Brazil for 16 communities were large and occupied about two hectares area an average. These gardens harbored a total of 86 plant species. Many of these species were cultivated by the farmers to suit to soil type and henceforth enabled the homegardens to become diverse from agro biodiversity. Urban and peri urban homegardens of Mexico were inventoried with 316 cultivated plant species representing 94 families including 103 species of shrubs, trees, palms, bananas and 213 herbaceous species. A total of 281 plant species was documented in Tehuacán valley, Mexico, of which 50 were medicinal plant species, 92 edible and 115 species ornamental [22]. Homegarden agrobiodiversity of Tabasco lowlands, México, was documented with 280 species represented by 28 genera and 84 families with a total of 38 functions. Among these, 14 were ecological, 12 were economic and 12 were socio-cultural. Nicaraguan homegardens were listed with 324 plant species. Thereby, majority of these were ornamentals followed by fruit trees, medicinal plants, food crops and spices. In a single homegarden at Java 60 plant species were reported. Similarly, from a small hamlet in west Java, 219 species were reported. The size of 278 homegardens at five villages in Beijing Municipality of China varied as 150–200 m^2. These homegardens were grown with 100 of edible species and 152 ornamental species.

In Thailand Karen, homegardens were listed with 268 plant species ranging from 100–146 species in the studied villages. Homegardens are prominent agroforestry systems in Sri Lanka ensuring nutritional and food security to the Lankan locals and harboured more than 400 woody species. In the western Sri Lanka, 289 plant species were documented from 106 suburban homegardens including 12% medicinal plants, 36% food plants and 51% ornamental plants [21]. Homegardens in Moneragala district located in the drier region of Sri Lanka harboured 70 plant species belonging to 30 families and 55 genera with a mean tree density of 23 and 80 trees ha^{-1} in medium and small sized homegardens. Tree density at inherited and settled homegardens in the urban area of Sri Lanka have been estimated as 54.5 and 40.3 trees per acre, respectively. There, 30% exotic species exist in settled homegardens in comparison with the 16% exotic species in inherited homegardens. In 402 homegardens in the six regions across southwestern Bangladesh, it was analyzed that 419 species existed (59% native) including six IUCN Red Listed plants were found.

Several studies also confirm that homegarden of India are highly diverse and also conserve biodiversity. The terai homegardens of West Bengal have been documented with 76–260 plant species [32]. On the other hand, Darjeeling Himalayan homegardens were inventoried with 262 plant species [40]. The documented species in Terai homegardens comprised of two bamboo species, three palms, 4 shrub species, 62 herb species and 71 tree species. These were used for various purposes such as food, medicinal, ornamental, aesthetic and cultural. In Barak valley, Assam, India, homegardens of 122 plant species were reported. Homegardens in the ethnic community in Manipur, India have been documented with 73 plant species representing 27 families and their economic, social and cultural values [44]. Homegardens of War Khasi community in Meghalaya have been listed with 197 plant species with an average of 89 plant species per homegarden of 750 m^2 average size. Shukla et al. [45] listed 116 plant species representing 50 families and 102 genera in 100 homegardens of Gumla district, Jharkhand. These included 31 species of fruits, 25 species of flowers, 51 species of vegetables and 31 species of traditional medicines. In 19 homegardens and 10 home-forest-gardens of South Andaman, 28 ligneous species were documented where plantation crops (coconut, areca nut), fruit plants (banana, mango, banana and papaya) and spices (nutmeg, clove and cinnamon) were primary species. In Kumaun Himalaya, 120 homegardens, the garden landscape with 70 plant species and 35 families dominated the garden biodiversity. These diversified as vegetable (50.78%), fruits (22.57%), medicinal herbs (18.03%), fodder (12.21%), fuel wood (7.35%) and ornamental commodity (1.43%) species. Homegardens located at Pachmarhi Biosphere Reserve, India were documented with 47 ethnobotanicals being deployed for food, medicine, vegetables, neutraceutical, fodder and cultural purposes.

Shastri et al. [43] documented 68 plant species in homegardens of Karnataka. In 50 urban gardens of Bangaluru, Jaganmohan et al. [17] found 117 species of herbs and 25 species of trees. In 66 rural and 62 urban homegardens of Kanyakumari district, Tamil Nadu India, 89 plant species comprising of 80 genus and 45 families were reported with 5–35 species per homegarden (average of 13.86). Homegardens of Kerala were found to be dominant with just 10 tree species which accounted 74% of the total homestead growing stock and 85% of the total wood volume in Kerala. However, 127 plant species were inventoried in the Kerala homegardens. Floristic and structural diversity of the 104 homegardens of indigenous communities in Kerala were assessed and 182 plant species under 160 genera and 67 families were documented including further classification as 39% to be edible, 24% to be ornamental and 25% to be medicinal type.

16.5 Functions of Homegardens

Indigenous and ethnic communities have traditional knowledge of cultivation and assurance of maintaining diverse food and medicinal plants in gardens cultivated near to their livelihood homes for support in both rural and urban regions [44]. Ever increasing demands for goods and services due to population rise forced the human society to find alternate ways from agroecosystems as well as family farming like homegardens in the context of supplementing the natural ecosystems sustainable development and livelihood [36]. Due to their flexibility and resiliency, the traditional farming systems such as homegardens are now recognized as the best option to sustain production and also maintain the ecosystem functions of the landscape. Thereby, improving rural livelihoods, food security and well-being is also targeted and achieved.

16.5.1 Social Capital and Sustainable Production System

Households with homegardens have been reported to share their garden products with their relatives, friends and neighbours and thereby deepen their friendship, neighbour and family relationship. In other words, they serve as Social Capital and ensure the home gardeners to become resilient to climate change and social conflicts. Agroforestry systems including homegardens have been also reported as a better land use option to generate income by the farmers and also ensure the realization of carbon-credits [42]. Studies have been also reported to infer that introduction or enhancement of species in homegardens do offer few ecological benefits. These particularly refer to the greater potential to deliver ecosystem service such as carbon sequestration.

Homegarden farming practice guided by the indigenous knowledge system of the area has been reported to cause overall sustainable development of the practicing community through ecological and socio-economic sustainability. This was demonstrated as the homegardens mediated ecological socio-economic sustainable development at Nandom traditional area in Ghana. Homegardens aid in food assurance for the rural community. Thereby, the needy farmers have been assured of their livelihood. Thus, they also provide them resilience against uncertainties of high-input agriculture. Homegardens in Maya have been reported to be important for semi-intensive production of crops such as vanilla, cacao and annatto.

16.5.2 Ecosystem Services

The significance of an of ecosystem facilities in human well-being and ecosystem functioning by the natural ecosystems has to be recognized. Fortunately, this was realized since the release of Millennium Ecosystem Assessment Report. This document envisioned support, provision, cultural regulation or protective services and thereby indicated both tangible and intangible benefits to the human society. In an agricultural landscape, production can be sustained through the resilience and flexibility of ecosystems through genetic variation within agricultural crops and thereby boost and enhance ecosystem functions [14].

There is growing attention in the study of ecosystem services offered by agroecosystems or traditional farming systems, including homegardens. This is due to the environmental, social and economic aspects that they ensure [4]. Ecosystem services provided by homegardens need to be understood and quantified for landscape planning, management and decision making. Very few studies till date devoted to the homegardens at a micro-landscape level and in the perspective of sustainable ecosystem services. Especially, research directions involving low energy input sustainable production s are to be targeted. Thus, ongoing research on their existence role to improve human well being should be somehow overcome with holistic pedogogies. Homegardens have been also reported to aid making cities more sustainable while providing rural folks with working opportunities, healthier food, cultural identity, aesthetic environment and dignified life [2].

Ecosystem services in larger perspectives being delivered by homegardens to agricultural and allied systems and also at micro-level of a household have been widely reported. These include protection and favourable environments for varieties that may be vulnerable to biotic and abiotic stress in fields; agro-biodiversity conservation; protection for micro and macro-fauna and allowance of gene-flow between plant populations inside and out of the garden area; provisional, regulating or supporting services in the form of food, clothing, shelter, materials and well-being; pollination, soil amelioration, soil conservation, effective nutrient cycling, soil fertility improvement, soil carbon build-up, water retention, air purification, reducing CO_2 emissions, microclimate modification, temperature regulation, providing livelihood opportunities as well as social-cultural preservation, aesthetic or recreational needs and empowerment or social position of women [24].

In 138 homegardens of Sangay, Ecuador, 484 plant species were reported to be associated with 20 ecosystem services. Most of these were cultural and provisional like ornamentation, food and medicine [4]. These Ecuadorian homegardens were found to be prominent for cultural ecosystem services (358 species i.e., 74%) followed by provisioning services (237 species i.e., 49%) and least for regulatory services (18 species i.e., 4%). In terms of individual species, 173 offered ornamental services, 145 medicinal, 142 foods, 42 construction, 19 rituals, 11 veterinaries, 10 shade, 9 for commercial and rest for other uses. Alnus-cardamom based traditional agroforestry system of Sikkim Himalayas offered provisional, supporting, and regulating types of services in the form of food, fodder, fuel wood, nitrogen fixation, soil erosion reduction as well as improvement of the aesthetic (beauty) of the areas. Homegardens have been reported to contribute 7–16% of the household income to its owner through sale of homegarden produce [8].

16.5.3 Biodiversity Conservation

Homegardens were considered to be highly relevant for conservation purposes as they signify agro-biodiversity at numerous levels in social and in other small avenues [2]. Homegardens constitute diverse and richer complex micro-production systems for the development and conservation at the genetic, ecosystem and species level. Thereby, they are effective to link natural ecosystems with agroecosystems and can as well serve as a focal point of the household's social interaction. These refer to maintaining varieties and sustaining diversity across generations [2].

However, more comprehensive and interdisciplinary studies at intra- and inter-specific levels within diversified cultural and environmental contexts is required for the better understanding of a homegarden's role in management and conservation of in situ genetic diversity [2]. Homegardens can be best understood for its conservation values from the targeted diversity holistic studies distribution within and across gardens. The resilience and evolution with time shall be targeted. Thereby, the minimum size for conservation of viable populations of the target species need to be defined and monitored for any pertinent evolutionary changes [2].

Harboured diversity and management techniques ensure effective strategies for regional biodiversity conservation in homegardens. Italian homegardens have many landraces of *Phaseolus* spp., a common bean, while, at least 10 native potatoes and cassava varieties were maintained in every Andean and Javanese homegardens, respectively. Homegardens were reported to maintain more than a few hundred plants of variable population size. This is dependent on the importance of the species and size of the garden. It was believed that to adequately represent the overall diversity of any given species by considering its inter- and intra-specific diversity, a number of homegardens in diverse agro-ecological need framework are to be defined. The relevance of homegardens in biodiversity conservation is possible only when the gardens represent biodiversity spatially across a specific ecological zone over small spaces [13].

Homegardens were also used to test new or indigenous crops and traditional varieties; nurseries for plantlets to be planted in open fields for weeds and wild plants and for complementing or restoring germplasm collections crucial for breeding which lead to a significant intra-specific diversity, and thereby enhance chances of gene flow. Thereby, opportunity for a species to adapt and survive over time and utilization of wild germplasm by the farmers to create and improve crops by trial and error methodology can be ensured.

However, it is not necessary that the homegardens dominantly represented with the species of higher frequency are indicators of efficient and viable systems of biodiversity conservation. Rather, it was suggested to analyse by considering their intrinsic pattern of multi-strata vegetation with predominant herbaceous strata [13] and thereby ensure that they are heterogeneous and dynamic on a temporal basis. Homegardens are maintained for production purpose based on owner's choice. Hence, species may disappear or reappear later and can be introduced in the same garden or elsewhere [13]. Thus, conservation function is only a by-product of production function but not a well defined objective. Thus, homegardens in a true sense as an efficient conservation system is questionable [13]. The intangibility of provisioning services with homegardens perhaps undermine its regulatory services. This is due to owner's perception of provisions (food or medicines) provided by homegardens as the most vital for his/her wellbeing and livelihood even if it is providing many services other than only provisional services [4]. Utilization of species depends on their availability. Hence, food and medicinal plants were mostly maintained in the vicinity of homegardens.

Homegardens were recommended as a better agroforestry practice for the rehabilitation of Gununo Watershed at Wolayitta zone, Ethiopia due to improvement of total nitrogen, good electrical conductivity and cation exchange capacity at all elevation gradients. The electrical conductivity, pH, cation exchange capacity and total nitrogen of all soil layers was significantly higher in homegardens than in the parkland and woodlot. On the other hand, electrical conductivity and cation exchange capacity of surface soil were significantly higher on lower elevation than those in the upper and middle elevation. Incidentally for all cases, no differences existed in the sub soil layers.

16.5.4 Provisioning Services

Provisioning particularly food production is the primary function of homegardens across the world [19]. Generally, homegardens satisfy more than half of the nutritional and medicinal needs of its owner. Hence, they were maintained as a primary source to provide dietary calorie, carbohydrates and minerals through the production of diverse species of vegetables, tubers, rhizomes, fruits and herbs. Thus, these are indispensably crucial for sustaining health and wellbeing of indigenous and rural communities. Foods species grown in the homegardens were mostly introduced domestic crops and were mostly reported as exotics in most of the places they were grown [4].

The farming of crop plants in the homestead of southern Edo State Nigeria fulfilled the need of 2–9 members of the household. Study of non-market food provisioning services via homegardens and community sharing in Satoyama socio-ecological production landscapes on Japan's Noto peninsula conveyed that the rural households in inland and coastal communities consume greater varieties and quantities of food grown at home and/or received from others than households in semi-urban community. The households with more connections with other households have been reported to consume more food varieties and quantities due to sharing of these communities among the households. This was prominent among rural households but not urban communities. The Japanese study reported that approximately 12% of all vegetables consumed by households were grown at home. However, a higher proportion was consumed in rural and agricultural communities (15%) in comparison to urban municipalities (10%).

Similarly, in Slovakia, home grown or produced food items were less purchased but commonly shared among the members of the communities in provincial areas in comparison to the urban locality. Fruits and vegetables have been reported as one of the important provisions provided by the homegardens. These were 51 species in Benin; 21 species in southwestern Ethiopia [24]; 31 fruit species in northern Thailand; 13 fruit species in Bangladesh. The Barak Valley homegardens of Assam, India, provided dominant fruit species and ensured nutritional, food and income security [24].

In Karen homegardens, Thailand, the most common fruit species were mango, jackfruit, guava and taro. Michon et al. [27] reported 250 crop species from the Javanese home gardens. Homegarden agrobiodiversity of Tabasco lowlands, México, was documented with a total of 38 functions. Among these, 14 were ecological, 12 economic and 12 socio-cultural. Home products and mutual non-market food interactions were more frequent in rural areas of USA. In developing countries, the underutilized and wild edible plant species being grown in the homegardens promote nutritional security. They as well prevent infections and diseases and even contribute significantly to the dietary diversity of HIV-positive rural households. Homegarden grown traditional fruits were reported to nourish poor families in Bangladesh.

16.5.5 Carbon Sequestration

Maintenance of multispecies and multistrata agroforests like traditional farming systems of homegardens have been recommended to be worthwhile for the offsetting of carbon emissions by the Kyoto Protocol under article 3.3 (Watson and Eyzaguirre, 2002; Smith et al., 2007). This is due to their financial feasibility that ensures releasing pressure from the forests and improving soil productivity through storing atmospheric carbon in its biomass for longer duration in an agricultural landscape along with the grower's sustainability [7]. Compared to any other land uses in an agricultural landscape, the homegarden agroforestry with tree-based land use systems have higher carbon content. They henceforth have higher net gains in carbon stocks. The relationship between biomass carbon stocks of homegardens to its soil organic carbon stocks across a landscape has been inferred to aid in the efficient management homegarden carbon pools.

Fundamentally, carbon sequestration potential of homegardens is achieved through the biological/ecological processes of photosynthesis, respiration, and decomposition. Net primary productivity of the homegardens is higher as majority of the plant materials produced are either accumulated in the biomass or returned back to the soil in form of organic carbon. Homegarden based agroforestry systems in the tropics have been reported with carbon sequestration rate of 1.5–3.5 Mg C ha^{-1} $year^{-1}$ and ensured upon the storage of 0.7–6.3 Gt carbon per year. At this rate, carbon stocks can be tripled in twenty years to 70 Mg C ha^{-1}. Greater plant biodiversity and incomplete biomass removal ensure homegardens to become more permanent, stable, resilient and productive agroecosystems.

The efficiency of homegardens to become carbon sink depends on the size, natural site qualities, choice of species and management practices along with the owner's survival strategies with his garden. Alternately, carbon sequestered by a homegarden is a function of its structure and composition modified by environmental and socio-economic factors. Different tree species have different amount of carbon stock. Also, soil carbon as well varies under different tree species in terms of its litter input and chemistry. Tree species vary in their influence to root: shoot ratio, litter quality and soil organic carbon [16]. Other factors reported to influence the carbon content of tree biomass were diameter of the tree, stand age, stand structure and diversity of the system.

Higher root biomass and efficient distribution of organic matter across soil layers aid in higher soil organic matter production in homegardens. Diverse species in homegardens aid in permanent continuous carbon build up in its soil even at deeper layers while preventing net emission. Moreover, drilosphere activity (i.e., association of earthworms and soil bacteria) in homegardens has been also reported higher due to higher organic matter input and optimum soil moisture relations making the garden soil productive to produce more biomass. Unfortunately, the influence of tree species to soil carbon along with its quantity, mechanism and duration of storage in the soil before the carbon is sequestered as an option for mitigation is not still clearly understood. This is also due to the perception that quantitative estimate of tree species effect on soil is highly complex.

Bamboo based homegardens were also reported as significant carbon sink due to their fast growth and productivity. Soil organic carbon stock in the homegardens of Terai zone of West Bengal up to 40 cm depth was highest in small sized gardens and affirmed about 46.85 Mg ha^{-1}. On the other hand, plant biomass carbon was the highest in large homegardens with an amount of 60.38 Mg ha^{-1}. The total tree biomass of 45 investigated homegardens in West Bengal, India has been quantified as 7482.67 Mg. This could offset upto 507.94 Mg CO_2. Large cardamom based traditional agroforestry systems in Sikkim Himalayas have been reported to be rich in soil organic carbon but lesser than natural forest cardamom agroforestry [42]. Overall, within the one-meter soil profile, carbon stock in traditional homegardens of Mizoram has been estimated in the range of 108.83–258.43 Mg ha^{-1}. Incidentally, older gardens stored more carbon than the younger ones. Biomass carbon in Mizoram homegardens has been estimated as 59.0–140.0 Mg ha^{-1} with CO_2 mitigation potential rate of 4.86–22.89 Mg ha^{-1} yr^{-1}.

Homegardens in Kerala have been reported to sequester higher amount of carbon in comparison to natural forests [33]. The above-ground homegarden carbon stock in Kerala has been estimated as 16–36 Mg ha^{-1} with smaller gardens reportedly storing more than the larger ones. Homegardens with different species composition in Kashmir Himalayas varied significantly in terms of carbon sequestration as homegardens with Salix, Poplar, Beans, Kale and Apple accumulated biomass of 104.86 Mg/ha in comparison with 44.53 Mg/ha by homegardens with Poplar, Kale, Beans and Apple.

Homegardens in dry zones of Sri Lanka were estimated with lesser amount of carbon (10–55 Mg ha^{-1}) than the homegardens in wet zone (48–145 Mg ha^{-1}).

This was due to lesser tree density in the former than in the latter. Estimated carbon storage of 145 Mg ha^{-1} in Panamanian traditional agroforestry systems was lesser than managed forest (335 Mg ha^{-1}) but higher than pastures (46 Mg ha^{-1}). In Ethiopia, homegardens were estimated with higher amount of ecosystem carbon stock (100.4 Mg ha^{-1}) in comparison to 72.90 Mg C ha^{-1} being estimated in woodlot agroforestry systems (Semere, 2019). Studies from northern Ethiopia found that the total carbon stock of 148.32 Mg ha^{-1} in homegardens was comparable to that of forests having carbon stock of 157.27 Mg ha^{-1}. However, wide variation in the carbon stock of homegardens and forest was reported from southern Ethiopia. Here, the mean above and below ground carbon stock in forest has been 205.14 and 51.29 Mg ha^{-1}, respectively. Thereby, 59.05 and 18.59 Mg C ha^{-1} has been estimated to be sequestered, respectively by the homegardens.

Total biomass carbon stock of 27.4 Mg ha^{-1} being quantified in homegardens was significantly lesser than the total carbon stock of 63.1 Mg ha^{-1} in adjacent coffee-based agroforestry systems in Mana district, southwestern Ethiopia. Even urban homegardens in Ethiopia have been quantified with a considerable amount of total biomass carbon of 884.2 Mg ha^{-1}. Bajigo et al. [3] quantified 86.4 Mg ha^{-1} carbon in biomass of homegardens at Gununo watershed, Ethiopia. Total carbon stock of natural forest, shade grown coffee agroforestry, homegarden and crop field at Dallo Mena districts in Bale zone, southeastern Ethiopia have been estimated as 426.54, 266.61, 185.26 and 97.56 Mg ha^{-1}, respectively. The above ground biomass of each tree in homegarden, cropland and pastureland at Sokuru district, Jimma zone of southwest Ethiopia has been determined as 5.54, 9.0 and 3.47 Mg, respectively.

16.6 Factors Affecting the Structure and Composition of Homegardens

Human utility of ecosystem service has been reported to vary with geographical location and social structure. This is due to alterations in income, gender, class and education. Rural households with low income rely more on natural resources due to easy access, lack of markets and limited alternatives. Households with higher income have better access to natural resources and markets. Hence, they have been reported to utilize more and variety of ecosystem services. Strategies on conservation of ecosystem service and biodiversity have been thus recommended and formulated based on societal needs [10]. In Catalan Pyrenees, Spain female valuation on ecosystem services provided by the homegardens was higher than male evaluation.

The structure and composition of homegardens was primarily influenced by its provisioning functions [4] and with several interacting factors. These are often dependent on the owner's needs and preferences. A garden's physical (size, age, ethnicity, location or accessibility i.e. distance from main road), and socio-personal features of the owner's (education, occupation, gender, customs, traditions, market access and aesthetic preferences) and other socio-economic attributes profoundly influence the owner's choice of species and hence the associated bio-diversity [4, 24]. Species richness, plant diversity, size, shape, plant density, composition of species and management in homegardens vary from region to region. These even vary within a locality and within the same community. This is due to cultural, socio-economic, edaphic (moisture, nutrients) and climatic conditions along with the owner's preference. However, relevant information on these in India are mostly anecdotal.

Owners in urban areas of South Africa due to their better socioeconomic conditions preferred to grow more exotics in their diverse homegardens than the owners with lower socioeconomic status who preferred mainly natives in their less diverse homegardens. In contrast, studies also have reported lesser species in homegardens near urban areas but with more ornamental and commercial plants than in the rural areas that emphasize upon provisioning services. Attraction to urbanized environments tends to neglect traditional values and knowledge along with the preference for exotics. This also catalyzes the disappearance of traditionally cultivated species.

Studies from Ecuador, China and Europe found that homegarden owners preferred ornamental plants or plants providing regulatory services over edibles due to higher incomes, better education and easier market accessibility. Italian homegarden owners prefer species with better taste suitable for a certain valued cuisine and those that satisfy their aesthetic needs. Women were reported to be instrumental in maintaining wellbeing of their family, preparing planting materials, planting, practicing and transmitting traditional knowledge. Economically backward families with less land holdings in Ethiopia were found with smaller homegardens and thus with lesser species richness. Large sized homegarden owners in Ethiopia, Bangladesh, Thailand and Costa Rica preferred to grow more fruits species in their gardens than their counterparts with smaller homegarden who preferred more annual vegetables [24].

16.7 Conclusions and Recommendations

A homegarden is a classical example of sustainable ecological plant bio-diversity. It provides huge scope and hope for the scientific understanding and maturity of biodiversity conservation. The homegarden structure and functions are highly complex and are not well understood. Studies found it difficult to evaluate upon the implications of the threats to the homegardens. Thereby, human ecology based conceptual approaches for better insight into complex socio-ecological interactions have been recommended. Studies further recommend responsible institutions with adequate socio-ecological framework and need based policy support for the sustainability of healthy homegardens. This can be achieved with participatory and biocultural conservation approaches along with the recognition incorporation of traditional knowledge and practices into site specific planning methodologies [6, 44]. Further, Castro et al. [4] and Shimrah et al. [44] suggested awareness drive and extension support providing ecological, nutritional, and economic importance of homegardens in the community. Besides these, the folk knowledge of homegarden management and biodiversity must be acknowledged and conserved through an interaction mechanism. This is due to the fact that the folk knowledge has a better say towards the complex interaction of cultivated and wild species in the homegardens in comparison to the limited rational objectives set through scientific rationalism.

In summary, subsistence of economy, consumption of traditional foods that fosters conservation of bio-cultural diversity, holistic yet effective strategies for food security can be together better achieved through the cultivation of homegardens and as a priority in the vast landscape of North-east India.

References

1. Abdoellah, O.S., Hadikusumah, H.K., Takeuchi, K., Okubo, S.: Parikesit: commercialization of homegardens in an Indonesian village: vegetation composition and functional changes. Agrofor. Syst. **68**, 1–13 (2006)
2. Agbogidi, O.M., Adolor, E.B.: Homegardens in the maintenance of biological diversity. Appl. Sci. Rep. **1**, 19–25 (2013)
3. Bajigo, A., Tadesse, M., Moges, Y., Anjulo, A.: Monitoring of seasonal variation in physico-chemical water parameters in Nalasopara Region. J. Ecosyst. Ecography **5**, 157 (2015). https://doi.org/10.4172/2157-7625.1000157
4. Caballero-Serrano, V., Onaindia, M., Alday, J.G., Caballero, D., Carrasco, J.C., McLaren, B.: Plant diversity and ecosystem services in Amazonian homegardens of Ecuador. Agr. Ecosyst. Environ. **225**, 116–125 (2016)
5. Calvet-Mir, L., Gómez-Baggethun, E., Reyes-García, V.: Beyond food production: ecosystem services provided by Homegardens. A case study in Vall Fosca, Catalan Pyrenees, Northeastern Spain. Ecol. Econ. **74**, 153–160 (2012)
6. Castro, A., Lascurain-Rangel, M., Gómez-Díaz, J.A., Sosa, V.: Mayan homegardens in decline: the case of the Pitahaya (*Hylocereus undatus*), a Vine Cactus with Edible Fruit. Trop. Conserv. Sci. **11**, 1–10 (2018)
7. Chakravarty, S., Puri, A., Abha, A.M., Rai, P., Lepcha, U., Vineeta, P.N.A., Shukla, G.: Linking social dimensions of climate change: transforming vulnerable smallholder producers for empowering and resiliency. In: Raj, A., Jhariya, M.K., Yadav, M.K., Banerjee, A. (eds.) Climate Change and Agroforestry Systems. Apple Academy Press (2020). https://doi.org/10.1201/9780429286759-7
8. Das, T., Das, A.K.: Inventorying plant biodiversity in homegarden: a case study in Barak valley, Assam, north east India. Curr. Sci. **98**, 155–163 (2005)
9. Das, T., Das, A.K.: Conservation of plant diversity in rural homegardens with cultural and geographical variation in three districts of Barak Valley, Northeast India. Econ. Botany **69**, 57–71 (2015)
10. Daw, T., Katrina, B., Rosendo, S., Pomeroy, R.: Applying the ecosystem services concept to poverty alleviation: the need to disaggregate human well-being. Environ. Conserv. **38**, 370–379 (2011)
11. Egodawatta, C., Warnasooriya, R.: Diversity of tree flora in homegardens in urban proximity of Anuradhapura city, Sri Lanka. Agric. For. Fish. **3**, 34–39 (2014)
12. Folke, C., Carpenter, S.R., Walker, B., Scheffer, M., Chapin, T., Rockström, J.: Resilience thinking: integrating resilience, adaptability and transformability. Ecol. Soc. **15**, 20 (2010)
13. Gbedomon, R.S., Salako, V.K., Adomou, A.C., Kakaï, R.G., Assogbadjo: Plants in traditional home gardens: richness, composition, conservation and implications for native biodiversity in Benin. Biodivers. Conserv **26**, 3307–3327 (2017)
14. Gebrewahid, Y., Meressa, E.: Tree species diversity and its relationship with carbon stock in the parkland agroforestry of northern Ethiopia. Cogent Biol. **6**, 17728945 (2020). https://doi.org/10.1080/23312025.2020.1728945

15. George, M.V., Christopher, G.: Structure, diversity and utilization of plant species in tribal homegardens of Kerala, India. Agrofor. Syst. **94**, 297–307 (2020)
16. Guo, L.B., Cowie, A.L., Montagu, K.D., Gifford, R.M.: Carbon and nitrogen stocks in a native pasture and an adjacent 16 year old *Pinus radiate* d. Don. Plantation in Australia. Agric. Ecosyst. Environ. **124**, 205–218 (2008)
17. Jaganmohan, M., Vailshery, L.S., Nagendra, H.: Patterns of insect abundance and distribution in urban domestic gardens in Bangalore, India. Diversity **5**, 767–778 (2013)
18. Kahane, R., Hodgkin, T., Jaenicke, H., Hoogendoorn, C., Hermann, M., Keatinge, J.D.H.(D), Hughes, J.d'A., Padulosi, S., Looney, N.: Agrobiodiversity for food security, health and income. Agron. Sustain. Dev. **33**, 671–693 (2013). https://doi.org/10.1007/s13593-013-0147-8
19. Kumar, B.M., Nair, P.K.R.: The enigma of tropical homegardens. Agrofor. Syst. **61**, 135–152 (2004)
20. Kumar, B.M., Nair, P.K.R.: Tropical Homegardens: A Time Tested Example of Sustainable Agroforestry. Environment Expert, India (2006)
21. Kumari, M.A.S., Kansuntisukmongkol, K., Brockelman, W.Y.: Plant diversity in Homegardens and its contribution to household economy in suburban areas in Sri Lanka. Environ. Nat. Resour. J. **7**, 12–30 (2009)
22. Larios, C., Casas, A., Vallejo, M., Moreno-Calles, A.I., Blancas, J.: Plant management and biodiversity conservation in Náhuatl homegardens of the Tehuacán Valley, Mexico. J. Ethnobiol. Ethnomedicine **9**, 74–89 (2013)
23. Linger, E.: Agro-ecosystem and socio-economic role of homegarden agroforestry in Jabithenan District, North-Western Ethiopia: implication for climate change adaptation. Springer Plus **3**, 154–162 (2014)
24. Mathewos, M., Hundera, K., Biber-Freudenberger, L.: Planting fruits and vegetable in home gardens as a way to improve livelihoods and conserve plant biodiversity. Agriculture 8 (2018). https://doi.org/10.3390/agriculture8120190
25. Mekonnen, E.L., Asfaw, Z., Zewudie, S.: Plant species diversity of Homegarden agroforestry in Jabithenan District, North-Western Ethiopia. Int. J. Biodivers. Conserv. **6**, 301–307 (2014)
26. Mengistu, F., Alemayehu, M.: Species assortment and biodiversity conservation in homegardens of Bahir Dar city, Ethiopiea. Ethiop. J. Agric. Sci. **27**, 31–48 (2017)
27. Michon, G., Bompard, J., Hecketseiler, P., Ducatillion, C.: Tropical forest architectural analysis as applied to agroforests in the humid tropics: the example of traditional village agroforests in West Java. Agrofor. Syst. **1**, 117–129 (1983)
28. Mohri, H., Lahoti, S., Saito, O., Mahalingam, A., Gunatilleke, N., Irham Hoang V.T., Hitinayake, G., Takeuchi, K., Herath, S.: Assessment of ecosystem services in homegarden systems in Indonesia, Sri Lanka, and Vietnam. Ecosyst. Ser. **5**, 124–136 (2013). https://doi.org/10.1016/j.ecoser.2013.07.006
29. Nath, A.J., Das, A.K.: Carbon storage and sequestration in bamboo-based small holder homegardens of Barak Valley, Assam. Curr. Sci. **100**, 229–233 (2011)
30. Niñez, V.K.: Household gardens: theoretical and policy considerations. Agrofor. Syst. **23**, 167–186 (1987)
31. Nguyen, M.C., Winters, P.: The impact of migration on food consumption patterns: the case of Vietnam. Food Policy **36**, 71–87 (2011)
32. Panwar, P., Chakravarty, S.: Floristic structure and ecological function of homegardens in humid tropics of West Bengal, India. Indian J. Agrofor. **12**, 69–78 (2010)
33. Peyre, A., Guidal, A., Wiersum, K.F., Bongers, F.: Dynamics of homegarden structure and function in Kerala, India. Agrofor. Syst. **66**, 101–115 (2006)
34. Poot-Pool, W.S., Van Der Wal, H., Flores-Guido, S., Pat-Fernández, J.M., Esparza-Olguín, L.: Economic stratification differentiates home gardens in the Maya village of Pomuch, México. Econ. Botany 66**,** 264-275 (2012)
35. Pushpakumara, D.K.N.G., Wijesekara, A., Hunter, D.G.: Kandyan homegardens: a promising land management system in Sri Lanka. Sustainable use of biological diversity in Socio-ecological production landscapes. Background to the Satoyama Initiative for the benefit of biodiversity and human well-being. CBD Technical Series **52**, 102–108 (2010)

36. Rana, K., Goyal, N., Sharma, G.P.: Staging stewards of agro-ecosystems in the ecosystem services framework. Ecosyst. Serv. 33**,** 89–101
37. Rivera, L.F., Armbrecht, I., Calle, Z.: Silvopastoral systems and ant diversity conservation in a cattle-dominated landscape of the Colombian Andes. Agr. Ecosyst. Environ. **181**, 188–194 (2013)
38. Roy, M., Sarkar, B.C., Shukla, G., Vineeta Debnath, M.K., Nath, A.J., Bhat, J.A., Chakravarty, S.: Traditional homegardens and ethnomedicinal plants: Insights from the Indian Sub-Himalayan region. Trees For. People **8**, 100236 (2022). https://doi.org/10.1016/j.tfp.2022.100236
39. Sahoo, U.T., Rocky, P.: Species composition and plant Diversity as influenced by altitude and size of homegardens in Mizoram, North-East India. Ecol. Environ. Sci. **41**, 3–4 (2015)
40. Sarkar, B.C., Shukla, G., Debnath, M.K., Pala, N.A., Chakravarty, S.: Diversity and utilization pattern of homegardens: a case study from Eastern Himalayan region of West Bengal. Indian J. Agrofor. **22**, 31–36 (2020)
41. Seta, T., Demissew, S., Asfaw, Z.: Homegardens of Wolayta, Southern Ethiopia (2013)
42. Sharma, G., Hunsdorfer, B., Singh, K.K.: Comparative analysis on the socio-ecological and economic potentials of traditional agroforestry systems in the Sikkim Himalaya. Trop. Ecol. **57**, 751–764 (2016)
43. Shastri, C.M., Bhat, D.M., Nataraj, D.C., Murall, K.S., Ravindranath, N.S.: Tree species diversity in a village ecosystem in Uttara Kannada district in Western Ghats, Karnataka. Curr. Sci. **82**, 1080–1084 (2002)
44. Shimrah, T., Lungleng, P., Shimrah, C., Khuman, Y.S.C., Varah, F.: Role of traditional homegardens in biodiversity conservation and socioecological significance in Tangkhul community in Northeast India. Trop. Ecol. **59**, 533–539 (2018)
45. Shukla, G., Vineeta Kumari, A., Chakravarty, S.: Plant diversity, structure and uses of plants in homegardens of Jharkhand, India. Indian J. Trop. Biodivers. **25**, 40–50 (2017)
46. Taylor, J.R., Lovell, S.T.: Urban home food gardens in the Global North: research traditions and future directions. Agric. Hum. Values **31**, 285–305 (2014)
47. Vibhuti, B.K., Bargali, S.S.: Species composition, diversity and traditional uses of homegardens in Kumaun Himalaya, India. Indian J. Agric. Sci. **89**, 1415–1418 (2019)
48. Whitney, C.W., Tabuti, J.R.S., Hensel, O., Yeh, C.-H., Genauer, J., Luedeling, E.: Homegarden and the future of food and nutrition security in southwest Uganda. Agric. Syst. **154**, 133–144 (2017)
49. Wiersum, K.F.: Diversity and change in homegarden cultivation in Indonesia. In: Kumar, B.M., Nair, P.K.R. (eds.) Tropical Homegardens: A Time-Tested Example of Sustainable Agroforestry, pp. 13–24. Springer (2006)
50. Woldeyes, F., Asfaw, Z., Demissew, S., Roussel, B.: Homegardens (Aal-oos-gad) of the basket people of southwestern Ethiopia: sustainable agro-ecosystems characterizing a traditional landscape. Ethnobot. Res. Appl. **14**, 549–563 (2016)

The manufacturer's authorised representative in the EU is Springer Nature Customer Service Centre GmbH, Europaplatz 3, 69115 Heidelberg, Germany. If you have any concerns regarding our products, please contact ProductSafety@springernature.com

Printed and bound by CPI Group (UK) Ltd, Croydon, CR0 4YY
01/10/2026
02225851-0008